Methods in Molecular Biology

Series Editor
John M. Walker
School of Life and Medical Sciences
University of Hertfordshire
Hatfield, Hertfordshire, UK

For further volumes:
http://www.springer.com/series/7651

For over 35 years, biological scientists have come to rely on the research protocols and methodologies in the critically acclaimed *Methods in Molecular Biology* series. The series was the first to introduce the step-by-step protocols approach that has become the standard in all biomedical protocol publishing. Each protocol is provided in readily-reproducible step-by-step fashion, opening with an introductory overview, a list of the materials and reagents needed to complete the experiment, and followed by a detailed procedure that is supported with a helpful notes section offering tips and tricks of the trade as well as troubleshooting advice. These hallmark features were introduced by series editor Dr. John Walker and constitute the key ingredient in each and every volume of the *Methods in Molecular Biology* series. Tested and trusted, comprehensive and reliable, all protocols from the series are indexed in PubMed.

Natural Product Isolation and Identification

Methods and Protocols

Edited by

Young Hae Choi and Özlem Erol

Natural Products Lab, Institute of Biology, Leiden University, Leiden, The Netherlands

Sandra Irmisch

Plant Sciences, Institute of Biology, Leiden University, Leiden, The Netherlands

Humana Press

Editors
Young Hae Choi
Natural Products Lab
Institute of Biology
Leiden University
Leiden, The Netherlands

Özlem Erol
Natural Products Lab
Institute of Biology
Leiden University
Leiden, The Netherlands

Sandra Irmisch
Plant Sciences
Institute of Biology
Leiden University
Leiden, The Netherlands

ISSN 1064-3745 ISSN 1940-6029 (electronic)
Methods in Molecular Biology
ISBN 978-1-0716-4349-5 ISBN 978-1-0716-4350-1 (eBook)
https://doi.org/10.1007/978-1-0716-4350-1

This Humana imprint is published by the registered company Springer Science+Business Media, LLC, part of Springer Nature.
The registered company address is: 1 New York Plaza, New York, NY 10004, U.S.A.

Preface

Every natural science, in a broad sense, engages with the products of nature to some degree. Genes, proteins, metabolites, and even precursors of chemical synthesis originate from nature. The utilization of natural products began with human history, as all needs of life, including food, housing, clothing, and medicine, have relied on nature. In the case of medicines, over time, the sources of medicinal compounds have progressively expanded, beginning with the chemical constituents of terrestrial plants and extending to include microorganisms and marine organisms. This expansion aims to exploit the diverse range of chemicals known as metabolites. It is estimated that a single organism may contain over 30,000 compounds with a vast chemical diversity that has provided a rich source of bioactive molecules. In ancient times, this chemical diversity was an unparalleled advantage for the use of natural products for specific purposes. However, with increasing available technological, humans began to use natural products in purified forms such as extracts, refined fractions, and isolated compounds instead of fresh or dried materials. The advantage posed by this chemical diversity became a drawback for their practical utilization, given the numerous and tedious steps required to isolate desired chemicals from raw materials. Metabolites, which constitute a major target for natural products utilization, are particularly challenging in this sense. Among these, genes and proteins, although numerous, are an exception to this, as being based on 4 nucleosides and 20 amino acids, their extraction and purification is straightforward. Most metabolites, on the other hand, have unique chemical properties, making it impossible to develop a general or standard protocol for their extraction and purification. Each class of metabolites, including primary and secondary (specialized) metabolites like alkaloids, carbohydrates, flavonoids, lignans, lipids, phenylpropanoids, quinones, terpenoids, and others, requires specific processing protocols. Moreover, even within the same class, individual compounds may require distinct protocols.

As a result, numerous books and papers have been written on the extraction, isolation, and chemical confirmation (structure elucidation) of natural products. However, with the continual advances in the technical means used for these processes, protocols must be continually updated.

The purpose of this book is to provide a practical overview of how natural products can be extracted, isolated, purified, and elucidated for targeted chemicals. While a single book cannot cover all protocols for individual metabolites, it offers exemplified protocols for certain metabolic groups. Additionally, several profiling techniques based on chromatography and spectroscopy are included. In the field of natural products research, as in other life sciences, the implementation of a systems biology approach, based on untargeted multiomics techniques that cover as many metabolites as possible, is becoming increasingly popular. Reflecting this trend, this book includes methods for the extraction and chemical profiling of chemical components in natural products.

This protocol book is divided into two parts: (1) extraction and separation, including preanalytical methods, and (2) identification of individual groups of metabolites. The first part presents a general approach with various techniques (high performace liquid chromatography (HPLC), gas chromatography (GC), nuclear magnetic resonance spectroscopy (NMR), thin layer chromatography (TLC), and deep eutectic solvents, as well as sample preparation for marine organisms). The second part focuses on the chemical elucidation of individual groups of metabolites, including the chemical profiling of primary and secondary metabolites, phenolics, coumarins, lignans, saponins, and their integration with a biological model. When writing this book, while not aspiring to cover all topics and techniques, given their vastness, we included comprehensive background information on a substantial part of them, providing a useful reference guide for natural products researchers or researchers beginning practical work in the natural products field.

Leiden, The Netherlands

Young Hae Choi
Özlem Erol
Sandra Irmisch

Contents

Contributors

J. WILLIAM ALLWOOD • *Environmental and Biochemical Sciences, James Hutton Institute, Dundee, Scotland, UK*
VISHWA J. BARUAH • *Plant Biology Section, School of Integrative Plant Science, Cornell University, Ithaca, NY, USA*
CHINMAYEE BAWKAR • *Marine Natural Products Chemistry Laboratory, Korea Institute of Ocean Science and Technology, Busan, Republic of Korea; Department of Marine Technology and Convergence Engineering, University of Science and Technology, Daejeon, Republic of Korea*
ALEXANDRA A. BENNETT • *Department of Chemistry, Institute of Analytical Chemistry, University of Natural Resources and Life Sciences, ViennaVienna, Austria*
YOUNG HAE CHOI • *Natural Products Lab, Institute of Biology, Leiden University, Leiden, The Netherlands*
YUNTAO DAI • *Institute of Chinese Materia Medica, China Academy of Chinese Medical Sciences, Beijing, China*
THI KIEU TIÊN DO • *CAMAG, Muttenz, Switzerland; HPTLC Association, Rheinfelden, Switzerland*
ÖZLEM EROL • *Natural Products Lab, Institute of Biology, Leiden University, Leiden, The Netherlands*
HOCELAYNE PAULINO FERNANDES • *Natural Products Laboratory, Institute of Biology, Leiden University, BELeiden, The Netherlands*
MAXIMILIAN FREY • *Department of Cell and Metabolic Biology, Leibniz Institute of Plant Biochemistry, Halle, Germany*
DAVY GUILLARME • *School of Pharmaceutical Sciences, University of Geneva, Geneva, Switzerland; Institute of Pharmaceutical Sciences of Western Switzerland (ISPSO), University of Geneva, Geneva, Switzerland*
KARTHICK BABU SAI SANKAR GUPTA • *Department of NMR facility, Leiden Institute of Chemistry, Leiden University, Leiden, The Netherlands*
WON SOK HA • *Department of Integrated Drug Development and Natural Products, Graduate School, Kyung Hee University, Seoul, Republic of Korea*
SANG BEOM HAN • *College of Pharmacy, Chung-Ang University, Seoul, Republic of Korea*
MOHAMMAD IRFAN • *Plant Biology Section, School of Integrative Plant Science, Cornell University, Ithaca, NY, USA*
SANDRA IRMISCH • *Plant Sciences, Institute of Biology, Leiden University, Leiden, The Netherlands*
YOUNG PYO JANG • *Department of Biomedical and Pharmaceutical Sciences, Graduate School, Kyung Hee University, Seoul, Republic of Korea; Department of Integrated Drug Development and Natural Products, Graduate School, Kyung Hee University, Seoul, Republic of Korea*
YOOSEONG JEONG • *College of Pharmacy, Chung-Ang University, Seoul, Republic of Korea*
CHUL YOUNG KIM • *College of Pharmacy and Institute of Pharmaceutical Sciences and Technology, Hanyang University, Ansan, Gyeonggi-do, Republic of Korea*
HYE KYONG KIM • *Natural Products Lab, Institute of Biology, Leiden University, Leiden, The Netherlands; Fytagoras BV, Leiden, The Netherlands*

Hye Mi Kim • *College of Pharmacy and Institute of Pharmaceutical Sciences and Technology, Hanyang University, Ansan, Gyeonggi-do, Republic of Korea*
Kyu Won Kim • *Department of Biomedical and Pharmaceutical Sciences, Graduate School, Kyung Hee University, Seoul, Republic of Korea*
Seung Hyun Kim • *College of Pharmacy, Yonsei Institute of Pharmaceutical Sciences, Yonsei University, Incheon, Republic of Korea*
Yena Kim • *Department of Integrated Drug Development and Natural Products, Graduate School, Kyung Hee University, Seoul, Republic of Korea*
Henrie Korthout • *Fytagoras BV, Leiden, The Netherlands*
Sandra T. Krause • *Institute of Pharmacy, Martin Luther University Halle-Wittenberg, Halle, Germany*
Lars H. Kruse • *Michael Smith Laboratories, University of British Columbia, Vancouver, BC, Canada*
Dong-Kyu Lee • *College of Pharmacy, Chung-Ang University, Seoul, Republic of Korea*
Jaeyoun Lee • *College of Pharmacy, Yonsei Institute of Pharmaceutical Sciences, Yonsei University, Incheon, Republic of Korea*
Kyung Ho Lee • *Department of Integrated Drug Development and Natural Products, Graduate School, Kyung Hee University, Seoul, Republic of Korea*
Yeon-Ju Lee • *Marine Natural Products Chemistry Laboratory, Korea Institute of Ocean Science and Technology, Busan, Republic of Korea; Department of Marine Technology and Convergence Engineering, University of Science and Technology, Daejeon, Republic of Korea*
Gaurav D. Moghe • *Plant Biology Section, School of Integrative Plant Science, Cornell University, Ithaca, NY, USA*
Natali Rianika Mustafa • *Natural Products Laboratory, Institute of Biology Leiden, Leiden University, Leiden, The Netherlands*
Mira Oh • *Business Support Division, Korea Testing and Research Institute, Cheongju, Republic of Korea*
Simon D. A. Pont • *Environmental and Biochemical Sciences, James Hutton Institute, Dundee, Scotland, UK*
Emerson Ferreira Queiroz • *School of Pharmaceutical Sciences, University of Geneva, Geneva, Switzerland; Institute of Pharmaceutical Sciences of Western Switzerland (ISPSO), University of Geneva, Geneva, Switzerland*
Eike Reich • *HPTLC Association, Rheinfelden, Switzerland*
Luis Francisco Salomé-Abarca • *Laborotorio de Química de Products Naturales, Departmento de Biotecnologia y Bioquímica, Centro de Investigatión y de Estudios Avanzados de IPN, Guanajuato, Mexico*
Sumin Seo • *Natural Products Lab, Institute of Biology, Leiden University, Leiden, The Netherlands; College of Pharmacy, Chung-Ang University, Seoul, Republic of Korea*
Minki Shim • *College of Pharmacy, Chung-Ang University, Seoul, Republic of Korea*
Jesus G. Arellano Spadaro • *Vivaltes BV, Bunnik, The Netherlands*
Ilona Trettin • *CAMAG, Muttenz, Switzerland; HPTLC Association, Rheinfelden, Switzerland*
Soohyun Um • *College of Pharmacy, Yonsei Institute of Pharmaceutical Sciences, Yonsei University, Incheon, Republic of Korea*

ROBERT VERPOORTE • *Natural Products Lab, Institute of Biology, Leiden University, Leiden, The Netherlands*
MARJOLEIN WILDWATER • *Vivaltes BV, Bunnik, The Netherlands*
JEAN-LUC WOLFENDER • *School of Pharmaceutical Sciences, University of Geneva, Geneva, Switzerland; Institute of Pharmaceutical Sciences of Western Switzerland (ISPSO), University of Geneva, Geneva, Switzerland*

Part I

Extraction and Separation

Chapter 1

Natural Products Analysis Through Time: From Past Achievements to Future Prospects

Robert Verpoorte and Hye Kyong Kim

Abstract

This introductory chapter traces the evolution of (bio)chemical assays, emphasizing the critical role of robust protocols in ensuring reproducibility—a fundamental aspect of scientific research. With the advent of systems biology, the need for standardized methods has intensified, particularly for integrating vast datasets in open-access formats. The historical progression from basic plant morphology to advanced chromatographic and spectroscopic techniques in phytochemistry highlights the necessity for precise, reproducible protocols.

As metabolomics advances, there is a renewed focus on targeted approaches, shifting from broad, untargeted analyses to more specific, hypothesis-driven studies. This chapter also explores the future of analytical techniques, including cellomics and real-time metabolic flux measurements, which offer new insights into dynamic biochemical processes.

Ultimately, this introduction underscores the importance of innovation in developing new methods that address current scientific challenges, particularly in pharmacognosy and analytical phytochemistry. The chapter sets the stage for the broader discussion on the necessity of well-designed protocols that facilitate effective data sharing and collaboration across research disciplines.

Key words Introduction, Protocols, Chromatography, Spectroscopy, Necessity

1 Introduction

Protocols are fundamental to research, ensuring the reproducibility that is a cornerstone of scientific integrity. They are essential not only for replicating previous findings but also for laying the groundwork for future studies. Historically, scientific publications were self-contained, often centered on validating or refuting a hypothesis. However, the significance of well-defined protocols has grown, particularly with the advent of systems biology—a modern approach in life sciences that seeks to integrate vast amounts of raw data. This trend requires the use of chemometric methods and handling of big datasets, or synthesizing data from

Young Hae Choi et al. (eds.), *Natural Product Isolation and Identification: Methods and Protocols*, Methods in Molecular Biology, vol. 2895, https://doi.org/10.1007/978-1-0716-4350-1_1,

various sources. However, even in this case, raw data must be obtained in a reproducible manner.

Reproducibility remains critical for scientific progress, necessitating a thorough analysis and control of experimental parameters. Mastering the tools and techniques employed in experiments is, therefore, fundamental.

In the realm of natural products, particularly those with biomedical applications, the quality control (QC) of herbal medicine poses an undoubtable challenge. In the nineteenth century, the identification of plant material relied heavily on plant morphology. Microscopy emerged as a crucial tool and continues to be vital for QC today. As phytochemistry advanced, protocols evolved to include the identification of active compounds. Initially, these protocols involved simple color reactions, such as alkaloid-specific reagents, but the advent of chromatographic methods greatly expanded the possibilities. Mass spectrometry (MS) and nuclear magnetic resonance (NMR) have since become powerful tools in analytical phytochemistry.

This introduction does not aim to provide an exhaustive review of analytical phytochemistry. Instead, it is intended to encourage the reflection on the importance of protocols, using the historical progression of phytochemistry as a lens to examine how these protocols have evolved over time. We hope this discussion will inspire further efforts to create an open-access data format that enables the integration of experimental results from different laboratories.

2 Early Days of Chromatography

In the historical evolution of the study of medicinal plants, the isolation of morphine by Sertürner in 1806 [1] is an undoubtedly significant milestone, resulting from an increasing interest in the field in the 1900s. The initial isolation of compounds was based on their physicochemical properties, including pH, solubility in various solvents, crystallization techniques, and like wet chemistry. Although other alkaloids were subsequently isolated, considerable chemical research was required before their complete structures were elucidated in the 1940s. For instance, the full structure and stereochemistry of strychnine, first isolated between 1818 and 1819, required nearly 130 years of research, during which over 200 scientific papers were published on the topic [2].

Chromatography, which later became a crucial tool in chemical analysis, took some time to be integrated into the study of natural products. Liquid chromatography initially relied on the differential adsorption of compounds on a stationary phase, such as silica or aluminum oxide, packed into a column and eluted with an organic solvent. Although Tswett first reported this technique in 1906 [3], it was the work of Kuhn et al. in 1931 [4] that generated significant interest and led to the widespread adoption of column chromatography for the isolation of natural compounds.

A major advancement in chromatography occurred with the work of Martin and Synge in 1941 [5], who developed the theoretical foundation for liquid-liquid partition chromatography. This breakthrough also paved the way for the development of paper chromatography, the precursor to planar chromatography techniques such as thin-layer chromatography (TLC) and electrophoresis.

3 Planar Chromatography and Thin-Layer Chromatography (TLC)

Chromatographic methods gained prominence in biochemical and natural products research during the 1960s, initially revolutionizing the separation of extracts and leading to the development of preparative chromatography to isolate pure compounds. Given the complexity of biological extracts, chromatography also enabled novel applications, such as QC of food and medicines, identification and structure elucidation of natural compounds, and measurement of pharmacological activities in conjunction with chromatographic separation, including bioassay-guided isolation of biologically active compounds.

Egon Stahl played a pivotal role in establishing TLC as a major tool across various fields, including biosynthesis, chemotaxonomy, medicinal plant studies, QC, and biological activity analysis [6, 7]. This advancement was grounded on rigorous standardization and strict protocols. There are several advantages in the use of TLC: low cost, rapid processing (10–20 samples in parallel within 5–10 min), selective detection, and suitability for "dirty" extracts due to the single-use feature of the plates. The method can be implemented to be non-destructive and allows numerous both selective and non-selective detection techniques, such as UV absorbance, fluorescence, and chemical (color) derivatization reactions. Bioassays can also be conducted directly on TLC plates. For instance, antimicrobial compounds can be identified by observing no-growth zones in "biograms" on TLC plates covered with a microorganism-containing medium.

Morlock and colleagues [8] have developed numerous TLC-bioassays, as detailed in an excellent review highlighting the significance of such multidimensional assays. By scaling up the size and number of TLC plates and using consistent separation methods, small quantities of active compounds can be isolated for identification via spectroscopic methods. The integration of TLC with MS has further improved the identification of spots of interest for the study of bioactive compounds. Given that a large number of samples can be analyzed under uniform conditions in a short time (typically 5–15 min), the combination of chromatographic separation and bioassays represents a highly effective strategy for screening various biological samples for novel drug development candidates.

4 From Low-Pressure Liquid Chromatography to High-Performance Liquid Chromatography (HPLC)

The 1970s marked the introduction of high-performance liquid chromatography (HPLC), which revolutionized the field. Prior to HPLC, single-use columns filled with adsorbents like silica or aluminum oxide were employed to separate crude extracts, aiming to isolate and identify the compounds within. These columns were eluted with gradients of organic solvents with increasing polarity, and separations could take days or even weeks, often leading to significant decomposition of the compounds. While such decomposition could enhance chemodiversity, it also posed challenges for studying biosynthetic pathways in plants.

The development of column chromatography theory revealed that reducing the size of stationary phase particles and applying pressure to the mobile phase resulted in high-resolution separations, giving rise to high-pressure/performance liquid chromatography (HPLC). A key innovation was the introduction of reversed-phase or chemically bonded materials that resulted in increasingly efficient and diverse stationary phases. In 1976, we successfully published the separation of atropine and scopolamine from various medicinal plants [9]. By 2008, a search on "Web of Science" revealed over 80 papers on the GC or HPLC analysis of these compounds [10]. Today, it is likely that hundreds of papers have been published using new protocols, though few compare them with existing methods, making it difficult to draw general conclusions from comparisons with older techniques.

Another major development over the past 50 years was the hyphenation of HPLC with spectroscopic methods, especially MS; coupling with NMR is still in its early days, at least in a commercially accessible format. While NMR has low sensitivity and is typically used offline, i.e., fractions collected from the HPLC column are analyzed separately—liquid chromatography mass spectrometry (LCMS) has evolved into a highly sensitive detection method, capable even of single-cell analysis [11]. Each compound generates a specific detector response, and absolute quantitation requires the creation of a calibration curve, although relative quantitation of a single compound is feasible.

One limitation of HPLC is that the commonly used ultraviolet–visible spectroscopy (UV) and diode array detector (DAD) detection methods can only detect compounds with a chromophore. For compounds lacking a chromophore, derivatization is necessary, or alternatively, a light scattering detector or an MS (MS/MS) detec tor can be connected to the HPLC.

5 Development of the Analysis of Volatiles

Gas chromatography (GC) was developed in the 1960s. Packed columns were used initially, but these were replaced almost completely by capillary columns. The detection of analytes in gas liquid chromatography (GLC) can be achieved using flame ionization detection (FID) that provides a general fingerprint of a sample, though limited to volatile compounds. Derivatization methods were developed for non-volatile or thermally unstable compounds. Furthermore, the use of MS detectors, achieved by coupling a gas chromatograph with a mass spectrometer, enhanced the sensitivity but also provided the possibility of identifying the analytes. However, the sensitivity of MS can vary significantly between individual compounds, making the technique unsuitable for the absolute quantification of specific compounds while allowing relative quantification. It is important to bear in mind that while MS aids in putative identifications, the complete identification of a compound requires additional comprehensive spectral data, including ^{1}H-NMR, ^{13}C-NMR, various 2D NMR techniques, ultraviolet–visible spectroscopy (UV), infrared spectroscopy (IR), circular dichroism spectroscopy (CD), or optical rotatory dispersion spectroscopy (ORD) according to the type of compound, and polarimetry to establish absolute stereochemistry when needed. Many studies report the identification of compounds based solely on retention behavior, molecular mass, fragmentation patterns, and sometimes direct comparison with reference compounds. Although such an identification may be labeled as "putative" in the original studies, this qualifier is often omitted when cited by others. Mariott and Nolvachai [12] made a provocative statement: "If you are not using GC×GC, you will never know what you are missing!" The response to this is straightforward: "All the non-volatiles are missing." Ultimately, the focus should be on the quality of data rather than the quantity.

6 Systems Biology and the Integration of Multi-data Sets

Why delve into the history of chromatographic tools? To understand the future, it's essential to learn from the past. Over the past 60 years, there has been a consistent advancement in chemical, phytochemical, biochemical, and biological tools, particularly in terms of selectivity and sensitivity. Today's state-of-the-art techniques are not set in stone; they are documented digitally and on paper, which means their relevance is often short-lived compared to ancient artifacts like the Rosetta Stone. When developing a new protocol, it is crucial to consider previous methodologies and clearly articulate what new advantages the updated protocol offers.

Table 1
Time scale various biological processes

Level	Example	Processing time (s)
Molecular	Ion channel gating	10^{-6}
Cellular	Mitosis	10^{3}
Plant-elicitor	pH Expression of regulatory gene Expression of structural gene Biosynthesis proteins Biosynthesis phytoalexins	10^{2} 10^{3} 10^{4} 5×10^{4} 10^{5}
Plant development	Aging	10^{7}

Restriction of most protocols is that they give an outcome that is static, which is fine in the QC of biological samples. For studying fast and localized dynamic biochemical events, one needs to be able to measure fluxes and (sub)cellular localization rather than concentrations of the compound(s) of interest in crude extracts. The biological process has four dimensions, and one of them is time

The ever-increasing sensitivity and selectivity of analytical methods in the last two decades have driven the development of techniques that aim to capture the full metabolome of an organism, transitioning from targeted methods to holistic analyses. This marks a shift toward a systems biology approach, where the goal is to measure entire metabolomes and gain an insight into the roles of various metabolites within an organism, such as identifying markers for plant resistance against pests and diseases.

However, the time scale of biological reactions must be considered, as some processes occur too rapidly to be captured in real-time with existing tools (Table 1). Chromatography-based tools, particularly, have proven useful for studies conducted under different experimental conditions. While early chromatographic methods were predominantly targeted, advancements in sensitivity have increased the number of detectable features in chromatograms and mass spectra, making metabolomics a popular approach. Initially seen as a non-targeted strategy, metabolomics, when combined with chemometric methods, was believed to be capable of identifying disease markers [10, 13, 14].

Today, multivariate data analysis and artificial intelligence are common tools for exploring various biological processes [15]. Yet, the complexity of the challenge posed by non-targeted metabolomics has resulted in only a few examples of metabolites linked to human diseases or plant resistance against pests and diseases. Thus, there is now a renewed focus on returning to targeted methods to better understand these biochemical changes.

7 Necessity for New Protocols

The development of new protocols is essential for several reasons, including:

- Designing more environmentally friendly procedures.
- Ensuring reproducibility of experiments.
- Managing the extensive number of parameters that need to be controlled.
- Handling large volumes of data generated.
- Identifying bioactive compounds.
- Accurately identifying and quantifying unknown features in chromatograms or spectra, which may obscure the true findings.
- Addressing the significant variability in the physicochemical properties of metabolites within living systems, as well as their wide concentration ranges.
- Mitigating false signals arising from the processing and analysis of biological samples.
- Avoiding saturation issues with poorly soluble compounds, which may consistently appear at the same level due to saturation during extraction.
- Detecting potential synergistic effects.

To address environmental concerns, it is critical to replace toxic and environmentally harmful solvents with safer alternatives [16]. When utilizing liquid chromatography-mass spectrometry/mass spectrometry^n (LC-MS/MS^n), many signals are detected; however, da Silva et al. [17] estimated that only about 2% of these can be identified, with the remaining signals potentially being mere noise. For instance, a recent study reported that in an in-source fragmentation MS detector (ISF), over 70% of the observed peaks are likely fragments formed within the detector [18], indicating that most detected signals lack biological relevance.

In situ determination of biological activity through techniques such as TLC [8, 19] or post-column enzymatic reactions provides an elegant solution. A specific example includes the identification of acetylcholine esterase (AChE) inhibitors [20, 21]. The study by Rhee et al. [21] also highlights the critical issue of false positives, underscoring the need for proper controls. Unfortunately, this need is not widely acknowledged, as revealed by the fact that approximately half of the published studies on natural AChE inhibitors lacking appropriate controls. Consequently, these studies often report compounds that chemically react with the dye in the Ellman assay without demonstrating actual pharmacological AChE inhibition.

Another instructive example is the targeted analysis of salicylate, where the literature reports highly variable recoveries for this compound [22]. This variability was attributed to the volatility of salicylic acid (SA). As the solvent used for SA extraction evaporates, the results can vary greatly depending on the duration of the drying process. This issue can be resolved by adding small amounts of NaOH to increase the pH, thereby stabilizing SA during analysis.

In studies of medicinal plants, the potential involvement of synergism is significant—a factor often overlooked by the reductionist approach commonly applied in traditional medicine studies. However, by employing a systems biology approach with well-established study designs and protocols, it is possible to identify multiple chromatographic features that correlate with biological activity [14, 23, 24].

8 Back to the Future: Revisiting Targeted Approaches in Metabolomics

Following the rapid transition from targeted methodologies to comprehensive metabolomics, there is now an emerging trend of returning to targeted approaches [25]. For many research purposes, particularly those involving open-access data sharing, a targeted approach based on robust and standardized protocols may be more appropriate. This "back to the future" trend is evident throughout this book. One of the significant challenges in achieving effective open-access data sharing is the integration of results from various sources. Often, comparing results from different experiments is hindered by reproducibility issues in biological assays [26]. Ioannidis [27] highlighted this concern in his seminal paper, "Why Most Published Research Findings Are False," identifying the lack of control over numerous experimental parameters as a critical issue. The protocols used in metabolomics are particularly susceptible to artifact formation, which may contribute to these reproducibility problems [13, 17, 28, 29].

A potential solution to these challenges lies in the advancement of "cellomics," where large-scale omics techniques are employed to analyze individual cells selected through sophisticated three-dimensional imaging [30]. This single-cell metabolomic analysis could pave the way for subsequent subcellular analyses [31]. The ultimate goal would be to achieve real-time measurements that reveal how life sustains itself under conditions such as low temperature or drought and the molecular interactions that occur during recovery from complete senescence. Achieving this would require real-time monitoring of metabolic fluxes across inter- and intracellular networks.

9 Final Conclusions

The purpose of this introduction is to provide a concise overview of the evolution of (bio)chemical assays and to highlight key considerations for the development of novel protocols. While this is not an exhaustive review, it reflects our experience and perspectives on future advancements in phytochemical analysis.

There is an urgent need for well-designed, robust protocols that are also suitable for data-sharing initiatives. Metabolomics has evolved into a powerful tool capable of analyzing single cells, similar to the capabilities seen in transcriptomics and proteomics. Effective experimental planning using design of experiment (DoE) software is essential to ensure an adequate number of data points for statistical analysis. Additionally, it is crucial to identify and control relevant parameters throughout the experimental process.

To gain insights into rapid and localized dynamic biochemical events, it is necessary to measure fluxes and (sub)cellular localization rather than merely focusing on the concentration of compounds in crude extracts. The overarching conclusion is that targeted approaches will be essential to fully harness the potential of metabolomics. Using DoE software to develop a comprehensive research plan that manages all variables is crucial. From the onset, it is important to plan for open access to experimental data, ensuring that all findings can be integrated with existing datasets for further analysis.

This introduction aims to raise awareness of the vast opportunities within the fields of pharmacognosy and analytical phytochemistry. Innovation and creativity will be key to developing novel methods and applications in various sectors, including medicine, cosmetics, food, nutraceuticals, dyes, and agrochemicals. Significant challenges lie ahead, such as developing new antibiotics and treatments for infectious and parasitic diseases, designing strategies to combat pests and diseases that threaten food production, and assessing the potential of widely used medicinal plants.

References

1. Sertürner FWA (1806) Darstellung der reinen Mohnsäure (Opiumsäure) nebst einer chemischen untersuchung des Opiums mit vorzüglicher Hinsicht auf einen darin neu entdeckten Stoff und die dahin gehörigen Bemerkungen. J der Pharmacie fuer Aerzte und Apotheker 14:47–93
2. Verpoorte R (1976) Pharmacognostical studies of some African strychnos species. PhD thesis, Leiden University
3. Tswett MS (1906) Physikalisch-chemische Studienüber das chlorophyll. Die Adsorbtionen Ber Dtsch Bot Ges 24:316–323
4. Kuhn R, Winterstein A, Lederer E (1931) Zur Kenntnis der Xanthophylle. Z Physiol Chem 197:141–160
5. Martin AJP, Synge RLM (1941) A new form of chromatogram employing two liquid phases. Biochem J 35(12):1358–1368. https://doi.org/10.1042/bj0351358

6. Stahl E (1962) Dünnschicht-chromatographie. Springer, Berlin/Heidelberg
7. Stahl E (1990) Thin layer chromatography: a laboratory handbook (Übersetzt von M. R. F. Ashworth), 2. revidierte und erweiterte Ausgabe. Springer, Berlin, XXIV und 1041 Seiten ISBN 3-540-04736-0
8. Morlock GE (2023) Planar chromatographic super-hyphenations for rapid dereplication. Phytochem Rev. https://doi.org/10.1007/s11101-022-09844-x
9. Verpoorte R, Svendsen AB (1976) High-performance liquid chromatography of some tropane alkaloids. J Chromatogr 120:203–205. https://doi.org/10.1016/S0021-9673(01)99014-4
10. Verpoorte R, Choi YH, Mustafa RN, Kim HK (2008) Metabolomics: back to basics. Phytochem Rev 7:525–538. https://doi.org/10.1007/s11101-008-9091-7
11. Li CX, Wood JC, Vu AH, Hamilton JP et al (2023) Single-cell multi-omics in the medicinal plant *Catharanthus roseus*. Nat Chem Biol 19:1031–1041. https://doi.org/10.1038/s41589-023-01327-0
12. Marriott PJ, Nolvachai Y (2021) Separation multidimensionality for improved sample characterization – is it worth the effort? J Agric Food Chem 69:11535–11537. https://doi.org/10.1021/acs.jafc.1c05513
13. Kim HK, Choi YH, Verpoorte R (2023) Natural products drug discovery: on silica or in-silico? Handb Exp Pharmacol 277:117–141. https://doi.org/10.1007/164_2022_611
14. Wang M, Lamers RJAN, Korthout HAAJ, van Nesselrooij JHJ et al (2005) Metabolomics in the context of systems biology: bridging traditional Chinese medicine and molecular pharmacology. Phytother Res 19:173–182. https://doi.org/10.1002/ptr.1624
15. Chen Y, Li E-M, Xu L-Y (2022) Guide to metabolomics analysis: a bioinformatics workflow. Metabolites 12(4):357. https://doi.org/10.3390/metabo12040357
16. Funari CS, Rinaldo D, Bolzani VS, Verpoorte R (2023) Reaction of the phytochemistry community to green chemistry: insights obtained since 1990. J Nat Prod 86(2):440–459. https://doi.org/10.1021/acs.jnatprod.2c00501
17. da Silva RR, Dorrestein PC, Quinn RA (2015) Illuminating the dark matter in metabolomics. Proc Natl Acad Sci USA 112:12549–12550. https://doi.org/10.1073/pnas.1516878112
18. Giera M, Aisporna A, Uritboonthai W et al (2024) The hidden impact of in-source fragmentation in metabolic and chemical mass spectrometry data interpretation. Nat Metab. https://doi.org/10.1038/s42255-024-01076-x
19. Rhee IK, van de Meent M, Ingkaninan K, Verpoorte R (2001) Screening for acetylcholinesterase inhibitors from Amaryllidaceae using silica gel thin-layer chromatography in combination with bioactivity staining. J Chromatogr A 915:217–223. https://doi.org/10.1016/s0021-9673(01)00624-0. PMID: 11358251
20. Ingkaninan K, de Best CM, van der Heijden R, Hofte AJP et al (2000) HPLC with on-line coupled UV, mass spectrometric and biochemical detection for identification of acetylcholinesterase inhibitors from natural products. J Chromatogr A 872:61–73. https://doi.org/10.1016/S0021-9673(99)01292-3
21. Rhee IK, van Rijn RM, Verpoorte R (2003) Qualitative determination of false-positive effects in the acetylcholinesterase assays using thin layer chromatography. Phytochem Anal 14:127–131. https://doi.org/10.1002/pca.675
22. Verberne MC, Brouwer N, Delbianco F, Linthorst HJM et al (2002) Method for the extraction of the volatile compound salicylic acid from tobacco leaf material. Phytochem Anal 13:45–50. https://doi.org/10.1002/pca.615
23. Verpoorte R, Choi YH, Kim HK (2005) Ethnopharmacology and systems biology: a perfect holistic match. J Ethnopharmacol 100:53–56. https://doi.org/10.1016/j.jep.2005.05.033
24. Verpoorte R, Kim HK, Choi YH (2018) Synergy: easier to say than to prove. Synergy 7:34–35. https://doi.org/10.1016/j.synres.2018.10.004
25. Verpoorte R, Choi YH, Kim HK (2010) Metabolomics: will it stay? Phytochem Anal 21:2–3
26. Baker M (2016) Is there a reproducibility crisis? Nature 533:452–454. https://doi.org/10.1038/533452a
27. Ioannidis JPA (2005) Why most published research findings are false. PLoS Med 2:696–701. https://doi.org/10.1371/journal.pmed.1004085
28. Maltese F, van der Kooy F, Verpoorte R (2009) Solvent derived artifacts in natural products chemistry. Nat Prod Commun 4:447–454. PMID: 19413130
29. Verpoorte R, Kim HK, Choi YH (2022) Trivialities in metabolomics: artifacts in extraction

and analysis. Front Mol Biosci 971. https://doi.org/10.3389/fmolb.2022.972190
30. Mitani TT, Susaki EA, Matsumoto K, Ueda HR (2024) Realization of cellomics to dive into the whole body or whole-organ cell cloud. Nat Methods 21:1138–1142. https://doi.org/10.1038/s41592-024-02307-5
31. Sahota A, Monteza Cabrejos A, Kwan Z, Nadappuram BP et al (2023) Recent advances in single-cell subcellular sampling. Chem Commun 59:5312. https://doi.org/10.1039/D3CC00573A

Chapter 2

Optimizing Extraction Solvents for Plant Chemical Analysis Using ^{1}H NMR and Principal Component Analysis

Sumin Seo, Sang Beom Han, Karthick Babu Sai Sankar Gupta, and Young Hae Choi

Abstract

Natural products, particularly plants, remain a vital source of bioactive compounds owing to their unparalleled metabolic diversity across pharmaceuticals, cosmetics, foods, and agriculture. However, this diversity, encompassing not only a multitude of compounds but also their varying chemical and physical properties, presents a challenge in their effective utilization. Targeted analysis of specific metabolites, as well as untargeted approaches covering a wide metabolite range, necessitate optimal extraction solvents tailored to meet diverse requirements. Achieving optimization requires two crucial components: analytical methods capable of capturing a broad spectrum of metabolites and effective data analysis to derive meaningful conclusions. In this regard, 1H nuclear magnetic resonance (NMR) spectroscopy combined with principal component analysis (PCA) emerges as a promising approach for optimization. In this protocol, employing two model plants, leaves of basil (*Ocimum basilicum*) and lettuce (*Lactura sativa*), we deduce optimal extraction solvents from mixtures of methanol-water, acetonitrile, and chloroform, leveraging ^{1}H NMR and PCA analyses.

Key words Extraction optimization, ^{1}H NMR, Principal component analysis, Targeted analysis, Primary metabolites, Secondary metabolites, Basil, Lettuce

1 Introduction

The examination of plant metabolites has attracted considerable interest owing to their diverse bioactive properties and potential applications across various industries [1]. In addition to primary metabolites such as sugars, amino acids, and lipids, plant extracts comprise a multitude of secondary metabolites, including alkaloids [2], flavonoids [3], terpenoids [4], and phenolic compounds [5], each demonstrating a wide range of biological activities. However, to fully exploit the potential of these plant compounds, it is essential to extract them efficiently and selectively from the plant matrix.

Investigating and comprehending the influence of extraction solvents on the analysis of target compounds in plant studies is

Young Hae Choi et al. (eds.), *Natural Product Isolation and Identification: Methods and Protocols*, Methods in Molecular Biology, vol. 2895, https://doi.org/10.1007/978-1-0716-4350-1_2,

paramount. The process of analyzing plant composition entails a systematic sequence of techniques, commencing with sample preparation involving drying, lyophilization, comminution, grinding, or homogenization. Subsequently, the material undergoes extraction using designated solvents or their combinations. The resultant extracts are purified by eliminating solids through methods such as filtration, ultrafiltration, or centrifugation [6]. The selection of extraction solvent critically governs the solubility and selectivity of target compounds. Variations in solvent polarities affect their capability to extract specific classes of plant compounds, thereby influencing yield, purity, and composition. Hence, comprehending the solvent's impact on quantification is essential for accurate assessment of plant metabolites [7].

Solvent selection is crucial, as the desired compounds must demonstrate high solubility and extractability in the chosen solvent. Ensuring congruent characteristics between solute and solvent is essential to streamline the extraction process, leading to reduced processing times [8, 9]. While attaining selectivity presents a notable challenge, the scientific community is actively involved in optimizing experimental conditions for the extraction of specific compounds.

In evaluating extracts using various solvents, two key considerations must be addressed: (1) how to detect all the extract metabolites, and (2) how to reduce the large number of signals to draw accurate conclusions regarding optimal extraction solvents. In this protocol, the assessment of extracts was conducted using ^{1}H nuclear magnetic resonance (NMR) spectroscopy, a method previously demonstrated to yield a diverse array of metabolites with highly precise quantitation and provide insights into chemical structures [10–13]. Consequently, its application offers valuable insights into the composition, purity, and structural characteristics of extracted plant compounds [14]. Given its pivotal role in plant metabolite analysis, selecting an appropriate extraction solvent significantly impacts compound extraction efficiency and subsequent NMR analysis [7, 15, 16].

The next concern is dealing with a multitude of data signal to draw conclusion. For this, a kind of multivariate data analysis, principal component analysis (PCA), was applied to the obtained ^{1}H NMR data. Based on the results of PCA, score plots show the differentiation of extracted metabolites between solvents; loading plots do the contribution to the differentiation, which eventually might be the cornerstone for optimizing extraction solvents depending on the targeted analysis.

As examples, two model plants, leaves of basil (*Ocimum basilicum*) and lettuce (*Lactura sativa*), were employed using such solvents as the mixtures of methanol and water, acetonitrile, and chloroform.

The next consideration involves managing the multitude of data signals to draw conclusions. To address this, a form of multivariate data analysis, PCA, was applied to the acquired ^{1}H NMR data. Based on the results of PCA, score plots demonstrate the differentiation of extracted metabolites between solvents, while loading plots indicate their contribution to this differentiation. These findings may serve as the foundation for optimizing extraction solvents tailored to targeted analysis.

As illustrative examples, two model plants, leaves of basil (*Ocimum basilicum*) and lettuce (*Lactuca sativa*), were utilized, employing solvents such as mixtures of methanol and water, acetonitrile, and chloroform.

2 Materials

2.1 Chemicals and Solvents (See Note 1)

1. Methanol (MeOH), chloroform ($CHCl_3$), and acetonitrile (AcCN) of analytical grade.
2. Deionized water (*see* **Note 1**).
3. Monopotassium phosphate (KH_2PO_4) of American Chemical Society (ACS) reagent grade.
4. 3-(trimethylsilyl)propionic-2,2,3,3-d_4 acid sodium salt TMSP-d_4, 99 atom % D).
5. Methanol-d_4 (CD_3OD) 99.8% (Tetradeuteromethanol).
6. Hexamethyldisiloxane (98%>).
7. Deuterium oxide (D_2O, >99.9 atom % D).
8. 40% NaOD (sodium deuteroxide, w/v).
9. Liquid nitrogen (*see* **Note 2**).

2.2 Solvent Preparation

1. To extract plant materials, prepare 20% (v/v), 50% (v/v), and 80% (v/v) methanol (MeOH) solutions by mixing deionized water and MeOH in the appropriate ratios.
2. Add 40% NaOD to 9 mL of D_2O in the glass bottle and mix them well to prepare 1.0 M NaOD.
3. For ^{1}H NMR measurements, prepare a phosphate buffer solution (90 mM, pH 6.0) containing 0.58 mM Trimethylsilylpropanoic acid sodium salt (TMSP) by dissolving 1.232 g of KH_2PO_4 and 10 mg of TMSP-d_4 in 100 mL of D_2O, and stir the mixture until complete dissolution is achieved, then adjust the pH using 1.0 M NaOD.
4. For ^{1}H NMR measurements, prepare CD_3OD containing HMDSO (0.418 mM) by combining 450 μL of CD_3OD with 50 μL of HMDSO in a 1.5 mL microtube, then add 100 μL of the prepared solution to 100 mL of CD_3OD (*see* **Note 3**).

2.3 Equipments and Softwares

1. A deep freezer (−80 °C) for sample storage (*see* **Note 4**).
2. A set of mortar and pestle, a mechanical sample grinder, or a ball mill machine for grinding samples.
3. A freeze-drier (lyophilize) for sample drying.
4. An ultrasonicator for the extraction of samples.
5. Vortexer for mixing samples and solvents.
6. A centrifuge for microtubes to operate at 17000 × *g*.
7. A speedvac vacuum concentrator for drying extracts.
8. An analytical balance to weigh 0.1 mg level.
9. A pH meter to adjust the pH of NMR solvents.
10. NMR spectrometer (>400 MHz).
11. An NMR data processing software to facilitate phasing, calibration, and baseline correction of ^{1}H NMR spectra.
12. An NMR data processing software to enable the binning or bucketing of ^{1}H NMR spectra, offering options to select normalization methods.
13. A software for multivariate data analysis is utilized to generate score- and loading plots of PCA, offering options to select scaling methods.

3 Methods

3.1 Extraction of Plant Samples

1. Grind plant samples (leaves of basil and lettuce used in this protocol) stored in liquid nitrogen using a mortar and pestle, which can be substituted with a mechanical grinder or ball mill machine. The ground samples are stored in a deep freezer (−80 °C) for a minimum of 12 h and subsequently dried using a freeze dryer for at least 2 days.
2. Transfer 50 mg of the ground samples into 5 mL microtubes using a spatula. For $CHCl_3$ extraction, a 1.5 mL glass vial was utilized to mitigate the risk of contamination from plastic microtubes.
3. Add 2 mL of each extraction solvent [deionized water, 20% (v/v) MeOH, 50% (v/v) MeOH, 80% MeOH, 100% MeOH, acetonitrile (AcCN), and chloroform ($CHCl_3$)] to the samples in microtubes or glass vials for $CHCl_3$ extracts. Triplicates are used for each extraction, followed by vortexing for 1 min at room temperature.
4. Ultrasonicate the microtubes containing the samples and solvents for 20 min at room temperature to facilitate extraction.

5. Centrifuge the extracts in microtubes at room temperature for 20 min using a centrifuge (17,000×*g*) to obtain a clear supernatant (*see* **Note 5**).
6. Transfer 1.0 mL of supernatants from centrifuged microtubes to another 1.5 mL microtube and dry the supernatants using a Speedvac vacuum concentrator.
7. Reconstitute dried extracts with 1.0 mL of NMR deuterium solvents [CD_3OD containing 0.418 mM HMDSO or CD_3OD-KH_2PO_4 buffer (pH 6.0, containing 0.58 mM TMSP-d_4) (1:1, v/v)] to achieve a final concentration of 5 mg/mL (*see* **Note 6**).
8. Vortex for 1 min at room temperature to dissolve extracts in NMR solvents.
9. Transfer 300 μL of supernatant to a 3 mm NMR tube.

3.2 ^{1}H NMR Measurements

NMR measurements were performed on a Bruker Avance-III 600 MHz standard bore liquid-state NMR spectrometer with a 14.1 Tesla magnetic field. In this field, ^{1}H resonates at 600.13 MHz. Cryoprobe of type triple channel inverse (TCI) H&F/C/N-D with Z gradient was used. 3 mm NMR tubes (Z112272) were purchased from Cortecnet and used for the experiments. The temperature was kept constant at 298 K. For internal locking, CH_3OH-d_4 was used. For each proton experiment, a 30-degree pulse of 2.64 μs at 5.5 W power with a fid resolution of 0.36 Hz, 64 scans with a relaxation delay of 1.5 s, and an acquisition time of 2.7 s, in total taking 5 min to complete the experiment. The water signal was suppressed using a pre-saturation method and low-power selective irradiation at 0.3 Hz H_2O at 4.87 ppm. Time domain data was transformed to the frequency domain by Fourier transformation with a window function of an exponential function and a line broadening parameter set to 0.3 Hz for smoothening. The generated spectra were manually phased, baseline corrected, and calibrated to TMSP-d_4 at 0.0 ppm or HMDSO at 0.06 ppm.

1. Load the NMR tube into the spectrometer.
2. Set the sample temperature to 298 K (25 °C) and allow for thermal equilibration for several minutes.
3. Tune and match the NMR tube manually or in automatic mode.
4. Lock the spectrometer frequency to the deuterium resonance in the NMR solvents (either CD_3OD or D_2O, preferably CD_3OD).
5. Shim the sample using either manual or automated methods.

6. Determine the frequency of the water resonance and center the spectrum at that frequency. While these steps are typically automated, the initial sample should be processed manually to ensure well-resolved spectra. A single sample can be utilized for preliminary testing.
7. Commence experiments in automation mode for the remaining samples once all settings are correctly established.

3.3 Principal Component Analysis of ^{1}H NMR Spectra of Plant Extracts of Different Solvents

1. Employing NMR data processing software facilitates binning or bucketing to normalize ^{1}H NMR spectra of plant extracts obtained from different solvents. This normalization relies on the intensity of internal standards (TMSP-d_4 for CD_3OD-KH_2PO_4 buffer (pH 6.0) (1:1, v/v) or HMDSO for CD_3OD) or total intensity (*see* **Note 7**).
2. Define the bin (bucket) and exclusion range for residual deuterium solvents and HDO (*see* **Note 8**).
3. Transfer binned (bucketed) ^{1}H NMR data to software for multivariate data analysis and scale the binned (bucketed) data using either unit-variance (UV) or Pareto scaling methods to reduce peak-to-peak variation (*see* **Note 9**).
4. Reduce the binned (bucketed) ^{1}H NMR data through PCA and calculate principal components (PCs) to generate score plots for each extract. Compare the position of each extract in the obtained score plots. Two binned (bucketed) ^{1}H NMR data sets [CD_3OD-KH_2PO_4 buffer (pH 6.0) (1:1, v/v) or CD_3OD] are separately analyzed for PCA. The examples of leaves of basil and lettuce samples are illustrated in Fig. 1, wherein the proximity of extracted metabolic profiles is indicated by the distance between PC scores.
5. Identify the PCs that contribute significantly to the separation. In the case of basil samples measured in a CD_3OD-KH_2PO_4 buffer (pH 6.0) at a ratio of 1:1 (v/v), as depicted in Fig. 1a, PC1 predominantly separates the extracts of AcCN and $CHCl_3$ from those of aqueous MeOH and MeOH. Conversely, in the scenario of basil samples measured in CD_3OD, as illustrated in Fig. 1b, PC2 primarily distinguishes $CHCl_3$ extracts from the others, notably differing from the pattern observed in the former basil case.
6. Utilizing multivariate data analysis software, calculate the loadings for each bin (bucket) of ^{1}H NMR data corresponding to the specified PCs. Figure 2a presents the loadings of each ^{1}H NMR variable for PC1 of basil samples measured in CD_3OD-KH_2PO_4 buffer (pH 6.0) at a ratio of 1:1 (v/v), while Fig. 2b delineates the loadings for PC2 of lettuce samples measured in CD_3OD.

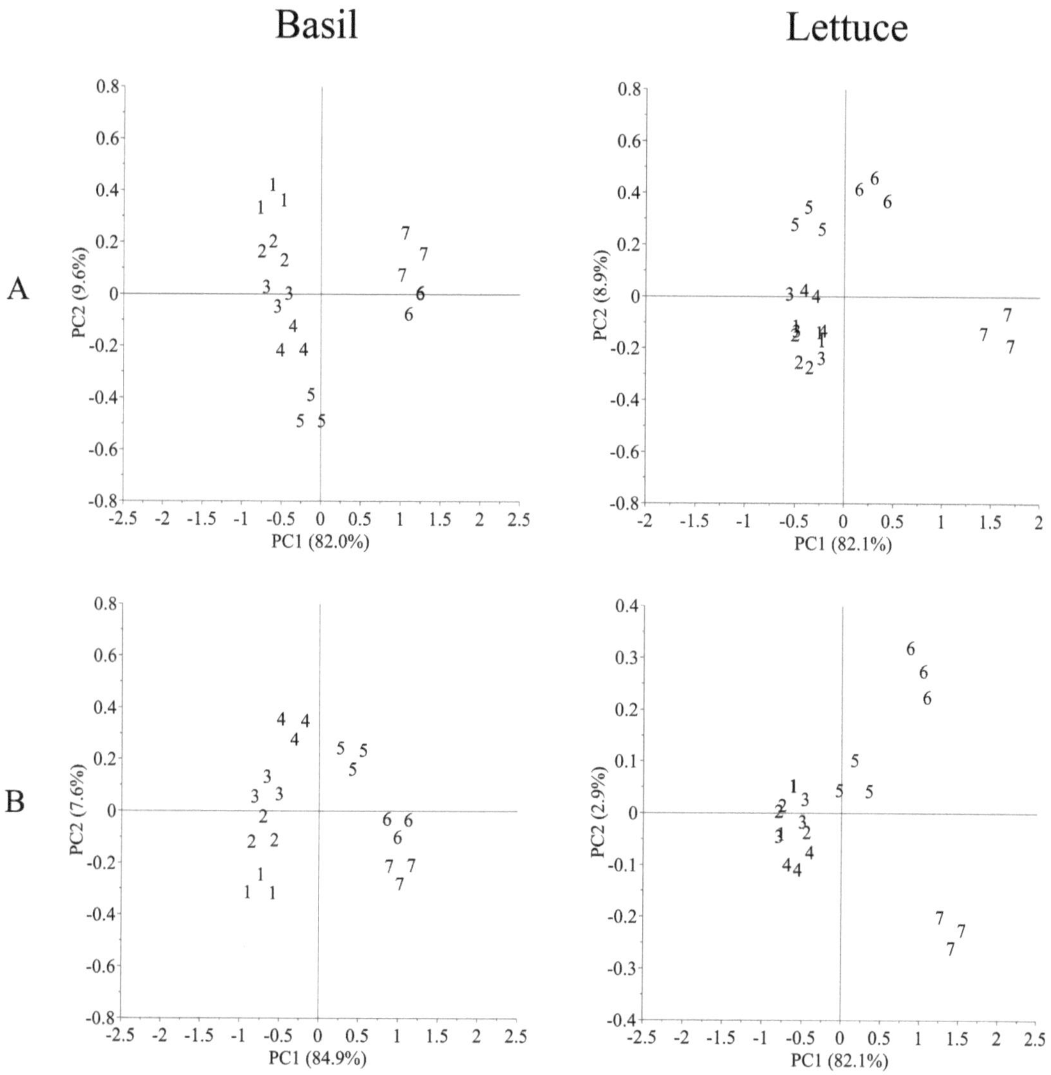

Fig. 1 Score plot [principal component (PC)1/2] of PCA of basil leaves (*Ocimum basilicum*) and lettuce leaves (*Lactura sativa*) based on ^{1}H NMR spectra measured in CD_3OD-KH_2PO_4 buffer (pH 6.0) containing 0.29 mM of TMSP-d_4 (**a**) and CD_3OD containing 0.418 mM of HMDSO (**b**). (*1*) water extracts, (*2*) 20% MeOH extracts, (*3*) 50% MeOH extracts, (*4*) 80% MeOH extracts, (*5*) 100% MeOH extracts, (*6*) AcCN extracts, (*7*) $CHCl_3$ extracts

3.4 Investigation of Loading Plots of Principal Component Analysis to Detect Extracted Metabolites in Each Solvent and Confirmation by ^{1}H NMR Spectra

1. Prepare a comprehensive list outlining the ^{1}H NMR bins (buckets) observed in the loading plots of PCA, presumably indicating heightened levels within each extraction solvent. Identify the specific signals in the original ^{1}H NMR spectra along with pertinent information (*see* **Note 10**).
2. To elucidate the ^{1}H NMR bins (buckets) selected in loading plots, two types of ^{1}H NMR spectra are prepared for each solvent extract, measured in CD_3OD-KH_2PO_4 buffer (pH 6.0) (1:1, v/v) or CD_3OD. Examples are shown in Fig. 3 for basil extracts and Fig. 4 for lettuce extracts.

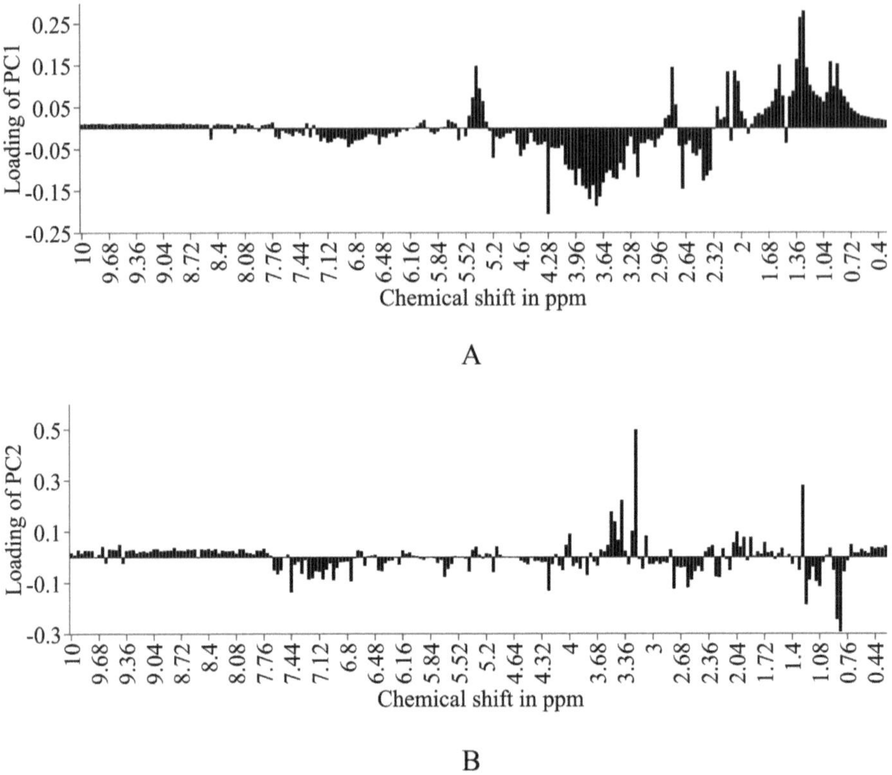

Fig. 2 PC1 loading plot of PCA of ^{1}H NMR spectra of basil leaves measured in CD_3OD-KH_2PO_4 buffer (pH 6.0) containing 0.29 mM of TMSP-d_4 (**a**) and PC2 loading plot of ^{1}H NMR spectra of lettuce samples measured in CD_3OD containing 0.418 mM of HMDSO (**b**)

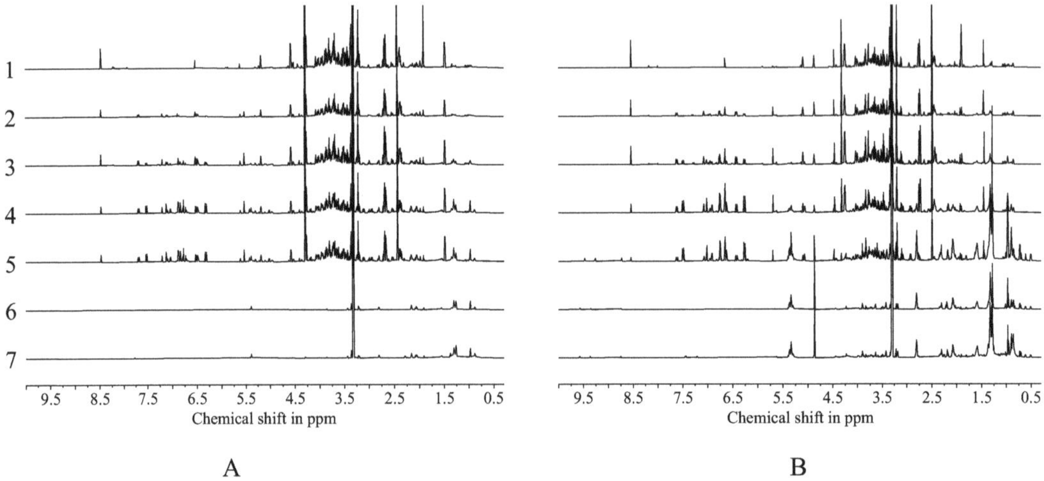

Fig. 3 Typical ^{1}H NMR spectra of basil (*Ocimum basilicum*) leaves extracted by water (*1*), 20% MeOH (*2*), 50% MeOH (*3*), 80% MeOH (*4*), 100% MeOH (*5*), AcCN (*6*), $CHCl_3$ (*7*) measured in CD_3OD-KH_2PO_4 buffer (pH 6.0) containing 0.29 mM of TMSP d_4 (**a**) and CD_3OD containing 0.418 mM of HMDSO (**b**) in the range of δ 0.3–δ 10.0

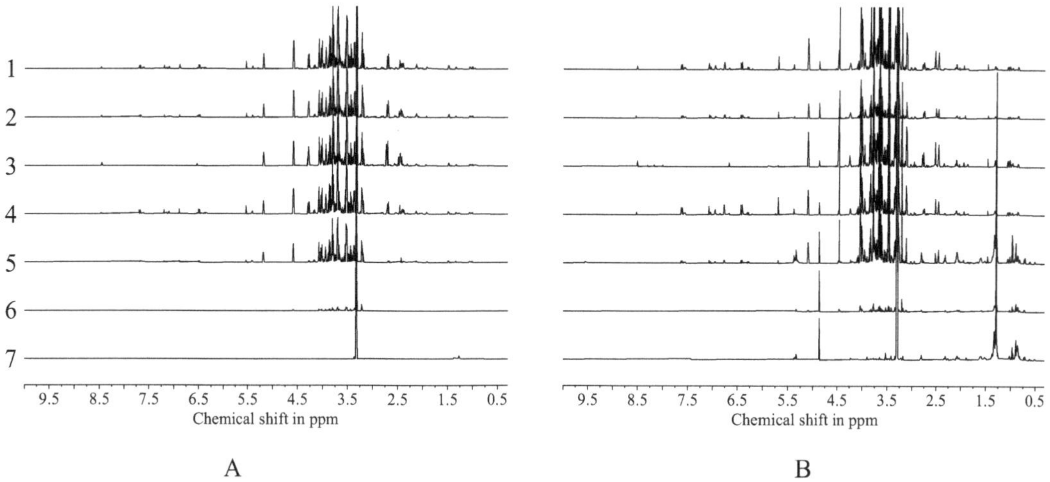

Fig. 4 Typical ^{1}H NMR spectra of lettuce (*Lactura sativa*) extracted by water (*1*), 20% MeOH (*2*), 50% MeOH (*3*), 80% MeOH (*4*), 100% MeOH (*5*), AcCN (*6*), $CHCl_3$ (*7*) measured in CD_3OD-KH_2PO_4 buffer (pH 6.0) containing 0.29 mM of TMSP-d_4 (**a**) and CD_3OD containing 0.418 mM of HMDSO (**b**) in the range of δ 0.3–δ 10.0

3. For the identification of sugars, employ the spectra measured in CD_3OD-KH_2PO_4 buffer (pH 6.0) (1:1, v/v) targeting on anomeric protons in the range of δ 4.5–δ 5.5 of doublet. The chemical shifts and coupling constants of common sugars are listed in the Chapter 16 of this book (*see* **Note 11**).
4. Confirm the presence of organic acids such as malic acid and citric acid within the range of δ 2.5–δ 3.0, distinguished by their characteristic signals. For fumaric acid, acetic acid, and formic acid, a singlet at δ 6.6, δ 1.9, and δ 8.5, respectively, is evident in the spectra of CD_3OD-KH_2PO_4 buffer (pH 6.0). For a comprehensive identification of organic acids, please consult Chapter 16 of this book (*see* **Note 12**).
5. Common amino acids such as leucine, isoleucine, valine, alanine, threonine, methionine, proline, arginine, glutamine, glutamic acid, asparagine, aspartic acid, phenylalanine, tyrosine, and histidine are observed in the spectra of CD_3OD-KH_2PO_4 buffer (pH 6.0) with their characteristic ^{1}H signals. For detailed identification of organic acids, please refer to Chapter 16 of this book.
6. Examine the broad ^{1}H signals around δ 0.96 (t, terminal CH_3) and δ 1.3 (brs, $-CH_2-$) indicative of total lipids, fatty acids, or fatty alcohols in the spectra acquired in CD_3OD, as demonstrated in MeOH extracts of basil (Fig. 5a) and lettuce (Fig. 5b) examples (*see* **Note 13**). For direct comparison of extraction yield, utilize the integral of the signal normalized to the internal standard, HMDSO signal at δ 0.06 in CD_3OD solvent.

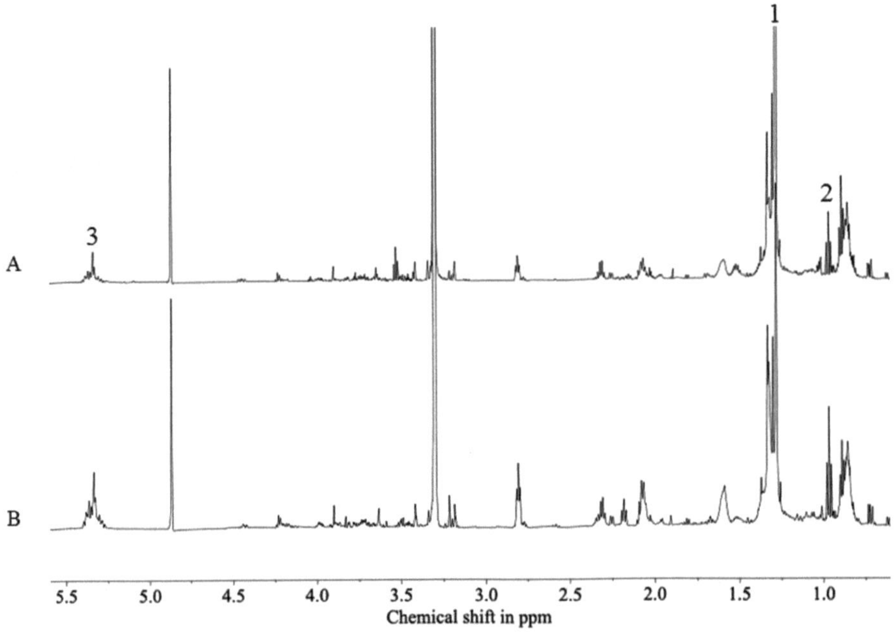

Fig. 5 Typical ^{1}H NMR spectra of MeOH extracts of leaves of basil (**a**) and lettuce (**b**) measure in CD_3OD in the range of δ 5.6–δ 0.6. (*1*) $-CH_2-$ of lipids, fatty acids, and fatty alcohols, (*2*) terminal CH_3 of lipids, fatty acids, and fatty alcohols, (*3*) olefinic protons of lipids, fatty acids, and fatty alcohols

7. Employ the signal of olefinic protons around δ 5.3 (m) to identify unsaturated fatty acids (*see* **Note 13**) (Fig. 5a, b).
8. For phenolics analysis, investigate the region spanning δ 5.8–δ 8.5, where flavonoids, phenylpropanoids, and simple phenolics are detected (*see* **Note 14**). Especially in the range of δ 6.0–δ 6.6 ^{1}H signals of H-6 and H-8 in flavonoids (doublet, $J = 1.8$–2.2 Hz) and H-8 (doublet, $J = 15.8$–16.2 Hz), they are utilized to identify common flavonoids possessing hydroxyl groups at positions H-5 and H-7 and phenylpropanoids, respectively (Fig. 6) (*see* **Note 15**).

4 Notes

1. Ensure all reagents are prepared and stored at room temperature unless explicitly stated otherwise. Ultrapure water was generated by filtering deionized water to attain a sensitivity of 18 MΩ-cm at 25 °C. When discarding waste materials, strictly adhere to all waste disposals.
2. Handling should be performed carefully while wearing gloves and safety glasses.
3. Typically, tetramethylsilane (TMS) serves as the internal standard for organic solvents, while TMSP-d_4 or sodium 2,2

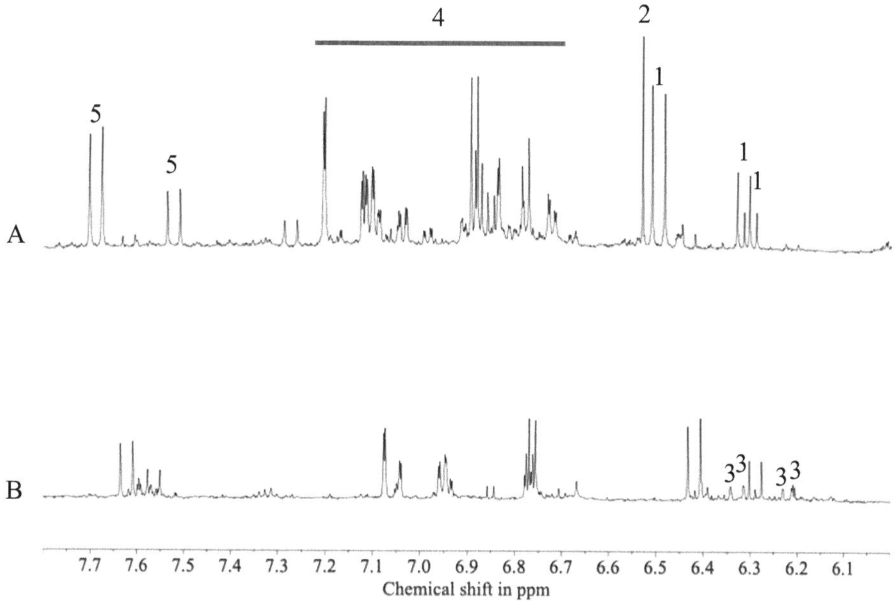

Fig. 6 Typical ^{1}H spectra of basil 50% MeOH extracts in CD_3OD-KH_2PO_4 buffer (pH 6.0) containing 0.29 mM of TMSP-d_4 (**a**) and lettuce 100% MeOH extracts in CD_3OD (**b**) in the range of δ 6.0–δ 7.8. (*1*) H-8 of phenylpropanoids, (*2*) fumaric acid, (*3*) H-6 or H-8 of flavonoids having OHs at C-5 and C-7, (*4*) the region of simple phenolics and phenolic moieties of phenylpropanoids and flavonoids, (*5*) H-7 of phenylpropanoids

dimethyl-2 silapenta-5-sulfonate-d_6 (DSS) are used for water. However, TMS is excessively volatile for quantitation purposes. Hence, HMDSO is chosen as the internal standard for CD_3OD solutions to calibrate chemical shifts and enable quantitation in ^{1}H NMR spectroscopy. HMDSO is less volatile than TMS, yet it exhibits a singlet near 0 ppm (calibrated to 0.06 ppm in this protocol), facilitating accurate measurements.

4. A deep freezer is used only for fresh samples before drying. Dries samples are stored in −4 °C.
5. In the case of $CHCl_3$ extracts, the supernatant is not completely separated from solid materials due to its heavy weight. Therefore, the extract solutions were filtered to obtain 1 mL of clear supernatant.
6. Sample amounts are normalized to the dried weight; 50 mg samples are extracted by 2 mL solvents, and 1 mL of each extract is reconstituted by 1 mL of NMR solvents to result in 25 mg/mL.
7. For comparing the relative quantities of extracted metabolites, it is recommended to utilize total intensity normalization as it diminishes variations between extracts. Regarding internal standard normalization, utilize the range of δ −0.2–δ 0.2 for TMSP-d_4 and δ 0.08 to δ 0.04 for HMDSO.

8. The recommended bucket (bin) range is δ 10.0 to δ 0.3, with a bin (bucket) width of 0.04 ppm. However, if signals are detected beyond this range, the range can be extended. The ranges of δ 3.28–δ 3.24 for residual CD_3OD and δ 5.0–δ 4.7 for residual HDO are excluded from the spectra.
9. Generally, UV and Pareto scaling methods are employed for scaling spectral data. The interpretation of loading plots is typically easier with the Pareto scaling method compared to UV.
10. In loading plots, higher loading values, whether positive or negative, signify a greater contribution to the selected PCs. Specifically, positively higher values in ^1H NMR bins indicate a more pronounced extraction, whereas negatively higher values suggest a diminished extraction.
11. Common plant sugars and sugar alcohols typically detected in the spectra of CD_3OD-KH_2PO_4 buffer (pH 6.0) include glucose, fructose, sucrose, and myo-inositol. However, some minor sugars and sugar alcohols, such as trehalose, maltose, raffinose, stachyose, and mannitol, are frequently observed in specific species. In general, two anomeric protons are detected whenever there is a hydroxyl group at the anomeric positions of aldoses: the α- and β-forms, distinguished by their chemical shifts and coupling constants. For identification of these sugars, please refer to Chapter 16 of this book.
12. Chemical shifts and coupling constants of organic acids vary significantly depending on concentration, even in buffer solvents. In the range of δ 2.5–δ 3.0, compounds such as aspartic acid, asparagine, malic acid, and citric acid are commonly detected. Citric acid is distinguished from the others by having two sets of doublet signals, whereas the rest exhibit two sets of doublet signals. Generally, the chemical shifts of malic acids are more upfield shifted.
13. ^1H signal around δ 0.96 (t, terminal CH_3), δ 1.3 of $-CH_2-$ (bs), and δ 5.3 (m, olefinic protons) of lipids, fatty acids, and fatty alcohols are only detected in CD_3OD because generally lipids, fatty acids, and fatty alcohols are not soluble in CD_3OD-KH_2PO_4 buffer (pH 6.0).
14. To analyze phenolics effectively, both spectra of CD_3OD-KH_2PO_4 buffer (pH 6.0) and CD3OD solvent must be examined. Generally, CD_3OD-KH_2PO_4 buffer (pH 6.0) exhibits a higher presence of glycosides, whereas CD_3OD predominantly displays aglycones.
15. In the range of δ 6.0–δ 6.6, H-6 and H-8 of flavonoids having OHs at C-5 and C-7 and H-8 of phenylpropanoids may overlap. To resolve this spectral region, J-resolved spectroscopy can be employed to discern splitting patterns accurately, with precise coupling constants ($J = 1.8$–2.2 Hz for H-6 and H-8 of

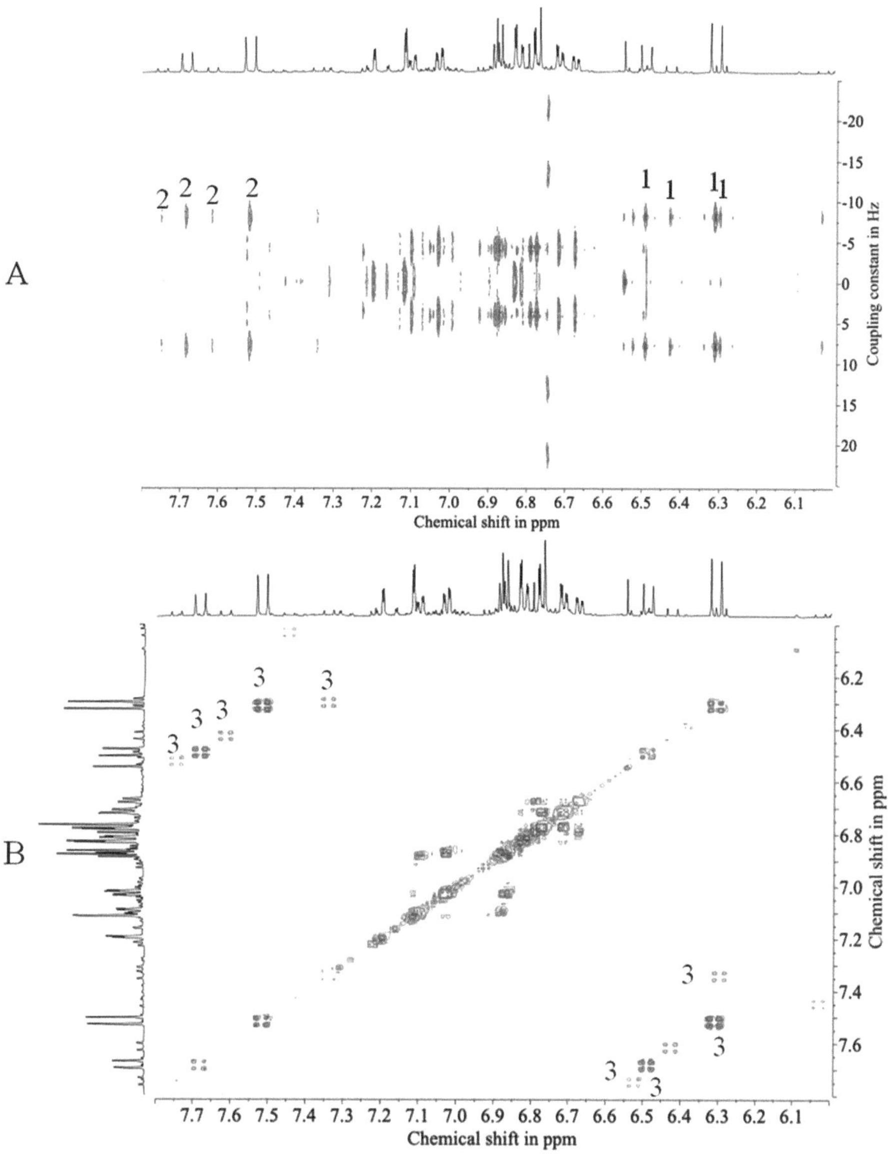

Fig. 7 Typical J-resolved (**a**) and ^{1}H-^{1}H-COSY spectra (**b**) of the extract of basil leaves (CD_3OD-KH_2PO_4 buffer (pH 6.0)) in the phenolic region (δ 6.0–δ 7.8). (*1*) H-8 of phenylpropanoids in J-resolved spectra, (*2*) H-7 of phenylpropanoids in J-resolved spectra, (*3*) cross peaks between H-7 and H-8 in ^{1}H-^{1}H-COSY spectra

the flavonoids, and $J = 15.8$–16.2 Hz for H-8 of phenylpropanoids). Additionally, ^{1}H-^{1}H-COSY spectroscopy can be utilized to detect cross peaks between adjacent protons, such as H-6 and H-8 of the flavonoids, and H-8 with H-7 at δ 7.5–δ 7.7. The examples of application of J-resolved and ^{1}H-^{1}H-COSY to the 50% MeOH extracts are shown in Fig. 7 (basil) and Fig. 8 (lettuce).

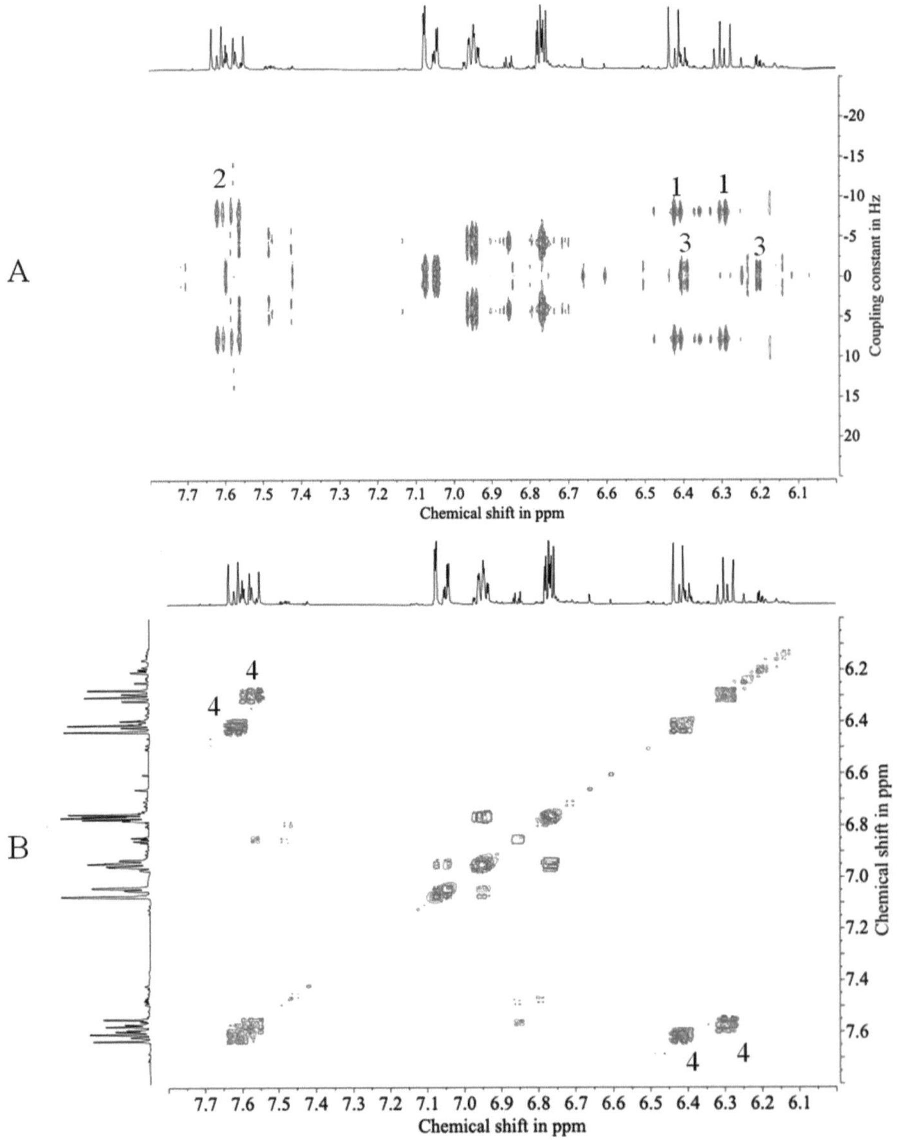

Fig. 8 Typical J-resolved (**a**) and ^{1}H-^{1}H-COSY spectra (**b**) of the extract of lettuce leaves (CD_3OD) in the phenolic region (δ 6.0–δ 7.8). (*1*) H-8 of phenylpropanoids in J-resolved spectra, (*2*) H-7 of phenylpropanoids in J-resolved spectra, (*3*) H-6 or H-8 of flavonoids having OHs at C-5 and C-7, (*4*) cross peaks between H-7 and H-8 in ^{1}H-^{1}H-COSY spectra

Acknowledgement

This research is part of the toegepaste en technische wegenschappen (TTW) Perspectief program "Sky High," which is supported by AMS Institute, Bayer, Bosman van Zaal, Certhon, Fresh Forward, Grodan, Growy, Own Greens/Vitroplus, Priva, Signify, Solynta, Unilever, van Bergen Kolpa Architects, and the Dutch Research Council (NWO).

References

1. Ingle KP et al (2017) Phytochemicals: extraction methods, identification and detection of bioactive compounds from plant extracts. J Pharmacogn Phytochem 6(1):32–36
2. Gutiérrez-Grijalva EP et al (2020) Plant alkaloids: structures and bioactive properties. In: Plant-derived bioactives: chemistry and mode of action. Springer, Singapore, pp 85–117
3. Wang T-Y, Li Q, Bi K-S (2018) Bioactive flavonoids in medicinal plants: structure, activity and biological fate. Asian J Pharm Sci 13(1): 12–23
4. Ludwiczuk A, Skalicka-Woźniak K, Georgiev M (2017) Terpenoids. In: Pharmacognosy. Elsevier, Amsterdam, pp 233–266
5. Khoddami A, Wilkes MA, Roberts TH (2013) Techniques for analysis of plant phenolic compounds. Molecules 18(2):2328–2375
6. Romanik G et al (2007) Techniques of preparing plant material for chromatographic separation and analysis. J Biochem Biophys Methods 70(2):253–261
7. Lefebvre T, Destandau E, Lesellier E (2021) Selective extraction of bioactive compounds from plants using recent extraction techniques: a review. J Chromatogr A 1635:461770
8. Azmir J et al (2013) Techniques for extraction of bioactive compounds from plant materials: a review. J Food Eng 117(4):426–436
9. Chaves JO et al (2020) Extraction of flavonoids from natural sources using modern techniques. Front Chem 8:507887
10. Simmler C et al (2014) Universal quantitative NMR analysis of complex natural samples. Curr Opin Biotechnol 25:51–59
11. Elyashberg M (2015) Identification and structure elucidation by NMR spectroscopy. TrAC Trends Anal Chem 69:88–97
12. Martinez I et al (2003) Destructive and non-destructive analytical techniques for authentication and composition analyses of foodstuffs. Trends Food Sci Technol 14(12): 489–498
13. Kikuchi J (2018) NMR analysis of molecular complexity. In: Experimental approaches of NMR spectroscopy: methodology and application to life science and materials science. Springer, Singapore, pp 461–489
14. Eisenreich W, Bacher A (2007) Advances of high-resolution NMR techniques in the structural and metabolic analysis of plant biochemistry. Phytochemistry 68(22–24):2799–2815
15. Ng ZX, Samsuri SN, Yong PH (2020) The antioxidant index and chemometric analysis of tannin, flavonoid, and total phenolic extracted from medicinal plant foods with the solvents of different polarities. J Food Process Preserv 44(9):e14680
16. Altemimi A et al (2017) Phytochemicals: extraction, isolation, and identification of bioactive compounds from plant extracts. Plan Theory 6(4):42

Chapter 3

High-Resolution Isolation of Natural Products: Efficient Combination of Dryload Injection and HPLC Gradient Transfer

Emerson Ferreira Queiroz, Davy Guillarme, and Jean-Luc Wolfender

Abstract

The isolation of pure compounds from complex extracts is a crucial step in natural products (NPs) research. Historically, this process has been recognized to be slow and laborious. However, significant advancements have been made in isolation methods. Efficient separation conditions can now be efficiently determined at the analytical scale using high- or ultra-high-performance liquid chromatography and transferred to the preparative scale by chromatographic calculations. This ensures consistent selectivity at both scales, offering precise separation predictions. High-resolution chromatographic conditions at the preparative scale can be achieved through optimized sample preparation and dry load sample introduction. Monitoring chromatographic traces issued from ultraviolet (UV), mass spectrometry (MS), and/or universal detectors such as evaporative light-scattering detectors (ELSD) can precisely guide the isolation or trigger the collection of specific NPs with diverse structural scaffolds. These approaches can be applied across different scales depending on the amounts of NPs targeted for isolation. This chapter presents a detailed description of an isolation protocol, using the aforementioned methodologies.

Key words Natural products, High-resolution isolation, Ultra-high-performance liquid chromatography, Gradient transfer, Dry load, Semi-preparative high-performance liquid chromatography

1 Introduction

In the classical bioguided natural product (NP) isolation procedure, the purification is usually carried out by preparative chromatographic methods such as open column chromatography, combined with low, medium, and high-performance liquid chromatography (HPLC) [1, 2]. Rudimentary methods are used to determine the best conditions for the separation at the preparative scale. The fractions obtained through this process are then submitted to biological assays to localize the active compounds. The identity of the compound is determined after the isolation. Although effective, this process must be repeated several times

Young Hae Choi et al. (eds.), *Natural Product Isolation and Identification: Methods and Protocols*, Methods in Molecular Biology, vol. 2895, https://doi.org/10.1007/978-1-0716-4350-1_3,

and is therefore considered long and tedious. Furthermore, it can lead to the identification of known compounds.

Today, this process is evolving, and sophisticated analytical methods are used for metabolite profiling of target extracts before any isolation efforts [3–6]. This procedure is referred to as dereplication, aiming to prevent the rediscovery of already-known bioactive compounds. In standard metabolomics investigations, the comparison of sample sets is achieved through metabolite profiling to identify biomarkers [3, 4]. In both scenarios, the precise identification of the compound of interest can be accomplished within the analytical profiles of the corresponding natural extract. For this purpose, it is essential to be able to reproduce the analytical separation conditions on a preparative scale.

The objective of this chapter is to describe in detail an isolation protocol that shows how to optimize a given extract separation on an analytical scale and then transfer it to a preparative scale without compromising the chromatographic resolution for the isolation of pure NP compounds in a single step.

In contemporary advanced isolation protocols, it is noteworthy that only microgram quantities of pure NPs are typically needed for unequivocal identification, owing to recent advancements in sensitive micro nuclear magnetic resonance (NMR) techniques [7] and X-ray diffraction methods [8, 9]. On the other hand, and despite the development of sensitive pharmacological assays, milligrams of pure NPs are however still typically needed for in-depth in vitro and in vivo bioactivity studies.

Preparative HPLC-UV is certainly the most widely used technique for the isolation of NPs at the μg/mg scale, [1, 2, 10]. The columns are typically packed with small particles (between 5 and 10 μm), in either normal or reversed phase modes, depending on the polarity of the compounds to be separated. The apparatus operates at high pressure and elevated solvent flow rates. Generally, this technique is used to purify fractions obtained by other preparative chromatographic techniques, such as open column chromatography (CC), medium-pressure liquid chromatography (MPLC), or liquid-liquid separation technique (centrifugal partition chromatography—CPC) [11–16]. However, with the improvements in sample treatment and sample injection (dry load injection) [17], it is possible to perform a one-step isolation of the compounds present in a complex matrix such as a natural extract [18–24].

When performing the scale-up from analytical to preparative scale in liquid chromatography (LC), some strict rules must be followed to maintain comparable chromatographic profiles (equivalent selectivity and resolution) at the preparative scale.

First, of all, a preparative LC column should be packed with the identical column packing chemistry that was used for the analytical column before scale-up. This ensures that there will be no change in retention factor and selectivity between the two columns. In the case where the same column chemistry is not available at a larger scale, it is possible to select an "equivalent" column from a different source. To find out the most appropriate preparative column chemistry for a given application, software for the comparison of column selectivity for more than 500 columns can be freely accessed at the US Pharmacopeia website [25]. Besides maintaining the same column chemistry, it is also recommended that the particle size remains similar for both the analytical and preparative scale columns so that the same column efficiency and resolution found for the small column will be duplicated for the large column. However, this rule is not always respected, and sometimes, scale-up from ultra-high-performance liquid chromatography (UHPLC) columns (sub-2 μm particles) to the preparative scale (in general 5 μm packings) can also be performed. To maintain resolution between the two dimensions, the ratio of column length (L) to particle size (d_p) must remain constant. In this context, an interesting approach consists of using a 50 mm, 1.7 μm UHPLC column at the analytical scale to rapidly develop a method, and a 150 mm, 5 μm column at the preparative scale, to maintain the same performance [26]. The latter is then transferred to the preparative scale [27]. Such an approach has for example been used for the separation of metabolites from marine organisms, plants, and biotransformation reactions [18, 19, 28].

In terms of mobile phase, it is also important to keep the same temperature and mobile phase composition (nature of the organic solvent, nature of buffers/additives, pH, etc.) to maintain the same chromatographic profile in analytical and preparative dimensions. However, flow rate, injected volume, and gradient profile must be adjusted based on the column dimensions. Indeed, the mobile phase flow rate must be scaled based on the ratio of column cross-sectional areas at the preparative and analytical scales, if particle size remains constant. If this is not the case, the mobile phase flow rate also must be increased in direct proportion to the inverse of particle size change [29]. To maintain a geometrically equivalent separation (same peak width and peak asymmetry), the injected volume must be scaled in direct proportion with the column volume (taking into account the length and internal diameter) [30]. However, preparative LC often involves mass overload conditions to collect a sufficient amount of purified compounds from a mixture. In such a case, detector and column overload can be observed. UV detector overload can be circumvented by using a short path length flow cell (preparative cell, 1 mm) or by changing the detection wavelength to decrease the absorbance of the compound of interest. Column overload can be handled by finding the conditions where the space between the peak of interest and its nearest neighbors just disappears, corresponding to baseline resolution (touching-peak separation) [31]. Obviously, an injection can be multiplied if the collected amount from one experiment is insufficient, or a larger column I.D. can be used to increase the amount injected while maintaining sufficient resolution. Besides flow rate and injected volume, the gradient profile also must be adjusted. It is indeed important to know that gradient elution is often used for analytical and small-scale preparative LC purification (using columns of 10–30 mm I. D.). This is however less convenient and more expensive for large-scale separations. In gradient mode, the duration of any isocratic step and slope parts have to be scaled in direct proportion with the change in column dead time from analytical to preparative scale [32]. In the case of gradient experiments, the differences in dwell volume (volume between the pump and column inlet, between the analytical and preparative LC instruments) will result in retention time shifts and must be considered during the method transfer [29]. Therefore, dwell volume differences must be compensated by introducing an initial isocratic hold time at the start of the preparative gradient. Calculations can be sometimes hard to perform for inexperienced users, and this is why an Excel tool to find out the geometrically equivalent conditions between analytical LC and preparative LC has been developed [33].

The classical procedure for the sample injection in a semi-preparative HPLC column uses a loop valve, where the sample is solubilized in an organic solvent. Since the solubility of crude natural extracts is limited, a high volume of solvent is required, which significantly compromises the final separation resolution. Moreover, when large amounts of extracts are injected, there is a risk of overpressure in the system [17].

Recently, it was demonstrated that a "dry load" injection can be an option to overcome such problems [17]. This approach can increase the sample amount introduced while maintaining a much better resolution than in the case of liquid injection. This approach was found particularly relevant when the extracts to purify were insoluble in AcCN or MeOH (due to the presence of highly lipophilic compounds) and can therefore not be injected under reverse phase liquid chromatography (RPLC) conditions using the classical loop injection [17]. In addition, the dry load setup was found to be relevant also when investigating relatively polar components such as the ones contained in methanolic plant extracts. Indeed, the significant difference in eluent strength between the sample diluent (pure MeOH) and the mobile phase (mostly constituted of water) leads to strong peak distortion and broadening with the classical loop injection. The dry load approach has been successfully used to purify a wide range of NPs [19–24, 34–37].

The methanolic extract of the leaves of the medicinal plant *Salvia fruticosa* is used as a model to demonstrate the isolation approach presented in Fig. 1. A specific compound at a retention time (Rt) of 19.7 min on the HPLC-UV-ELSD analysis was chosen to showcase the entire target isolation procedure (Fig. 1a). The

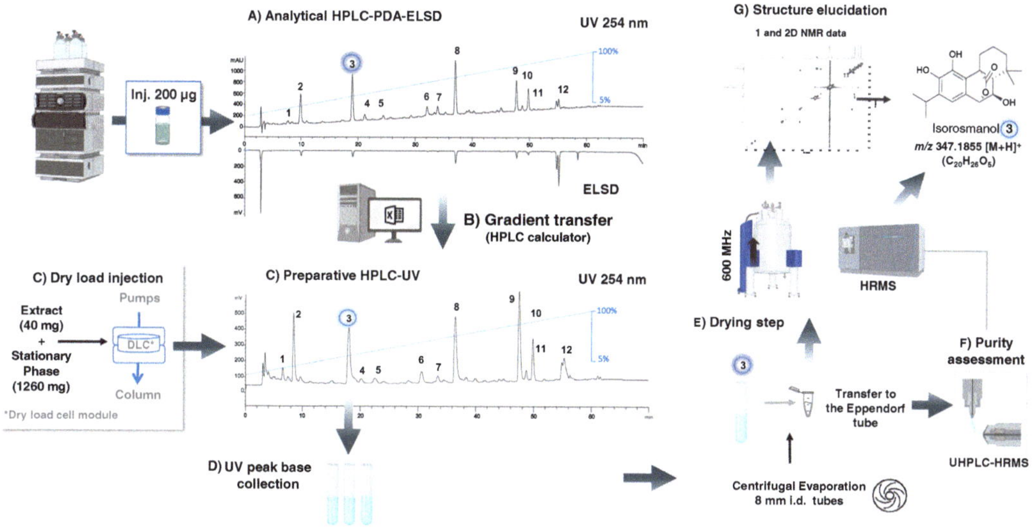

Fig. 1 Workflow used for the one-step targeted isolation of a compound from the methanolic extract of the leaves of *Salvia fruticosa*. (**a**) HPLC-UV-ELSD analysis of the extract (injection of 200 µg) using an analytical HPLC column (250 × 4.6 i.d, mm, 5 µm) at the flow rate of 1 mL/min. (**b**) Calculation of the gradient transfer using the "HPLC calculator" based on the analytical separation and in the dimensions of the preparative HPLC column. (**c**) Preparative HPLC-UV analysis using a column with the same stationary phase (250 × 19 i.d, mm, 5 µm) at 17 mL/min. Sample introduction by dry load to maximize resolution. (**d**) Fraction collection in 8 mm i.d. glass tubes evaporated by speedvac®. (**e**) Transfer to Eppendorfs and drying. (**f**) Purity assessment of the isolated compound by UHPLC-HRMS. (**g**) Solubilization of the sample in dimethylsulfoxide (DMSO)-d_6 (600 µL) and 1D and 2D NMR analysis using a 600 MHz NMR equipped with a cryoprobe. Analysis of an aliquot (10 µL) solubilized in 1 mL of MeOH (HPLC grade) by UHPLC-HRMS to determine its molecular formula. (This figure was adapted with permission from Wolfender et al. [3]. Copyright 2019 Elsevier)

conditions of the metabolites profiling HPLC-UV analysis of the extracts were transferred to the preparative HPLC-UV scale using the HPLC calculator developed at the University of Geneva and available at https://ispso.unige.ch/sciences-analytiques/hplc-calculator (Fig. 1b). The separation at the preparative scale was performed with a column packed with the same chemistry (C_{18} XBridge, Waters, Milford, MA, USA). The column dimensions provided the flow rate and gradient time needed to reproduce the analytical conditions (Fig. 1c). To maintain the resolution, the sample was injected into the column by dry load (Fig. 1c). The UV peaks were individually collected (Time-based peak collection) in 8 mm internal diameter (ID) glass tubes (Fig. 1d). The tube containing compound 3 (the same retention time observed in the analytical HPLC) was dried (Fig. 1e) and subjected to UHPLC-high-resolution mass spectrometry (HRMS) to define the molecular weight of the compound and assess its purity (Fig. 1f). The sample is analyzed by NMR to enable full de novo identification (Fig. 1g). In this case, it has been unambiguously identified as a diterpene, isorosmanol [38]. The same procedure can be repeated on any HPLC peak of interest.

2 Materials

The reader can find below the material required for one-step NP isolation from a complex plant extract. This involves instrument characterization for efficient gradient transfer.

3 Methods

3.1 Measurement of System Dwell Volume

1. The measurements must be performed on both the analytical HPLC and semi-preparative HPLC. Once the system dwell volume is found, this value can be taken as a reference for further method transfer experiments (no need to repeat the dwell volume measurement experiment).
2. Prepare the mobile phase for the HPLC analysis: add 1 mL of acetone in 1 L of HPLC-grade water to obtain a concentration of 0.1% (solvent B). Pure HPLC-grade water will be used as solvent A.
3. Replace the column with a zero dead volume (ZDV) connector, or ideally with a tubing that generates at least 10–20 bar, to ensure that the pump operates correctly.
4. Equilibrate the system with mobile phase A for about 10 min.
5. Run a gradient program from 0% B to 100% B in 20 min at 0.5 mL/min (analytical HPLC) and 3 mL/min (preparative HPLC), followed by 0% B at 20.1 min and equilibration during 1 min at 0% B at the end of the gradient.
6. Monitor the signal using UV detection at 265 nm.
7. The resulting chromatogram is shown in Fig. 2. To identify the midpoint of the gradient ($t_{50\%}$), which corresponds to the time at which the absorbance is half the maximal value, subtract the UV absorbance at the end of the gradient (A_{max}) from the UV

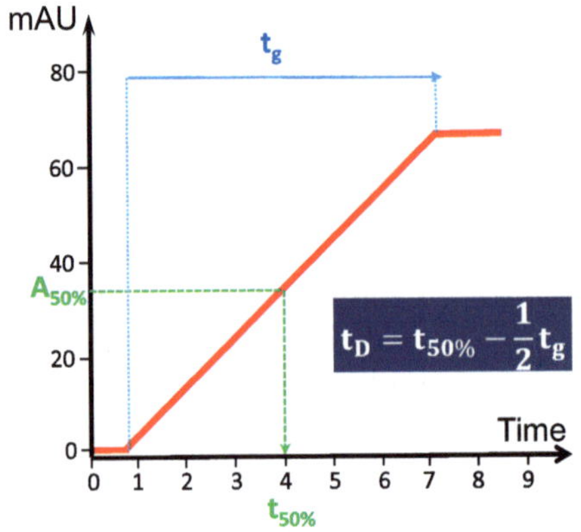

Fig. 2 Example of chromatogram produced to calculate dwell volume

absorbance at the beginning of the gradient, divide this value by 2 and find the corresponding time on the chromatogram.

8. The system dwell time can be calculated from the $t_{50\%}$ simply by using the equation provided in Fig. 2. The dwell time (t_D) corresponds to the difference between the $t_{50\%}$ and half the gradient time ($1/2\ t_G$).
9. Finally, by considering the mobile phase flow rate, dwell time can be easily transformed into dwell volume.
10. The dwell volume value is used to correctly transfer the method from analytical to preparative scale, using the Excel calculator.

3.2 Workflow

The various steps required to isolate a selected compound using a high-resolution chromatographic method are illustrated in Fig. 3. This starts with the plant extraction and characterization of the analytical and preparative instruments, followed by the determination of the best analytical conditions to separate the targeted compounds. The separation conditions are geometric gradient transfer to the preparative HPLC scale using the HPC calculator. The samples are then injected using the dry load approach, and the

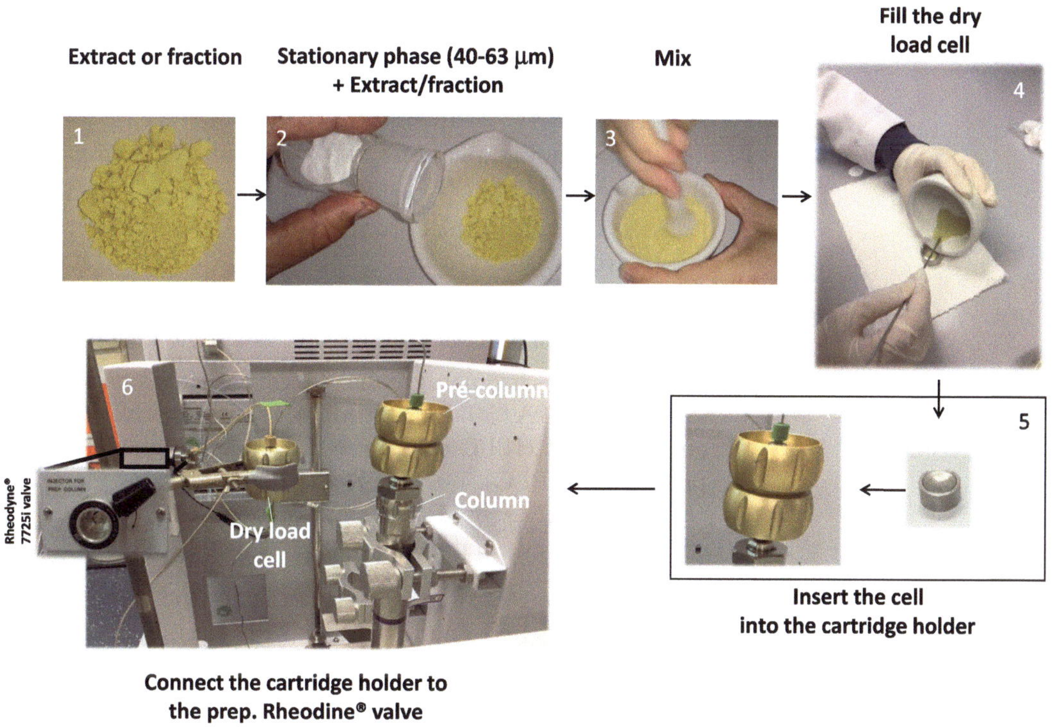

Fig. 3 Scheme illustrating the preparation of the dry load cell. (*1*) to (*3*) the extract or fraction is mixed to the stationary phase (40–63 μm), (*4*) the mixture is placed inside the dry load cell, (*5*) the dry load cell is introduced inside the cartridge holder. (*6*) The cartridge holder is connected to an injection valve (Rheodyne® 7725i) in replacement of the classical injection loop. (This figure was adapted with permission from Queiroz et al. [17]. Copyright 2019 Elsevier)

peak of interest will be collected based on the UV detector. The samples are dried, weighed, and submitted to an NMR and HRMS analysis for their structural elucidation.

3.3 Plant Extraction

1. The ground leaves of *Salvia fruticosa* (10 g) are acquired from Hänseler (Art.09-2741-02), Switzerland.
2. Extract 10 g of the plant material using a Dionex™ ASE™ 350 accelerated solvent extractor.
3. The grounded plant material was introduced into a 34 mL cartridge and the extraction was performed at 40 °C.
4. The plant material is first defatted with hexane (3 × 15 mL), the same plant material is then subsequently extracted with dichloromethane (3 × 15 mL), and finally extracted with MeOH (3 × 15 mL).
5. The dichloromethane and methanolic extracts obtained are dried under pressure by rotary evaporation to yield 440 mg and 1.1 g, respectively.
6. To improve purification yields, the extract can be enriched in compounds of interest. (*see* **Note 1**).

3.4 Measurement of Column Dead Time

1. Measurements must be performed on both the analytical HPLC and preparative HPLC. The column dead time is unique to each column thus, it is crucial to measure the dead time each time a column is changed at any scale.
2. Prepare the mobile phase for the HPLC analysis: Pure water and pure MeOH of HPLC grade are used as the mobile phase.
3. Place the column in the chromatographic system (analytical/preparative LC).
4. Prepare a solution of uracil at 0.1 mg/mL, in a volume of 1 mL, and use a 50% MeOH in water as sample diluent.
5. Equilibrate the chromatographic system with an isocratic composition of 50% A/50% B (50% MeOH) at a flow rate of 1 mL/min (analytical HPLC of 4.6 mm i.d.) and a higher flow rate (preparative HPLC).
6. Launch an injection of uracil (injection volume of 5 μL in analytical HPLC and higher value in preparative HPLC).
7. The elution time of uracil directly provides the column dead time as uracil is non-retained under RPLC conditions using a hydrophobic stationary phase (such as C18).
8. The column dead time will be used in the calculator to correctly transfer the method from the analytical to the preparative scale.

3.5 HPLC Analysis of the Sample

1. In the illustrated example, HPLC-UV analyses are conducted on an HPLC 1260 system (Agilent) equipped with a photodiode array detector and an evaporative light-scattering detector (ELSD). The system was controlled by the LabSolutions software.
2. Prepare the mobile phase for the HPLC analysis: add 1 mL of formic acid (HPLC grade) into 1 L of MeOH to obtain a concentration of 0.1% (solvent B). Add 1 mL of formic acid (HPLC grade) in 1 L of water HPLC grade to similarly obtain a concentration of 0.1% (Solvent A).
3. Install the HPLC column Waters XBridge C18 column (250 × 4.6 mm i.d., 5 μm) (Waters, Milford, MA, USA) equipped with a Waters C18 pre-column cartridge holder (20 × 4.6 mm i.d., 5 μm).
4. Set the UV detector at 254 nm and the UV-PDA spectra between 190 and 600 nm.
5. Set the ELSD detector (Sedere, France) at a temperature of 45 °C, gain of 9, and a nebulization pressure of 3.5 bar.
6. Wash the column with 100% of solvent B for 10 min with a flow rate of 1 mL/min.
7. Equilibrate the column with 5% of solvent B for 10 min with a flow rate of 1 mL/min.
8. Prepare the sample for the HPLC analysis: 10 mg of the methanolic extract should be weighed and solubilized on 1 mL of HPLC-grade MeOH.
9. Inject 20 μL of sample using the injector module.
10. Perform the separation in gradient mode using the following conditions: 5–100% B in 60 min, followed by 10 min of washing at 100% B with a flow rate of 1 mL/min.
11. In the illustrated example this resulted in the HPLC-ELSD-PDA chromatogram obtained is presented in Fig. 1.

3.6 Gradient Transfer of the Conditions to Preparative Scale

The gradient transfer was carried out using the HPLC calculator (Excel spreadsheet) freely available on the University of Geneva website (https://ispso.unige.ch/sciences-analytiques/hplc-calculator).

1. In the first instance, the analytical scale separation is optimized.
2. The gradient conditions, mobile phase flow rate, and injection volume of the original method (analytical scale separation) are introduced into the calculator. The column dead volume and system dwell volume of the analytical system, previously calculated, are indicated in the calculator.
3. The column dead volume and system dwell volume of the preparative system are indicated in the calculator.

4. The calculator automatically calculates the new gradient conditions to use at the geometrically equivalent flow rate under preparative conditions. For this purpose, some basic equations of chromatography are used [32]. The user also has the option to use a different flow rate that can be indicated. In the case of transfer from analytical to preparative scale, the injection volume should not be considered, as dry load injection should be preferred to keep sharp peaks and high resolution.
5. The suggested gradient experiment can then be performed. It will be important to check if the overall separation is maintained (*see* **Note 2**).

3.7 Fractionation of the Sample at Preparative HPLC Preparative Scale

1. In the illustrated example, Semi-preparative HPLC-UV analyses were conducted on a Shimadzu system equipped with LC-20A module pumps, an SPD-20A UV/VIS, a 7725I Rheodyne® valve, and a FRC-10A fraction collector (Shimadzu, Kyoto, Japan). The system was controlled by the LabSolutions software from Shimadzu.
2. Prepare the mobile phase for the HPLC analysis: add 1 mL of formic acid (HPLC grade) in 1 L of MeOH to obtain a concentration of 0.1% (solvent B). Add 1 mL of formic acid (HPLC grade) in 1 L of HPLC water grade to similarly obtain a concentration of 0.1% (Solvent A).
3. Install the HPLC column of the semi-preparative HPLC instrument: Waters XBridge C18 column (250 × 19 mm i.d., 5 μm) equipped with a Waters C18 pre-column cartridge holder (10 × 19 mm i.d.).
4. Mix 40 mg of the methanolic extract of the leaves of *Salvia fruticosa* with the same amount of stationary phase (C_{18} Zeoprep® 40–63 μm), transfer this mixture to the dry load cell (modified pre-column) as a thin layer in the bottom of the cell. The large size of particles used in the dry load (40–63 μm) cartridge reduces the risk of system backpressure (Fig. 4).
5. After this step, this thin layer is covered with the stationary phase until reaching a final mass of 1300 mg (capacity of the cell). Then, the metallic insert initially removed, is used to close the cell. The cell is introduced on the cartridge holder.
6. The cartridge containing the cell with the sample is directly connected to the Rheodyne® valve and replaced with the loop position between the pump and the column inlet [17].
7. Install 8 mm i.d. glass tubes on the fraction collector.
8. Wash the column with 100% of solvent B for 10 min with a flow rate of 17 mL/min.
9. Equilibrate the column with 5% of solvent B for 15 min with a flow rate of 17 mL/min.

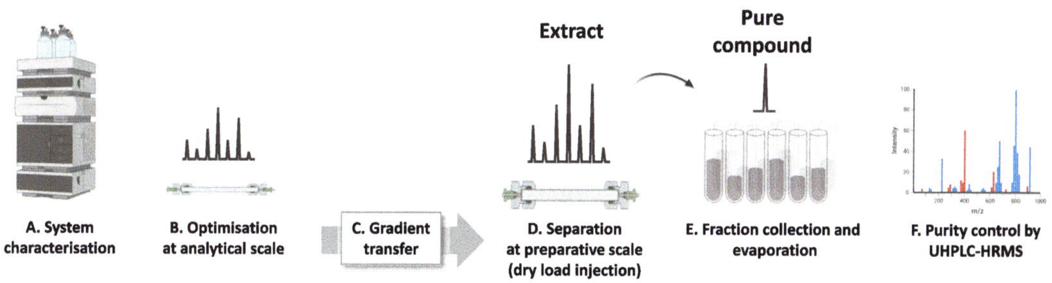

Fig. 4 Workflow presenting the process for the isolation of a NP using high-resolution chromatography. (**a**) Characterization of the analytical and preparative HPLC, (**b**) HPLC analysis of the sample, (**c**) Gradient transfer of the analytical conditions to the semi-preparative HPLC, (**d**) Separation at the semi-preparative scale using the dry load injection, (**e**) Fraction collection and evaporation, (**f**) Purity control using UHPLC-HRMS

10. Inject the sample using the Rheodyne® valve (*see* **Note 3**).
11. Trigger the peak collection using the UV trace at 254 nm and the retention time. Thanks to the gradient transfer method, the compound of interest presents the same retention time observed on an analytical scale. If an ELSD detector is used, a split system must be installed (*see* **Note 4**).
12. Wash the column after the purification time with 100% of solvent B for 10 min.
13. The semi-preparative HPLC-UV chromatogram obtained is presented in Fig. 1.

3.8 Fraction Collection and Evaporation

1. In the illustrated example, the UV peaks collected on the 8 mm i.d. glass tubes were transferred to the Speedvac® (EZ-2 Elite Genevac®, Stone Ridge, NY, United States) instrument where they are dried using the following program: step 1: 40 °C, 40 mbar for 1 h; step 2: 40 °C, 8 mbar for 8 h.
2. Transfer the volume of the sample to a pre-weighed Eppendorf tube. Re-dry the sample on the speedvac® instrument (step 1: 40 °C, 40 mbar for 2 h). Weigh the Eppendorf tube and calculate the purification yields.
3. The separation yielded 3.2 mg of compound 3.

3.9 Purity Assessment by UHPLC-UV-MS

1. Solubilize the samples in 1 mL of HPLC-grade MeOH. Take 10 μL of this solution and solubilize in MS grade MeOH for UHPLC-MS to check the purity of the compound. Transfer the remaining volume of the sample to a pre-weighed Eppendorf tube. Re-dry the sample on a speedvac instrument.
2. Prepare the mobile phase for the HPLC analysis: add 1 mL of formic acid (HPLC-MS grade) in 1 L of MeOH to obtain a concentration of 0.1% (solvent B). Add 1 mL of formic acid (HPLC-MS grade) in 1 L of HPLC-grade water to similarly obtain a concentration of 0.1% (Solvent A).

3. The analysis was performed on a Waters Acquity UHPLC system equipped with a Q-Exactive Focus mass spectrometer (Thermo Scientific, Bremen, Germany), using a heated electrospray ionization source (HESI-II).
4. Install the Acquity ultra-performance liquid chromatography (UPLC) ethylene bridged hybrid (BEH) C_{18} column (50 × 2.1 mm i.d., 1.7 μm; Waters).
5. Wash the column for 5 min with 100% of solvent B with a flow rate of 0.6 mL/min at 40 °C.
6. Equilibrate the column for 5 min with 100% of solvent B with a flow rate of 0.6 mL/min at 40 °C.
7. The gradient was carried out as follows: 5%–100% B in 7 min, 1 min at 100% B, and an equilibration step at 5% B in 2 min.
8. Inject the sample (1 μL) at a concentration of 100 μg/mL.
9. The MS ionization parameters were the same as the ones used by Rutz et al. [39].

3.10 NMR and HRMS Analysis

1. For the NMR and HRMS analysis of the isolated compound: Solubilize the sample on 600 μL of deuterated DMSO-d_6 for the NMR analysis. Take 10 μL of this solution and complete it to 1 mL for UHPLC-HRMS analysis.

4 Notes

1. To improve isolation yields, plant extracts should be treated to remove primary metabolites such as saccharides from polar extracts (such as aqueous, methanolic, or ethanolic) and lipids from apolar extracts (such as ethyl acetate or dichloromethane) [40]. In the case of isolation from a microbial solid culture, the compounds used for microbial growth (culture medium) should be removed before the isolation step [34, 36, 37].
2. Check if the gradient transfer works by using a standard (e.g., a commercially available NP of similar polarity of the target compounds).
3. Inject around 5 mg of the extract on a preparative scale (250 × 19 mm i.d. column) to check if the gradient transfer is performed correctly before injecting a large quantity of extract. This is particularly important when only a few amounts of extract are available.
4. If an ELSD detector is used on a preparative scale, optimize the split flow rate to avoid signal overload when a large quantity of extract is injected. Signal saturation could make it difficult to collect peaks.

Acknowledgements

The authors acknowledge Biorender (https://app.biorender.com/) for the icons used in the figures.

References

1. Bucar F, Wube A, Schmid M (2013) Natural product isolation – how to get from biological material to pure compounds. Nat Prod Rep 30(4):525–545. https://doi.org/10.1039/c3np20106f
2. Sticher O (2008) Natural product isolation. Nat Prod Rep 25(3):517–554. https://doi.org/10.1039/b700306b
3. Wolfender JL, Litaudon M, Touboul D, Queiroz EF (2019) Innovative omics-based approaches for prioritisation and targeted isolation of natural products – new strategies for drug discovery. Nat Prod Rep 36(6):855–868. https://doi.org/10.1039/c9np00004f
4. Bayona LM, de Voogd NJ, Choi YH (2022) Metabolomics on the study of marine organisms. Metabolomics 18(3). https://doi.org/10.1007/s11306-022-01874-y
5. Kind T, Fiehn O (2017) Strategies for dereplication of natural compounds using high-resolution tandem mass spectrometry. Phytochem Lett 21:313–319. https://doi.org/10.1016/j.phytol.2016.11.006
6. Li D, Gaquerel E (2021) Next-generation mass spectrometry metabolomics revives the functional analysis of plant metabolic diversity. Annu Rev Plant Biol 72:867–891. https://doi.org/10.1146/annurev-arplant-071720-114836
7. Emwas AH, Roy R, McKay RT, Tenori L, Saccenti E, Gowda GAN et al (2019) NMR spectroscopy for metabolomics research. Metabolites 9:7. https://doi.org/10.3390/metabo9070123
8. Gruene T, Wennmacher JTC, Zaubitzer C, Holstein JJ, Heidler J, Fecteau-Lefebvre A et al (2018) Rapid structure determination of microcrystalline molecular compounds using electron diffraction. Angew Chem Int Edit 57(50):16313–16317. https://doi.org/10.1002/anie.201811318
9. Kunde T, Schmidt BM (2019) Microcrystal electron diffraction (microed) for small-molecule structure determination. Angew Chem Int Edit 58(3):666–668. https://doi.org/10.1002/anie.201813215
10. Atanasov AG, Waltenberger B, Pferschy-Wenzig EM, Linder T, Wawrosch C, Uhrin P et al (2015) Discovery and resupply of pharmacologically active plant-derived natural products: a review. Biotechnol Adv 33(8):1582–1614. https://doi.org/10.1016/j.biotechadv.2015.08.001
11. Human C, Danton O, De Beer D, Maruyama T, Alexander L, Malherbe C et al (2021) Identification of a novel di-C-glycosyl dihydrochalcone and the thermal stability of polyphenols in model ready-to-drink beverage solutions with *Cyclopia subternata* extract as functional ingredient. Food Chem 351. https://doi.org/10.1016/j.foodchem.2021.129273
12. Keller M, Fankhauser S, Giezendanner N, Konig M, Keresztes F, Danton O et al (2021) Saponins from saffron corms inhibit the gene expression and secretion of pro-inflammatory cytokines. J Nat Prod 84(3):630–645. https://doi.org/10.1021/acs.jnatprod.0c01220
13. Jing SX, Alania Y, Reis M, McAlpine JB, Chen SN, Bedran-Russo AK, Pauli GF (2022) Proanthocyanidin tetramers and pentamers from *Cinnamomum verum* bark and their dentin biomodification bioactivities. J Nat Prod 85(2):391–404. https://doi.org/10.1021/acs.jnatprod.1c00972
14. Zhou B, Alania Y, Reis M, Jing SX, McAlpine JB, Bedran-Russo AK et al (2022) Seco B-type oligomers from *Pinus massoniana* expand the procyanidin chemical space and exhibit dental bioactivity. J Nat Prod 85(12):2753–2768. https://doi.org/10.1021/acs.jnatprod.2c00664
15. Angelis A, Mavros P, Nikolaou PE, Mitakou S, Halabalaki M, Skaltsounis L (2020) Phytochemical analysis of olive flowers' hydroalcoholic extract and in vitro evaluation of tyrosinase, elastase and collagenase inhibition activity. Fitoterapia 143:104602. https://doi.org/10.1016/j.fitote.2020.104602
16. Darme P, Escotte-Binet S, Cordonnier J, Remy S, Hubert J, Sayagh C et al (2022) Anti-toxoplasma gondii effect of lupane-type triterpenes from the bark of black alder (*Alnus glutinosa*) and identification of a potential target by reverse docking. Parasite 29(7). https://doi.org/10.1051/parasite/2022008

17. Queiroz EF, Alfattani A, Afzan A, Marcourt L, Guillarme D, Wolfender JL (2019) Utility of dry load injection for an efficient natural products isolation at the semi-preparative chromatographic scale. J Chromatogr A 1598:85–91. https://doi.org/10.1016/j.chroma.2019.03.042
18. Righi D, Huber R, Koval A, Marcourt L, Schnee S, Le Floch A et al (2020) Generation of stilbene antimicrobials against multiresistant strains of *Staphylococcus aureus* through biotransformation by the enzymatic secretome of *Botrytis cinerea*. J Nat Prod 83(8):2347–2356. https://doi.org/10.1021/acs.jnatprod.0c00071
19. Klimenko A, Huber R, Marcourt L, Chardonnens E, Koval A, Khotimchenko YS et al (2021) A cytotoxic porphyrin from north pacific brittle star *Ophiura sarsii*. Mar Drugs 19(1). https://doi.org/10.3390/md19010011
20. Huber R, Marcourt L, Koval A, Schnee S, Righi D, Michellod E et al (2021) Chemoenzymatic synthesis of complex phenylpropanoid derivatives by the *Botrytis cinerea* secretome and evaluation of their wnt inhibition activity. Front Plant Sci 12:805610. https://doi.org/10.3389/fpls.2021.805610
21. Huber R, Koval A, Marcourt L, Heritier M, Schnee S, Michellod E et al (2022) Chemoenzymatic synthesis of original stilbene dimers possessing wnt inhibition activity in triple-negative breast cancer cells using the enzymatic secretome of *Botrytis cinerea* Pers. Front Chem 10. https://doi.org/10.3389/fchem.2022.881298
22. Klimenko A, Huber R, Marcourt L, Tabakaev D, Koval A, Dautov SS et al (2023) Shallow- and deep-water Ophiura species produce a panel of chlorin compounds with potent photodynamic anticancer activities. Antioxidants 12(2). https://doi.org/10.3390/antiox12020386
23. Ferreira Queiroz MM, Huber R, Marcourt L, Guénée L, Allard PM, Rutz A et al (2023) Phytochemical study of *Alzatea verticillata*, the sole species belonging to the Alzateaceae family. Front Nat Prod 2. https://doi.org/10.3389/fntpr.2023.1147195
24. Huber R, Marcourt L, Heritier M, Luscher A, Guebey L, Schnee S et al (2023) Generation of potent antibacterial compounds through enzymatic and chemical modifications of the trans-delta-viniferin scaffold. Sci Rep 13:15986. https://doi.org/10.1038/s41598-023-43000-5
25. USP: chromatographic database (2021) https://www.usp.org/resources/chromatographic-columns. Accessed 16 Mar 2021
26. Waters: method migration from UPLC technology to preparative HPLC (2021) https://www.waters.com/waters/library.htm?cid=511436&lid=1512004&locale=en_US. Accessed 16 Mar 2021
27. Novakova L, Veuthey JL, Guillarme D (2011) Practical method transfer from high performance liquid chromatography to ultra-high performance liquid chromatography: the importance of frictional heating. J Chromatogr A 1218(44):7971–7981. https://doi.org/10.1016/j.chroma.2011.08.096
28. Brillatz T, Kubo M, Takahashi S, Jozukuri N, Takechi K, Queiroz EF et al (2020) Metabolite profiling of javanese ginger *Zingiber purpureum* and identification of antiseizure metabolites via a low-cost open-source zebrafish bioassay-guided isolation. J Agric Food Chem 68(30):7904–7915. https://doi.org/10.1021/acs.jafc.0c02641
29. Agilent: aspects of preparative liquid chromatography 5994-1016EN (2019) https://www.agilent.com/cs/library/primers/public/primer-preparative-liquid-chromatography-5994-1016EN-agilent.pdf. Accessed 16 Mar 2021
30. Guillarme D, Nguyen DTT, Rudaz S, Veuthey JL (2007) Method transfer for fast liquid chromatography in pharmaceutical analysis: application to short columns packed with small particle. Part I: isocratic separation. Eur J Pharm Biopharm 66(3):475–482. https://doi.org/10.1016/j.ejpb.2006.11.027
31. Snyder LR, Dolan JW (2007) High-performance gradient elution: the practical application of the linear-solvent-strength model. Wiley, Hoboken
32. Guillarme D, Nguyen DTT, Rudaz S, Veuthey JL (2008) Method transfer for fast liquid chromatography in pharmaceutical analysis: application to short columns packed with small particle. Part II: gradient experiments. Eur J Pharm Biopharm 68(2):430–440. https://doi.org/10.1016/j.ejpb.2007.06.018
33. Guillarme D (2020) HPLC calculator: software for chromatographic performance evaluation and HPLC method transfer. https://ispso.unige.ch/labs/fanal/hplc_calculator:en. Accessed 16 Mar 2021
34. Pellissier L, Koval A, Marcourt L, Ferreira Queiroz E, Lecoultre N, Leoni S et al (2021) Isolation and identification of isocoumarin derivatives with specific inhibitory activity

against wnt pathway and metabolome characterization of *Lasiodiplodia venezuelensis*. Front Chem 9. https://doi.org/10.3389/fchem.2021.664489

35. Quiros-Guerrero LM, Nothias LF, Gaudry A, Marcourt L, Allard PM, Rutz A et al (2022) Inventa: a computational tool to discover structural novelty in natural extracts libraries. Front Mol Biosci 9. https://doi.org/10.3389/fmolb.2022.1028334
36. Alfattani A, Marcourt L, Hofstetter V, Queiroz EF, Leoni S, Allard PM et al (2021) Combination of pseudo-LC-NMR and HRMS/MS-based molecular networking for the rapid identification of antimicrobial metabolites from *Fusarium petroliphilum*. Front Mol Biosci 8: 725691. https://doi.org/10.3389/fmolb.2021.725691
37. Alfattani A, Queiroz EF, Marcourt L, Leoni S, Stien D, Hofstetter V et al (2023) One step bio-guided isolation of secondary metabolites from the endophytic fungus *Penicillium crustosum* using high-resolution semi-preparative HPLC. Comb Chem High Throughput Screen. https://doi.org/10.2174/1386207326666230707110651
38. Sallam A, Mira A, Ashour A, Shimizu K (2016) Acetylcholine esterase inhibitors and melanin synthesis inhibitors from. Phytomedicine 23(10):1005–1011. https://doi.org/10.1016/j.phymed.2016.06.014
39. Rutz A, Dounoue-Kubo M, Ollivier S, Bisson J, Bagheri M, Saesong T et al (2019) Taxonomically informed scoring enhances confidence in natural products annotation. Front Plant Sci 10:1329. https://doi.org/10.3389/fpls.2019.01329
40. Houriet J, Allard PM, Queiroz EF, Marcourt L, Gaudry A, Vallin L et al (2020) A mass spectrometry based metabolite profiling workflow for selecting abundant specific markers and their structurally related multi-component signatures in traditional chinese medicine multi-herb formulae. Front Pharmacol 11. https://doi.org/10.3389/fphar.2020.578346

Chapter 4

Standardized HPTLC for Reproducible Chemical Fingerprinting

Thi Kieu Tiên Do, Ilona Trettin, and Eike Reich

Abstract

High-performance thin-layer chromatography (HPTLC) plays a crucial role in establishing the chemical fingerprint of natural products (NPs) during analysis. This technique involves standardized procedures for each step, ensuring the reproducibility of results. In this context, we present a comprehensive standard operating procedure (SOP) encompassing both manual instruments and suitable devices. The methods detailed herein adopt a non-targeted screening approach applicable to all unrelated samples, complemented by a two-step derivatization process capable of detecting a wide range of phytochemical classes. By following these procedures, researchers can obtain reliable and consistent results that could be used to build a database for NPs using HPTLC.

Key words Standardization, Complementary developing solvent (CDS), Universal HPTLC mixture (UHM), Fingerprinting, Standard operating procedure

1 Introduction

High-performance thin-layer chromatography (HPTLC) is widely recognized as a valuable technique in quality control for confirming the identity of plants and plant-derived products and quantifying chemical compounds [1, 2]. It involves a step-by-step offline process on an open-system plate, in which various factors can affect the analytical outcome. To ensure reproducibility and compliance with quality control standards, standardization of all parameters using a standard operating procedure (SOP) was proposed by the HPTLC Association [3].

This SOP served as the foundation for the incorporation of HPTLC as a product-neutral concept in the general chapters <203> of the United States Pharmacopoeia (USP) (2015) [4], 2.8.25 of the European Pharmacopoeia (Ph. Eur.) (2017) [5], and 2.4.17 of the Indian Pharmacopoeia (IP) (2022) [6]. As part of the SOP and as a crucial element of the validation process, a system

Young Hae Choi et al. (eds.), *Natural Product Isolation and Identification: Methods and Protocols*, Methods in Molecular Biology, vol. 2895, https://doi.org/10.1007/978-1-0716-4350-1_4,

suitability test (SST) is conducted using the universal HPTLC mixture (UHM) [7, 8] applied to each plate. This ensures that the chromatographic system is suitable for the intended analysis, enabling the results to be utilized in the subsequent evaluation step.

For researchers considering HPTLC outside of quality control, the initial steps and method selection may pose challenges. In response, we propose a simple and straightforward general approach tailored for non-targeted screening, wherein all unrelated samples are analyzed together. This approach involves the parallel use of three complementary developing solvents (CDS) with varying polarity and selectivity: low-polarity developing solvent (LPDS), medium-polarity developing solvent (MPDS), and high-polarity developing solvent (HPDS) [9]. In the realm of natural product (NP) research, CDS provides an excellent solution for separating individual molecules for subsequent identification, comparing diverse samples to discern similarities and differences, and performing pattern recognition. The resulting extensive datasets are then available for further multidimensional data analysis.

The following chapter describes the standardized HPTLC protocol following the SOP and using different levels of instrumentation ranging from manual to automated and fully automated systems. Additionally, it proposes various data treatments to enhance the analytical process.

2 Materials

Prepare all solutions using at least deionized water and analytical grade solvents and reagents. For cyclopentyl methyl ether, the solvent should contain 50 ppm BHT as inhibitor. Prepare and store (maximum 5 days) all developing solvents at room temperature and derivatization reagents in the fridge at 4 °C. Diligently follow all waste disposal regulations when disposing of waste materials. When not using manual instruments, employ appropriate devices and instruments capable of controlling and reproducing each HPTLC step. This may involve using an automated system with components such as an automatic TLC sampler, an automatic developing chamber, a visualizer, a TLC derivatizer, software, or a fully automatic system that eliminates environmental and human factors of influence.

2.1 Test and Standard Solutions

1. Test solution: 100 mg of a powdered botanical ingredient, 10 mg of a dry extract, or the amount of a dosage form containing the equivalent of the aforementioned quantities of the botanical ingredient are sonicated for 10 min with 1.0 mL of methanol. After filtration or centrifugation, the filtrate or supernatant is used as the test solution. For essential oils,

50 μL/mL solution in toluene is used as the test solution (*see* **Note 1**).

2. Standard preparation: reference standards of individual marker compounds are dissolved in methanol at a concentration of 1.0 mg/mL. The reference standard extracts are shaken and sonicated in methanol at a concentration of 10 mg/mL or, for essential oils, in toluene at 50 μL/mL.
3. Universal HPTLC Mixture (UHM) used as System Suitability Test (SST) [8]: prepared the UHM using 0.01 mg/mL for thioxanthen-9-one (CAS 492-22-8; ≥97%), 0.5 mg/mL for guanosine (CAS 118-00-3; ≥98%), 1.0 mg/mL for sulisobenzone (CAS 4065-45-6; ≥97%), thymidine (CAS 50-89-5; ≥99%), paracetamol (CAS 103-90-2; 98%), 9-hydroxyfluorene (CAS 1689-64-1; 96%) and 2-(2H-benzotriazol-2-yl)-4-(1,1,3,3-tetramethylbutyl)phenol (CAS 3147-75-9; 98%), and 2.0 mg/mL for phthalimide (CAS 85-41-6; ≥99%) in methanol (*see* **Note 2**).

2.2 Stationary Phase

Plate: use 20 × 10 cm glass plates coated with a uniform 200-μm layer of porous (60-Å pore size) silica gel having irregular particles of 2–10 μm and an average particle size of 5 μm, a polymeric binder, and a fluorescence indicator (F_{254}) (*see* **Note 3**).

2.3 Activation

1. Use a saturated salt solution in a desiccator or a suitable device for controlling the activity of the stationary phase by preconditioning the plate to a prescribed relative humidity after the samples have been applied.
2. Magnesium chloride salt solution for achieving a relative humidity of 33%: for manual instrumentation, fill the 500 mL humidity bottle with 200 mL of deionized water and mark the water level. Poor out the water and dry the bottle. Add 225 g of $MgCl_2$ into the humidity bottle. Add 50 mL of deionized water and heat it up to 40–50 °C under stirring. Add more deionized water to reach the 200 mL line that was marked earlier. Heat again at 40–50 °C for 10 min. Cool down the humidity bottle to room temperature. The final solution should contain about 2.5 cm of salt crystals on the bottom to ensure the saturation of the salt solution. Store at room temperature (*see* **Note 4**).

2.4 Development

1. Developing solvents: DS1: prepare 50.0 mL of solution containing 5.0 mL of ethyl acetate and 45.0 mL of toluene. DS2: prepare 66.0 mL of a solution containing 1.0 mL of formic acid, 40.0 mL of cyclopentyl methyl ether, 24.0 mL of tetrahydrofuran, and 1.0 mL of water. DS3: prepare 37.0 mL of a solution containing 16.0 mL of dichloromethane, 16.0 mL of ethanol, 1.0 mL of formic acid, and 4.0 mL of water. Store at room temperature.

2. Use a twin trough chamber (TTC) with a cover fitted and a saturation pad (for plates 2 and 3) for development to a specified distance (*see* **Note 5**), or a suitable device controlling the activation, the saturation, and the developing distance.
3. Use a hair dryer or a device suitable for reproducible and homogeneous drying of the developed plate (*see* **Note 6**).

2.5 Derivatization

1. Natural Product (NP) reagent: dissolve 1.0 g of 2-aminoethyl diphenylborinate in 100.0 mL of methanol. Store at 4 °C.
2. Anisaldehyde sulfuric acid (AS) reagent: slowly and carefully mix 170.0 mL of ice-cooled methanol with 20.0 mL of acetic acid and 10.0 mL of sulfuric acid. Allow the mixture to cool to room temperature, then add 1.0 mL of anisaldehyde (*p*-methoxy benzaldehyde). Store at 4°C.
3. Use a TLC sprayer for manual spraying or a device suitable for the treatment of the plate with derivatization reagents.
4. Use a plate heater to heat the plate prior to and/or after derivatization.

2.6 Documentation

1. Use a device suitable for illumination under shortwave UV (254 nm), longwave UV (350 nm broadband), and white light.
2. Utilize a suitable device to save all data with traceability.

3 Methods

3.1 Plate Layout

1. Open the box of plates, place the top plate back into the stack, face down, using it as protection. Retrieve the next three plates from the stack.
2. Place each plate with silica gel side facing up on, making sure that the short white edge of the plate (with silica) points to the right.
3. Label each plate in the upper right corner (*see* **Note 7**).

3.1.1 For Manual Instrumentation

1. With a ruler and a pencil, mark the application position (8.0 mm from the long, lower edge) and the expected position of the developing front (70.0 mm from the lower edge) on each plate.
2. Use the respective template in Fig. 1 to mark the plate or program a suitable application device.

3.1.2 For Automated Instrumentation

1. Samples are automatically sprayed-on as narrow bands of 8.0 mm length at a distance of 8.0 mm from the lower edge of the plate and the center of the nearest track edge at 20.0 mm

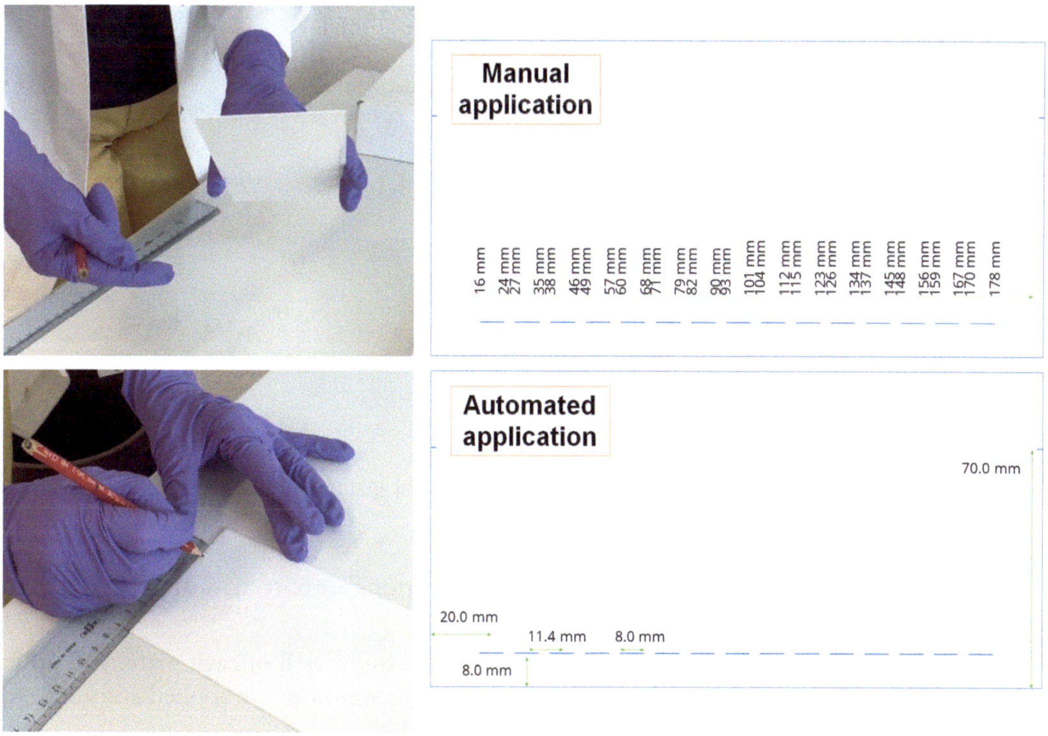

Fig. 1 Handling and layout of the HPTLC plate. (With permission from HPTLC Association)

from the left edge of the plate. The distance between tracks (center to center) is 11.4 mm.

2. The developing distance (70 mm from the lower edge) is marked with a pencil close to one of the edges of the plate before the development.

3.2 Application

Application: sample and standard solutions are quantitatively applied as bands using a capillary of the requested volume or an automated device suitable for application by spraying on of indicated volumes as band.

3.2.1 For Manual Instrumentation

1. Fill a 2.0 μL capillary with the UHM used as SST.
2. Gently apply the UHM on track 8 (2.0 μL) of the plate as a band, being careful not to damage the layer.
3. Discard the capillary.
4. Fill in a new capillary of specified volume with the sample solution or the reference and apply it gently according to the track assignment table (Fig. 2).

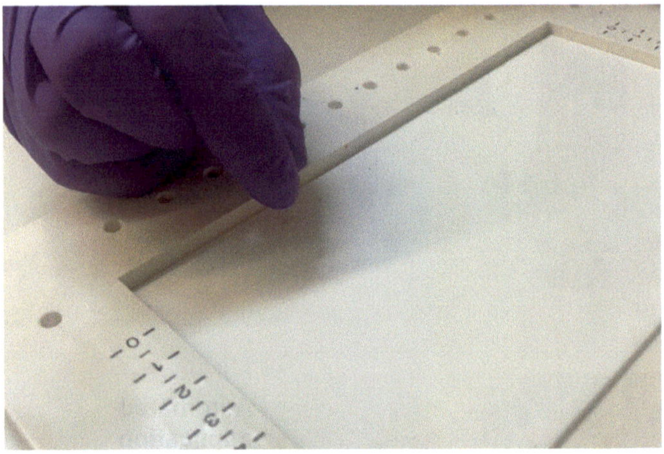

Fig. 2 Application step using manual instrumentation. (With permission from HPTLC Association)

3.2.2 For Automated Instrumentation

1. Samples and standard solutions are automatically sprayed onto HPTLC plates as bands.
2. The SST (UHM) is applied on track 8 of each plate, and other sample solutions or reference solutions are applied according to the track assignment table.

3.3 Plate Activation

3.3.1 For Manual Instrumentation (Fig. 3)

1. Shake vigorously and then pour the salt solution into a container entering the desiccator.
2. Position the container at the bottom of the desiccator.
3. Seal the desiccator.
4. Allow the humidity to equilibrate inside the desiccator. The duration of equilibration may vary depending on the size of the desiccator; it should be approximately 30 min.
5. Open the desiccator gently by sliding the lid.
6. Gently insert into the desiccator the plate with samples applied.
7. Allow the plate to equilibrate with the relative humidity inside the desiccator. Using manual instrumentation, the duration may vary depending on the size of the desiccator; it should be approximately 30 min.

3.3.2 For Automated Instrumentation

Using an automatic developing chamber, the activation is controlled by the instrument to ensure homogeneous diffusion on the layer, and it has a duration of 10 min.

3.4 Developments

3.4.1 For Manual Instrumentation

For Chamber Saturation for Plates 2 and 3 of the CDS (DS2 for MPDS, DS3 for HPDS) (Fig. 4)

1. 20 min prior to the end of plate activation, place a saturation pad in the rear trough of the TTC.

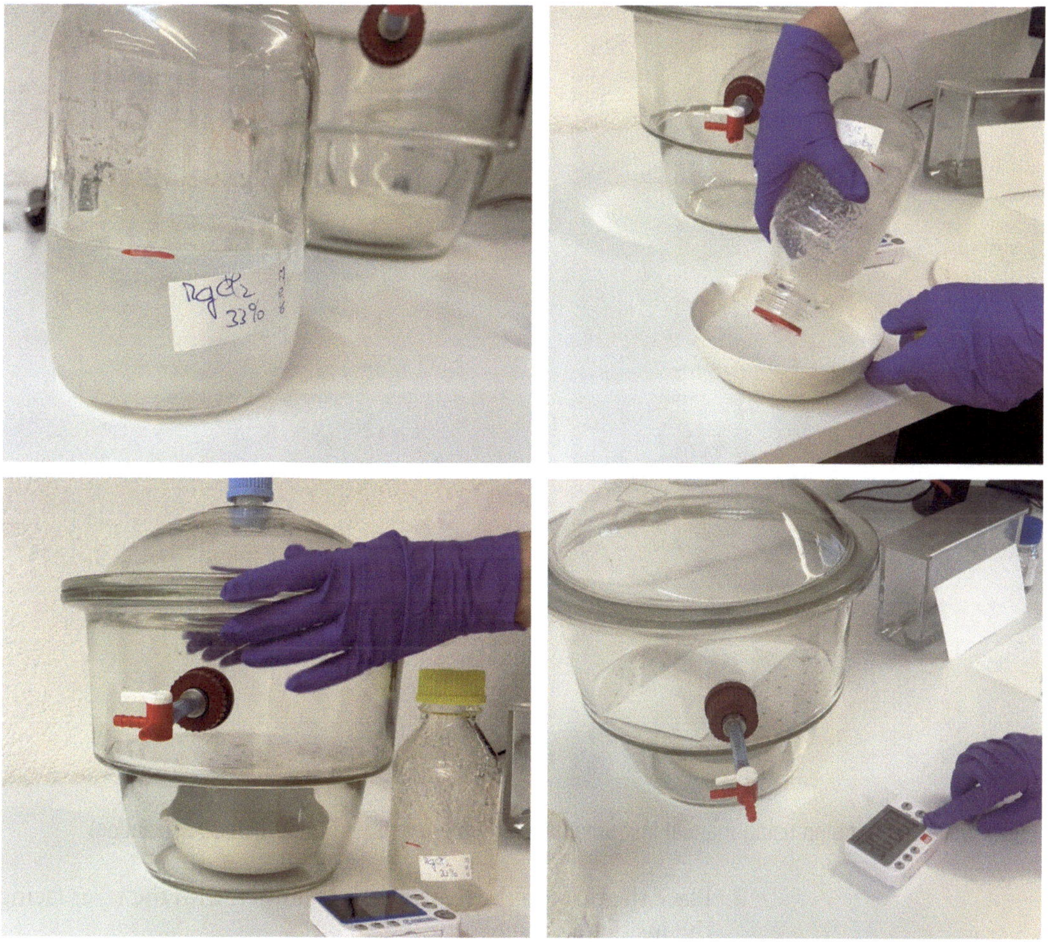

Fig. 3 Activation step using manual instrumentation. (With permission from HPTLC Association)

2. Pour 25 mL of the developing solvent (DS2 for MPDS or DS3 for HPDS) onto the saturation pad at the back of the TTC and 10 mL into the front trough.
3. Seal the TTC.
4. Allow the chamber saturation for 20 min.
5. At the end of plate activation and chamber saturation, gently open the TTC and position the plate in the front of the chamber, with the layer facing the saturation pad.
6. Replace the cover (*see* **Note 8**).

For Unsaturated Chamber for Plate 1 of the CDS (DS1 for LPDS)

1. At the end of the activation of the plate, pour 10 mL of the DS1 into the front trough of a TTC, leaving the rear trough empty.

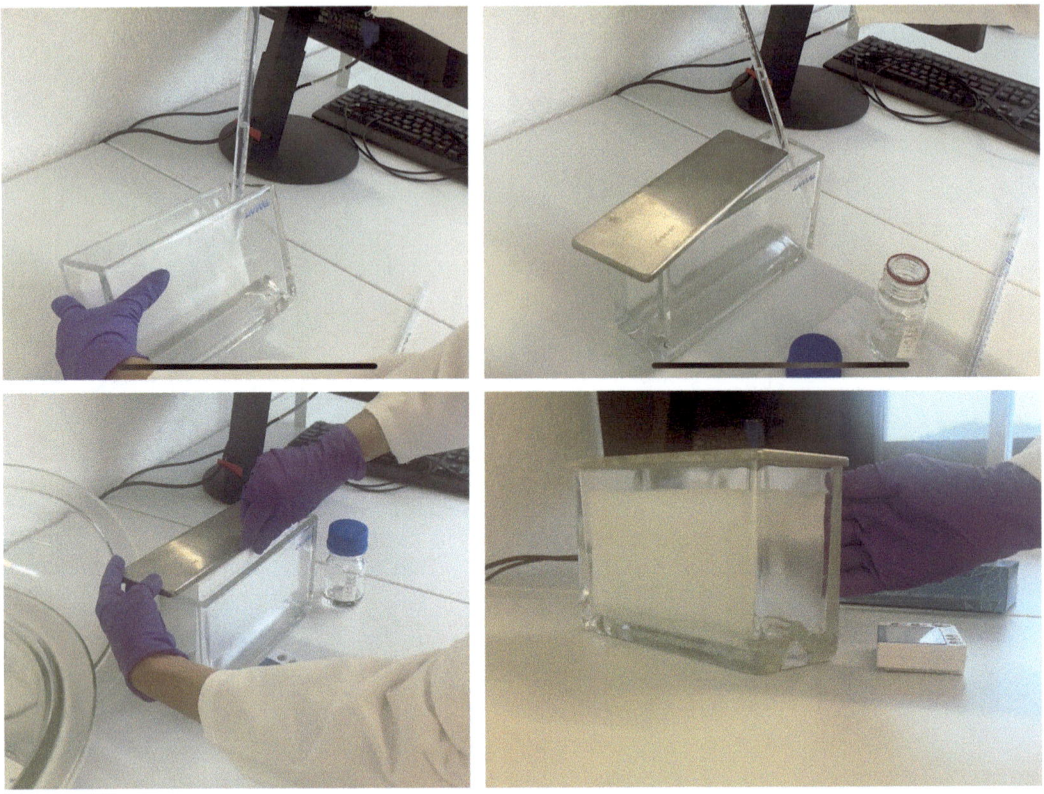

Fig. 4 Development step using manual instrumentation. (With permission from HPTLC Association)

2. Place the plate in the front of the chamber, with the layer facing the inside of the chamber.
3. Replace the cover (*see* **Note 8**).

For all Plates

Once the developing solvent reaches 70.0 mm of developing distance, remove the plate from the chamber and then dry it for 5 min in a vertical position plate (direction of chromatography facing up) in a flow of cold air.

3.4.2 For Automated Instrumentation

For Chamber Saturation for Plates 2 and 3 of the CDS (DS2 for MPDS, DS3 for HPDS)

1. Pour 10 mL of the developing solvent (DS2 for MPDS or DS3 for HPDS) for the development, and 25 mL of the developing solvent for the saturation.
2. Using an automatic developing chamber, the saturation is controlled by the instrument to ensure homogeneous diffusion through the saturation pad in the chamber, and it lasts 20 min.

For Unsaturated Chamber for Plate 1 of the CDS (DS1 for LPDS)

1. Pour only 10 mL of developing solvent for the development.
2. Using an automatic developing chamber, the development will start at the end of the activation.

For all Plates

Once the solvent reaches 70.0 mm of developing distance, the plate is automatically lifted and dried for 5 min using a stream of cold air.

3.5 Documentation Prior to Derivatization

1. Place the plate in a viewing cabinet with different sources of light, or in a TLC visualizer (Fig. 5).
2. Record for each plate pictures in shortwave UV, longwave UV, and white light.

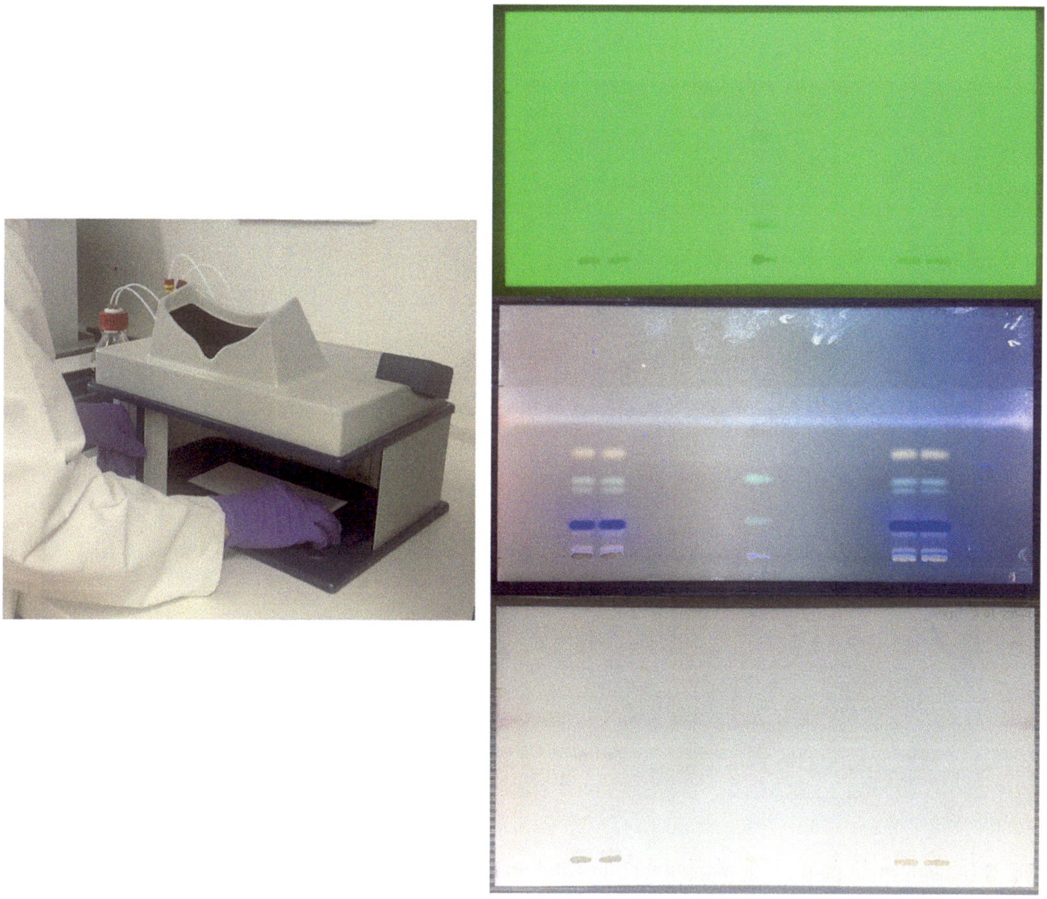

Fig. 5 Documentation step using manual instrumentation. (With permission from HPTLC Association)

3.6 Derivatization NP

3.6.1 For Manual Instrumentation (Fig. 6)

1. Put the plate on a plate heater for 3 min at 100 °C.
2. Position the plate vertically in a ventilated place.
3. Fill the sprayer with 2.0 mL of NP reagent.
4. Spray the plate homogenously with a regulated flow.
5. Dry the plate for 2 min with cold airflow.

3.6.2 For Automated Instrumentation

1. Put the plate on a plate heater for 3 min at 100 °C.
2. Position the plate in the TLC derivatizer.
3. Fill the green nozzle with the 3.0 mL of NP reagent.
4. Spray the plate homogenously using spraying level 3.
5. Dry the plate for 2 min with cold airflow.

3.7 Documentation After Derivatization

1. Place the derivatized plate in a viewing cabinet with different sources of light, or in a TLC visualizer.
2. Record pictures in longwave UV and white light.

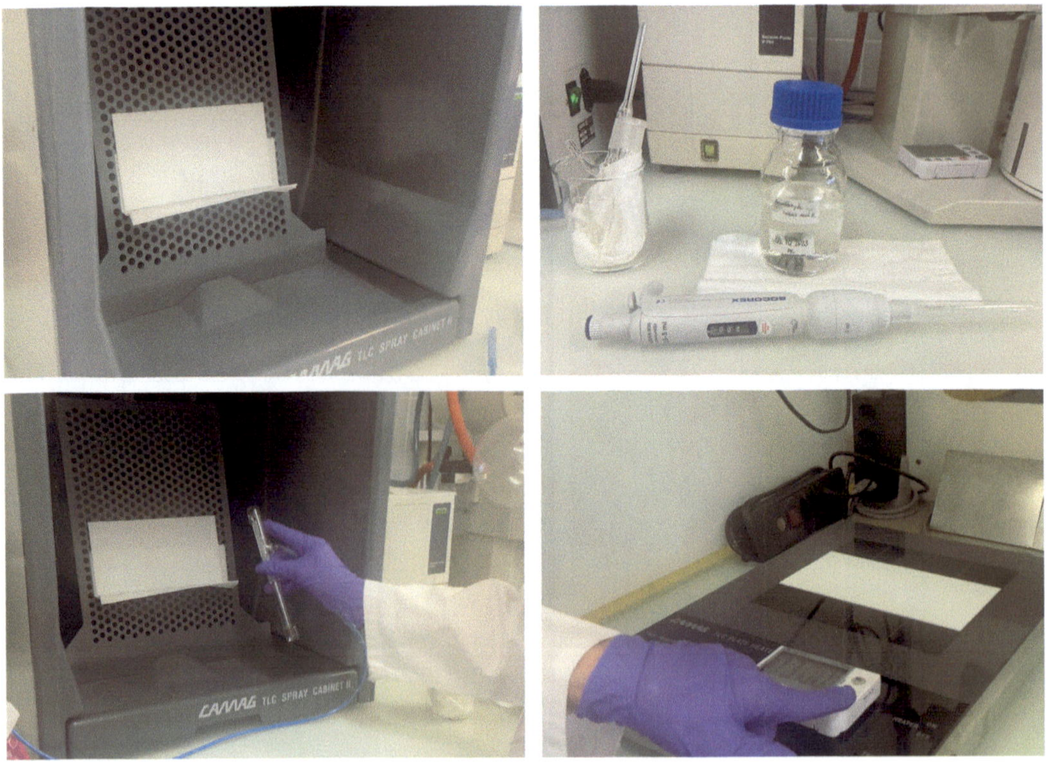

Fig. 6 Derivatization step using manual instrumentation. (With permission from HPTLC Association)

3.8 Derivatization AS

3.8.1 For Manual Instrumentation

1. Position the plate vertically in a ventilated place.
2. Fill the sprayer with the 2.0 mL of AS reagent.
3. Spray the plate homogenously with a regulated flow.
4. Put the plate on a plate heater for 3 min at 100 °C (*see* **Note 9**).

3.8.2 For Automated Instrumentation

1. Position the plate in the TLC derivatizer.
2. Fill the blue nozzle with the 3.0 mL of AS reagent.
3. Spray the plate homogenously using spraying level 3.
4. Put the plate on a plate heater for 3 min at 100 °C.

3.9 Documentation After Derivatization

1. Place the plate in a viewing cabinet with different sources of light, or in a TLC visualizer.
2. Record pictures in longwave UV and white light.

3.10 R_F Calculation

3.10.1 For Manual Instrumentation

1. On the electronic image, mark the positions of the application position and the solvent front with thin lines (Fig. 7; *see* **Note 10**).
2. Measure the distance of migration of the requested zone from the application position.
3. Calculate the R_F value based on the equation:

R_f = distance of the distance d_s/distance of the solvent front d_f

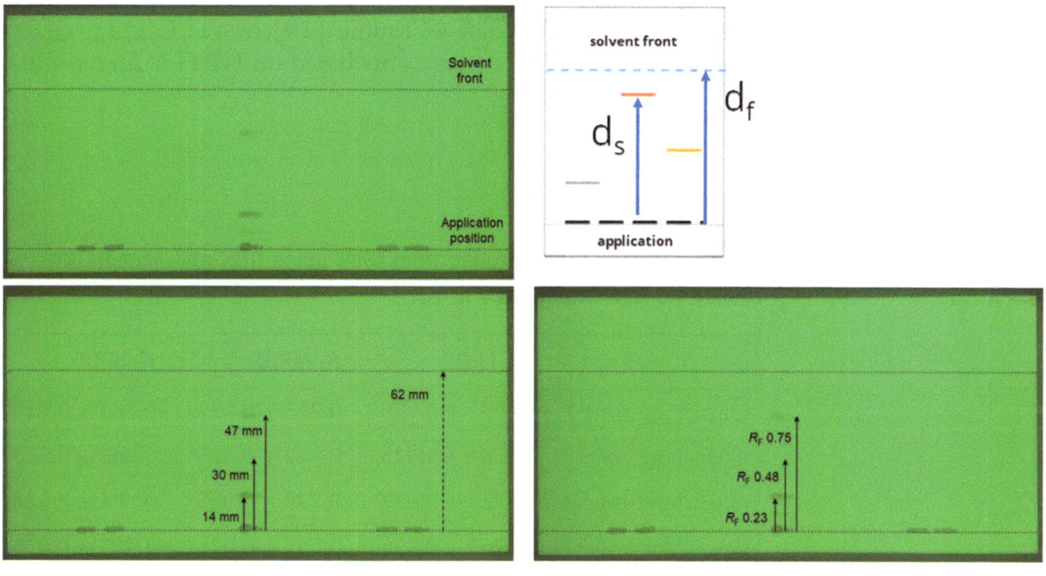

Fig. 7 R_F calculation. (With permission from HPTLC Association)

3.10.2 For Automated Instrumentation and Software

1. On the electronic image, check the position of the application and the solvent front.
2. R_F values are calculated automatically using the software.

3.11 SST Check

3.11.1 For Manual Instrumentation

1. Record the R_F values of the quenching zone obtained for the UHM on track 8 in shortwave UV.
2. Compare the values obtained to the values shown in Table 1.
3. The plates can be used for the evaluation step only if their SST are compliant.

3.11.2 For Automated Instrumentation and Software

1. The software incorporates an automated SST check.
2. In the track assignment table, track 8 is marked for SST.
3. In the SST tab, R_F values from Table 1 are pre-defined, and the evaluation is automatically computed.

3.12 Evaluation

1. Without appropriate instrumentation and software, data can only be compared visually to previous results.
2. With appropriate instrumentation and software, after SST compliance, data can be saved and compared visually or using profiles (generated from images, or from scanning densitometry) (*see* **Note 11**).

3.12.1 HPTLC Fingerprint

Following HPTLC analysis, each sample on the plate separates into distinct zones, forming a chromatogram starting at the application position ($R_F = 0.0$) and ending at the solvent front ($R_F = 1.0$). This chromatogram, known as the HPTLC fingerprint, can be visually assessed, or documented through an electronic image and/or a densitometric curve using software like *visionCATS*. Additional software, such as Knime [10] or rTLC [11], enable clustering or other data treatments based on HPTLC fingerprints.

Table 1
Values obtained for the zones from UHM using CDS and acceptance ranges for each zone

	R_F LPDS	R_F MPDS	R_F HPDS
Quenching zone 1	0.232 ± 0.025	0.213 ± 0.025	0.380 ± 0.025
Quenching zone 2	0.384 ± 0.025	0.442 ± 0.025	0.662 ± 0.025
Quenching zone 3	0.723 ± 0.025	0.735 ± 0.025	0.784 ± 0.025
Quenching zone 4	0.953 ± 0.025	0.741 ± 0.025	0.912 ± 0.025
Quenching zone 5	–	0.796 ± 0.025	0.977 ± 0.025

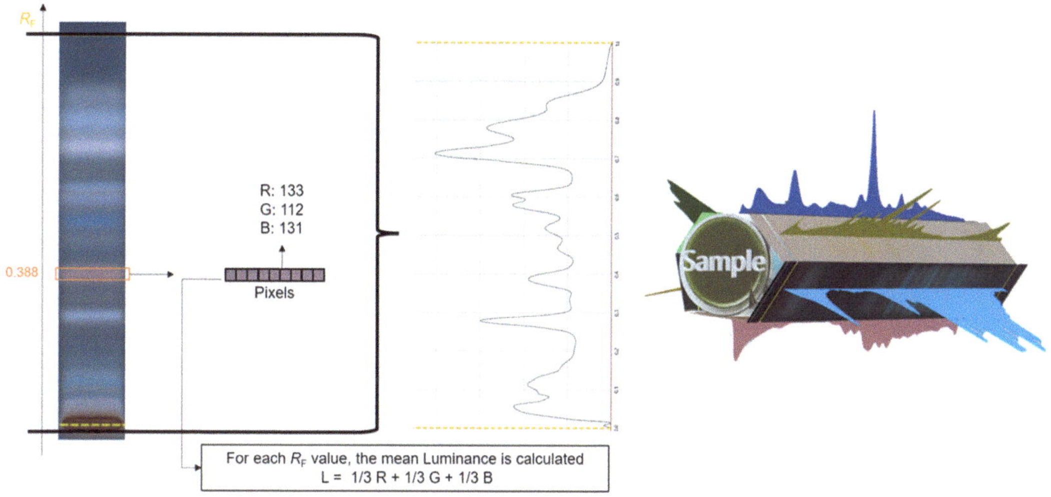

Fig. 8 Generation of PPI from the image used for comprehensive HPTLC fingerprinting

3.12.2 Comprehensive HPTLC Fingerprinting

The term "Comprehensive HPTLC Fingerprinting" (Fig. 8) highlights that each sample can offer qualitative and quantitative data stored in one analysis file. Images are made up of pixels, each with three channels: red, green, and blue (RGB). These RGB values help calculate luminance ($L = 1/3\ R + 1/3\ G + 1/3\ B$) for each pixel line in a chromatogram track. Plotting luminance against R_F values creates a peak profile from image (PPI), suitable for integration and quantitative assessment. Moreover, the same plate can be examined before and after derivatization through peak profile from scanning densitometry (PPSD) [12].

3.12.3 Searching for Similar Compounds in the Database

A molecule is characterized by its R_F values in LPDS, MPDS, and HPDS, generated by the software using the PPI. By utilizing the database created through the CDS approach analysis, a query can be assessed for a hypothesis that matches the specified similarity (*see* **Note 12**). The Euclidean distance can be employed for this purpose, and, for instance, the four closest neighbors (most similar) can be displayed [13] (*see* **Note 13**). To enhance specificity, additional spectral data, such as UV spectra and MS, may also be integrated into the database.

3.12.4 Predicting R_F Values

Utilizing machine learning algorithms to predict R_F values emerges as a practical approach. This becomes especially beneficial when there is access to a reference substance and a requirement to identify the best developing solvent among the three choices presented by the CDS. By training a model using the database, a predictive tool can be created to recommend the optimal developing solvent for a specific reference substance [13] (*see* **Note 14**).

4 Notes

1. Evaluate the stability of test solutions and standard solutions individually.
2. UHM remains stable in the fridge at 4 °C for 2 months.
3. Unless specified otherwise, use plates from Supelco.
4. The $MgCl_2$ salt solution remains stable for at least 1 year.
5. Before use, rinse and clean the TTC. Ensure it is dry and free of solvent vapor.
6. Dry it with cold airflow.
7. Label the plate with the project name, the date, and the plate number.
8. If a pronounced banana effect is observed on the plate, it could be attributed to a gas phase leak at the edges of the chamber. To prevent this, place a weight on the cover to ensure that the chamber is sealed.
9. Immediately after derivatization with AS, place the plate on the plate heater to prevent a pink background.
10. The solvent front position is usually slightly above 70 mm.
11. Electronic images are the most convenient way of conveying information contained in the fingerprint such as position, color, and intensity of zones.
12. Utilize the open-source software KNIME with Excel reader, concatenate, numeric distances, and similarity search nodes.
13. The efficiency of the identification process depends on the content and composition of the database. Successful compound identification relies on the similarity between the unknown zone and the chemical space represented in the database.
14. Use the open-source software KNIME along with its free complementary extensions "KNIME Base Chemistry Types" and "KNIME Chemistry Add-ons," as well as the "RDKit KNIME Integration." Employ Excel reader, partitioning, Random Forest learner (regression), and Random Forest prediction (regression) nodes.

Acknowledgement

This work was supported by the HPTLC Association (https://www.hptlc-association.org/home.cfm).

References

1. Reich E, Schibli A (2011) High-performance thin-layer chromatography for the analysis of medicinal plants. Thieme, New York. https://books.google.ch/books?id=ADXjekuY09gC
2. Bladt S, Rickl V (2009) Plant drug analysis: a thin layer chromatography atlas. Springer, Berlin/Heidelberg. https://books.google.ch/books?id=CdVKAAAAQBAJ
3. HPTLC Association (n.d.) https://www.hptlc-association.org/home.cfm. Accessed 24 Dec 2023
4. U.S. Pharmacopeia. General Chapter<203>, United States Pharmacopeia, USP 39–NF 34 second supplement. In: US Pharmacopeial convention: Rockville, MD, 2016
5. European Pharmacopoeia. General chapter 2.8.25, European Pharmacopeia 10.6. In: Council of Europe, Strasbourg, 2020
6. Indian Pharmacopoeia. General chapter 2.4.17, Indian Pharmacopiea, in: 2022
7. Schmid M, Do TKT, Trettin I, Reich E (2022) Applicability of the universal mixture for describing system suitability and quality of analytical data in routine normal phase high performance thin layer chromatography methods. J Chromatogr A 1666:462863
8. Do TKT, Schmid M, Phanse M, Charegaonkar A, Sprecher H, Obkircher M, Reich E (2021) Development of the first universal mixture for use in system suitability tests for high-performance thin layer chromatography. J Chromatogr A 1638:461830. https://doi.org/10.1016/j.chroma.2020.461830
9. Do TKT, Schmid M, Trettin I, Hänni M, Reich E (2022) Complementary developing solvents for simpler and more powerful routine analysis by high-performance thin-layer chromatography. JPC—J Planar Chromatogr—Mod TLC. https://doi.org/10.1007/s00764-022-00185-1
10. Software Knime used for data treatment (n.d.) https://www.knime.com/?utm_term=knime&utm_campaign=Knime-Brand-EMEA&utm_source=google&utm_medium=cpc&hsa_acc=2573270492&hsa_cam=20103126437&hsa_grp=148910434317&hsa_ad=666391965047&hsa_src=g&hsa_tgt=aud-2014909942526:kwd-298177681876&hsa_kw=knime&hsa_mt=e. Accessed 30 Dec 2023
11. Fichou D, Ristivojević P, Morlock GE (2016) Proof-of-principle of rTLC, an open-source software developed for image evaluation and multivariate analysis of planar chromatograms. Anal Chem 88:12494–12501. https://doi.org/10.1021/acs.analchem.6b04017
12. Frommenwiler DA, Trefzer D, Schmid M, Cañigueral S, Reich E (2020) Comprehensive HPTLC fingerprinting: a novel economic approach to evaluating the quality of Ganoderma lucidum fruiting body. J Liq Chromatogr Relat Technol 43:414–423. https://doi.org/10.1080/10826076.2020.1725560
13. Do TKT, Trettin I, Hänni M, Reich E (2023) Applying machine learning on high-performance thin-layer chromatography using the complementary developing solvents concept. J Liq Chromatogr Relat Technol:1–7. https://doi.org/10.1080/10826076.2023.2284707

Chapter 5

Extraction of Anthocyanins with Deep Eutectic Solvents and Chemical Evaluation by High-Performance Liquid Chromatography

Yuntao Dai, Robert Verpoorte, and Young Hae Choi

Abstract

Deep eutectic solvents (DES) and ionic liquids (ILs), specifically natural deep eutectic solvents (NADES), allow for the extraction of natural products using environmentally friendly solvents instead of organic solvents. Here we describe the extraction of anthocyanins from a medicinal plant using NADES prepared either by evaporating method or heating-and-stirring method with the help of ultrasound-assisted extraction (UAE). The NADES extract can be qualified by the high-performance liquid chromatography (HPLC) method, which can separate the component of NADES with target compounds from medicinal plant.

Key words Natural deep eutectic solvents (NADES), Anthocyanin, Extractability, Stability, Heating-and-stirring method, Ultrasound-assisted extraction

1 Introduction

Over the past decade, significant attention has been directed towards ionic liquids (ILs) and deep eutectic solvents (DES) derived from common primary metabolites, referred to as natural deep eutectic solvents (NADES). The aim is to position them as alternatives to organic solvents, serving the dual purpose of reducing environmental impact and minimizing health risks. NADES, characterized by their biocompatibility and exceptionally low toxicity, are formulated through diverse combinations of sugars, sugar alcohols, amino acids, organic acids, and bases [1]. These NADES exhibit notable applications, particularly in the eco-friendly extraction and separation of bioactive compounds within health-related fields such as food [2], pharmaceuticals [3], and cosmetics.

NADES, as indicated by their name, consist of natural primary metabolites held together by intermolecular interactions, primarily hydrogen or ionic bonds, resulting in liquid supramolecular systems [1, 4]. They offer numerous advantages over synthetic ILs,

Young Hae Choi et al. (eds.), *Natural Product Isolation and Identification: Methods and Protocols*, Methods in Molecular Biology, vol. 2895, https://doi.org/10.1007/978-1-0716-4350-1_5,

including cost-effectiveness, biodegradability, and simple preparation methods [4]. Furthermore, NADES exhibit exceptional physicochemical properties that are conducive to extraction purposes. Their inherent environmental friendliness, negligible vapor pressure, and ability to remain liquid even at temperatures well below 0 °C are notable characteristics. As extraction solvents, NADES display a wide polarity range and high solubilization capacity for various compounds [4, 5]. Additionally, their viscosity can be adjusted by manipulating water content [6]. Particularly in the mid-range polarity, where many bioactive compounds are found, NADES surpass classical solvents. Moreover, their composition of components abundant in our daily diet renders NADES inexpensive, sustainable, and safe commodities.

The exceptional properties of NADES have led to their application in producing ready-to-use extracts of blueberry phenolic compounds [4, 7]. Notably, certain NADES demonstrate outstanding solubilization capabilities for both nonpolar and polar compounds, surpassing those of water [4]. This highlights the significant potential for NADES as solvents in extracting valuable metabolites, applicable to the food, cosmetic, and pharmaceutical industries. The utilization of NADES in extracting bioactive compounds [8] and sample extraction has been extensively reviewed [9]. While initially regarded as general solvents, it is increasingly evident that NADES exhibit high selectivity. However, this realization also presents a challenge, necessitating the development of specific and highly selective methods for each application.

Herein, we present the extraction of anthocyanins, pivotal natural antioxidants, utilizing NADES via both evaporating method and heating-and-stirring methods, as well as ultrasound-assisted extraction (UAE) techniques. These methodologies yield extraction efficiencies comparable to those achieved with conventional organic solvents [10]. Moreover, NADES demonstrate improved thermal and storage stability for anthocyanins, thereby substantially enhancing the anthocyanin extraction process [10].

2 Materials

All chemicals and solvents utilized in the formulation of NADES and the extraction of plant materials adhere to analytical grade standards. All glassware employed in the procedures undergoes pre-cleaning protocols. The purity requirements for chemicals involved in NADES preparation, including malic acid (*see* **Note 1**), lactic acid, proline, glucose, fructose, sucrose, 1,2-propanediol, and choline chloride, mandate a minimum purity level of 95% or higher. Cyanidin, employed as a reference compound for anthocyanin, possesses a purity of 95%.

The flower petals sourced from two distinct colored varieties of *Catharanthus roseus*, specifically Pacifica XP Apricot (orange) and Pacifica Orchid Halo (purple), were obtained from Intratuin, Pijnacker (Heerhugowaard, The Netherlands), during the blooming season. Subsequently, the petals were ground into powder using a mortar and pestle filled with liquid nitrogen, followed by freeze-drying (*see* **Note 2**) [10].

2.1 Equipment

The methodology utilizes a magnetic stirrer and glass bottles for the synthesis of NADES. Ultrasonication baths were employed for the extraction of plant materials using NADES. Subsequently, the NADES extracts underwent analysis using an Agilent chromatographic system equipped with a Phenomenex Luna C18 column (4.6 mm × 250 mm, 5 μm particle size), maintained at 35 °C.

2.2 Preparation of Natural Deep Eutectic Solvents

Two methods (*see* **Note 3**) can be utilized for the preparation of NADES, as previously reported [4]. Either the evaporating method (Subheading 2.2.1) or the heating-and-stirring method (Subheading 2.2.2) can be employed for the preparation of NADES.

2.2.1 Evaporating Method

1. Dissolve two components in a defined molar ratio (*see* **Note 4**) in deionized water.
2. Employ a rotary evaporator to evaporate the water at 50 °C or below until the liquid volume stabilizes.
3. Transfer the resultant viscous liquid to a desiccator containing dry silica gel until a constant weight is achieved.

2.2.2 Heating-and-Stirring Method

1. Obtain NADES with a specified water content by adding the NADES ingredients in the desired ratio into a predetermined volume of water in a container.
2. Insert a magnetic stirring bar, seal the container, and heat it in a water bath at a temperature below 100 °C.
3. Allow approximately 30–90 min for the formation of a clear liquid.

2.2.3 Selection of NADES for Anthocyanin Analysis

In this protocol, the heating-and-stirring method is employed for the preparation of NADES. The following NADES were prepared using the heating-and-stirring method (*see* **Notes 3** and **5**): 1,2-propanediol-choline chloride-water (molar ratio, 1:1:1; designated as PCH); lactic acid-glucose-water (5:1:3; designated as LGH); proline-malic acid-water (1:1:3; designated as PMH); malic acid-choline chloride-water (1:1:2; designated as MCH); glucose-choline chloride-water (2:5:5; designated as GCH); and glucose-fructose-sucrose-water (1:1:1:11; designated as FGSH). All solvents were stored in colorless bottles with airtight caps at room temperature (*see* **Note 6**).

2.3 Preparation of HPLC Mobile Phase

1. High-performance liquid chromatography (HPLC) solvent A: HPLC-grade water containing 3% (v/v) mass spectrometry (MS)-grade formic acid. Using a clean 1 L measuring cylinder, accurately measure 1 L of HPLC-grade water, transfer it to a 1 L LC bottle, remove 30 mL of water, add 30 mL of MS-grade formic acid, and mix gently by swirling by hand.
2. HPLC solvent B: HPLC-grade acetonitrile containing 3% (v/v) MS-grade formic acid. Using a clean 1 L measuring cylinder, accurately measure 1 L of HPLC-grade acetonitrile, transfer it to a 1 L HPLC bottle, remove 30 mL of acetonitrile, add 30 mL of MS-grade formic acid, and mix gently by swirling by hand.
3. Degassing of solvents A and B is achieved by placing them in an asonicator set to the degas mode for 20 min.
4. HPLC wash solvent: 80% (v/v) AcCN and 20% (v/v) water. Using a clean 1 L measuring cylinder, accurately measure 800 mL of HPLC-grade acetonitrile and 200 mL of HPLC-grade water, transfer them to a 1 L HPLC bottle, and mix gently by swirling by hand.

3 Methods

3.1 Preparation of Organic Solvent Extracts as Controls Using Ultrasound-Assisted Extraction (UAE)

1. Weigh 100 mg (98–102 mg) of freeze-dried plant material powders into 10 mL glass bottles and add 3 mL of methanol/ethanol containing 3% formic acid.
2. Seal the bottles securely with airtight caps.
3. Transfer the sealed vials to an ultrasonicator bath.
4. Employ ultrasonication to extract the samples at room temperature for 30 min.
5. After extraction, transfer the resulting mixture from the glass bottles to 2 mL microcentrifuge tubes and centrifuge at 15,682×*g* for 20 min.
6. Retrieve the supernatant and filter it through a 0.4 μm membrane filter using a syringe into an HPLC vial for HPLC analysis.

3.2 Extraction of Plant Materials by Heating and Stirring Using NADES (See Notes 7 and 8)

1. Weigh 100 mg (98–102 mg) of freeze-dried plant material powders into 10 mL glass bottles and add 3 mL of NADES (*see* **Notes 9** and **10**).
2. Seal the tubes securely with airtight caps.
3. Stir and heat the mixture in a heating bath at 40 °C for 30 min.

4. Transfer the resulting mixture from the glass tubes to 2 mL microcentrifuge tubes, and subject it to centrifugation at 15,682×*g* for 20 min.
5. Retrieve the supernatant and filter it using a syringe through a 0.4 μm cellulose acetate membrane filter (PVDF).
6. Dilute the NADES extract with 3% aqueous formic acid in a 1: 2 ratio for HPLC analysis (*see* **Note 11**).

3.3 Ultrasound-Assisted Extraction (UAE) with and Without Heating

1. Weigh 100 mg (98–102 mg) of freeze-dried plant material powders into 10 mL glass tubes and add 3 mL of NADES.
2. Seal the tubes securely with caps.
3. Transfer the sealed tubes to an ultrasonicator bath.
4. For extractions without heating, maintain the samples at room temperature; for extractions with heating, set the bath temperature to 40 °C.
5. Utilize ultrasonication to extract the samples at room temperature for 30 min.
6. Follow the same centrifugation and filtering procedures as outlined in **steps 4–6** in Subheading 3.2.

3.4 Analyze Extracts Using HPLC

1. Gradient elution is employed with 3% aqueous formic acid (A) and methanol containing 3% formic acid (B) following this profile: 20–25% B (0–30 min), 25–45% B (30–35 min), 45–80% B (35–40 min) at a flow rate of 1.0 mL/min.
2. The injection volume is 10 μL.
3. Chromatograms were recorded at 520 nm.

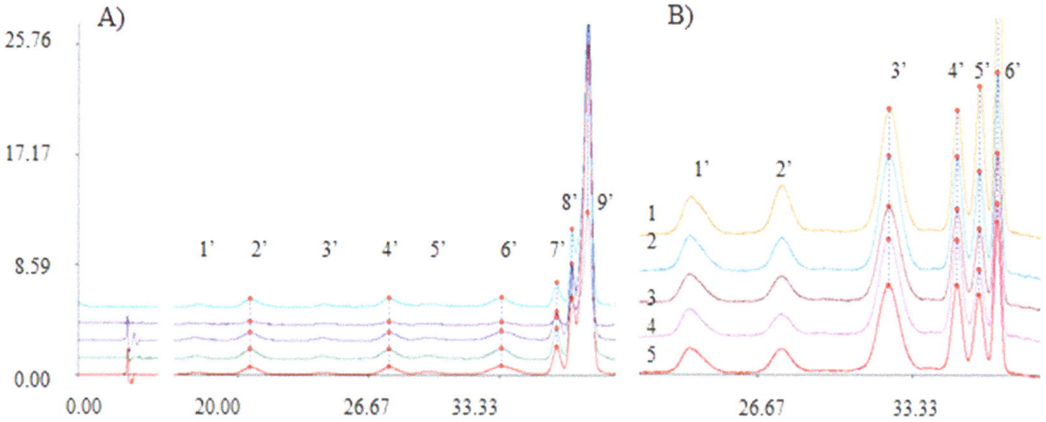

Fig. 1 HPLC-UV chromatograms depict anthocyanins extracted from *Catharanthus roseus* with (**a**) purple and (**b**) orange flower petals, utilizing various solvents, and detected at 520 nm. Solvents include (*1*) methanol with 3% formic acid; (*2*) LGH; (*3*) MCH; (*4*) PMH; (*5*) GCH (reproduced from ref. [10])

4. Compare HPLC chromatograms at 520 nm of the tested NADES and 3% methanolic formic acid extracts *C. roseus* as shown in Fig. 1.

4 Notes

1. In many cases, certain components of NADES possess stereoisomers, such as L-malic and D-malic acid. Unless there is a specific requirement, stereospecific forms of components in NADES are unnecessary.
2. The ground powder can be dried using a freeze-dryer. The drying time depends on the sample quantity, typically requiring 24 h.
3. There are two methods for the preparation of NADES: the evaporating method and the heating-and-stirring method. The evaporating method can produce NADES at temperatures lower than 50 °C, depending on vacuum power. It typically requires less time than the heating-and-stirring method for producing NADES with high density. However, the water content of the prepared NADES is unknown. On the other hand, the heating-and-stirring method is suitable for producing NADES with a known amount of water. It can be utilized at temperatures higher than 50 °C and is not suitable for making NADES containing volatile components.
4. It is advisable to desiccate the components of NADES in a vacuum desiccator at room temperature for 24 h prior to use. Certain components of NADES, such as choline chloride, have a propensity to absorb moisture, which may lead to inaccurate ratios of the NADES ingredients.
5. The extractability of NADES varies depending on the target compounds and the water content within the NADES. Different target compounds within NADES require varying water content for optimal extraction yields. Specifically, for a given target compound, the extractability of NADES fluctuates with its water content, and the highest extraction yield may be achieved with 10–50% water content in the NADES [11]. Compared to conventional organic solvents, NADES exhibit high viscosity, primarily suited for applications involving full extracts, such as cosmetics, or food and fodder applications. However, the high viscosity of NADES can be adjusted by incorporating water. For instance, the viscosity of PC (1,2-propanediol-choline chloride) notably decreased with increasing water content according to the function $y = 0.88x - 1.524$ ($r^2 = 0.99$) (where y represents viscosity and x denotes the water weight percentage in PCH). The elevated

viscosity of NADES is often attributed to the extensive hydrogen-bonding interactions among the components. Diluting NADES with water results in a significant reduction in viscosity due to the attenuation of hydrogen-bonding interactions among the components [4, 6]. Notably, NADES maintain their supramolecular characteristics even with less than 50% water dilution [6].

6. It is recommended to store NADES in a sealed container to maintain their integrity, as the water content within the mixture may fluctuate owing to the hygroscopic nature of NADES [4].
7. Different extraction methods can be selected based on the viscosity of the solvent, stability of the target compounds, and characteristics of the sample materials. When dealing with more viscous solvents, it is advisable to employ the stirring method or UAE at specific temperatures. Heating reduces the viscosity of NADES, while ultrasound enhances extraction efficiency by disrupting cell walls and inducing turbulence [10, 12]. In this experiment, ultrasonication at 25 °C was found to be the least efficient method. However, increasing the temperature to 40 °C resulted in a 2–20% improvement in yields. The highest extraction yield was achieved using stirring at 40 °C, which was 35–55% higher than that obtained with sonication at 25 °C (Table 1) [10].
8. For the stirring method, no significant difference in extraction yields was observed between 30 and 60 min of extraction time. However, at 90 min, the areas of all peaks decreased. Therefore, the optimal conditions, specifically stirring at 40 °C for 30 min, were selected for this study (Table 1).
9. Due to the high viscosity of NADES, it is preferable to measure its quantity by weight, with its volume calculated based on its density. However, for NADES with viscosity similar to that of water, its quantity can be measured directly in volume.
10. For NADES with high viscosity, they are unsuitable for use as extraction solvents, as their extracts can be challenging to separate from the sample. Additionally, the extractability of a particular target compound within NADES varies with its water content, with the optimal extraction yield typically achieved with 10–50% water content in the NADES [12]. In this protocol, most NADES or diluted NADES used exhibit low viscosity to mitigate the impact of high viscosity.
11. To assess the extractability of various NADES, the extract should be diluted with a solvent such as water or methanol [11, 12]. Otherwise, the NADES extract becomes too viscous for quantitative analysis. In this experiment, 3% aqueous formic acid was added in a 1:2 ratio.

Table 1
Effect of three distinct extraction methods (ultrasonication, ultrasonication with heating, heating-and-stirring) on the relative extraction efficiency of anthocyanins from *Catharanthus roseus* with NADES, featuring purple flowers, was investigated ($n = 3$)

Peaks[b]		Peak area ratio[a]		
		Ultrasonication (25 °C)	Ultrasonication (40 °C)	Heating and Stirring[c] (40 °C)
Peak 1	Unknown	0.6	0.8	1.0
Peak 2	Unknown	0.5	0.6	1.0
Peak 3	Unknown	0.6	0.4	1.0
Peak 4	Unknown	0.5	0.5	1.0
Peak 5	Unknown	0.4	0.3	1.0
Peak 6	Unknown	0.6	0.6	1.0
Peak 7	Petunidin 3-*O*-(6-*O*-*p*-coumaroyl) glucose	0.5	0.7	1.0
Peak 8	Malvidin 3-*O*-(6-*O*-*p*-coumaroyl) glucose	0.5	0.7	1.0
Peak 9	Hirsutidin 3-*O*-(6-*O*-*p*-coumaroyl) glucose	0.5	0.7	1.0

Reproduced from ref. 10
[a]The relative extraction yield is quantified by the peak area ratio observed in the chromatogram of the extract relative to that of the control group (stirred at 40 °C), which serves as the reference
[b]All the peaks are labeled in Fig. 1A
[c]Stirring with a magnetic stirrer

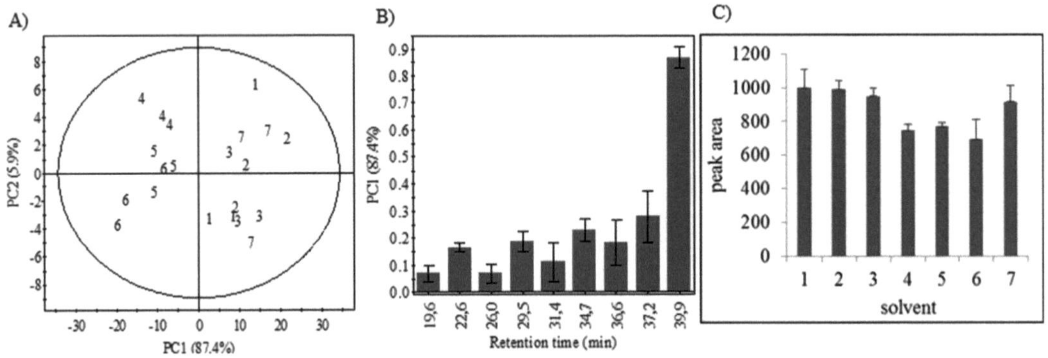

Fig. 2 Score plot (**A**) and loading plot (**B**) resulting from principal component analysis (PCA) of HPLC chromatograms derived from the extraction of *Catharanthus roseus* purple flowers using various solvents, including methanol (methanol 3% formic acid (*1*); PCH (*2*); LGH (*3*); PMH (*4*); MCH (*5*); GCH (*6*); 75% (v/v) FGSH (*7*)). (Reproduced from ref. 10) (*see* **Note 12**)

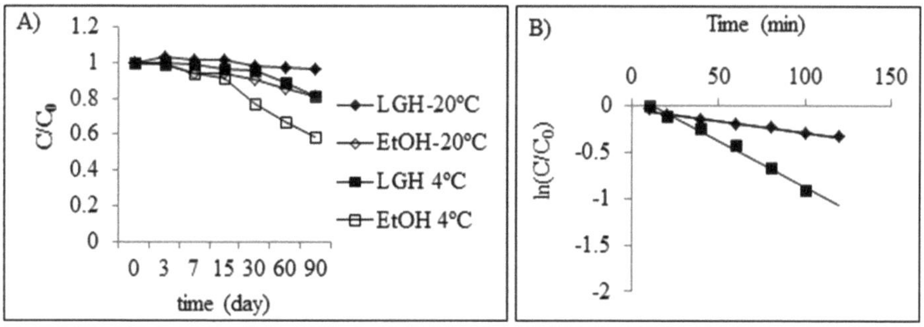

Fig. 3 Stability of cyanidin in LGH and ethanol with 3% formic acid. (**A**) Storage stability cyanin at −20 °C and 4 °C in the dark for 3 months. (**B**) Thermal stability of cyanidin at 60 °C in LGH (♦) and EtOH with 3% formic acid(■). (Reproduced from ref. 10) (*see* **Note 13)**

12. Different NADES exhibit varying extractabilities. LGH, PCH, and 75% FGSH demonstrate comparable efficiency to acidified methanol in extracting anthocyanins from *C. roseus* petal powder. Conversely, PMH, MCH, and GCH display lower extractability compared to acidified MeOH (Fig. 2). Among all the NADES tested, it can be inferred that the extraction efficiency of NADES is more likely correlated with the viscosity of the NADES rather than their polarity and pH.
13. Anthocyanin extracts or isolated anthocyanins are highly unstable and prone to degradation, which can be induced by factors such as temperature, light, and the solvent used for extraction. It is generally recommended to conduct anthocyanin extraction in the absence of light and at low temperatures to mitigate degradation. NADES offer improved thermal stability and storage stability for anthocyanins compared to conventional organic solvents. For instance, in LGH as a solvent, cyanidin demonstrated greater stability compared to ethanol containing 3% formic acid, particularly at high temperatures during extraction and low temperatures during storage (Fig. 3). Similarly, carthamin from safflower exhibited enhanced thermal stability and storage stability in sugar-based NADES relative to water or a 40% EtOH solution. This robust stabilization is attributed to the formation of strong hydrogen bonding interactions between the solutes and NADES molecules [5].

References

1. Choi YH, van Spronsen J, Dai Y, Verberne M, Hollmann F, Arends IW et al (2011) Are natural deep eutectic solvents the missing link in understanding cellular metabolism and physiology? Plant Physiol 156:1701–1705
2. Benvenutti L, Zielinski AAF, Ferreira SRS (2019) Which is the best food emerging solvent: IL, DES or NADES? Trends Food Sci Technol 90:133–146
3. Álvarez MS, Zhang Y (2019) Sketching neoteric solvents for boosting drugs bioavailability. J Control Release 311:225–232
4. Dai Y, van Spronsen J, Witkamp GJ, Verpoorte R, Choi YH (2013) Natural deep

eutectic solvents as new potential media for green technology. Anal Chim Acta 766:61–68

5. Dai Y, Verpoorte R, Choi YH (2014) Natural deep eutectic solvents providing enhanced stability of natural colorants from safflower (*Carthamus tinctorius*). Food Chem 159: 116–121
6. Dai Y, Witkamp GJ, Verpoorte R, Choi YH (2015) Tailoring properties of natural deep eutectic solvents with water to facilitate their applications. Food Chem 187:14–19
7. da Silva DT, Smaniotto FA, Costa IF, Baranzelli J, Muller A, Somacal S et al (2021) Natural deep eutectic solvent (NADES): a strategy to improve the bioavailability of blueberry phenolic compounds in a ready-to-use extract. Food Chem 364:130370
8. Zainal-Abidin MH, Hayyan M, Hayyan A, Jayakumar NS (2017) New horizons in the extraction of bioactive compounds using deep eutectic solvents: a review. Anal Chim Acta 979:1–23
9. Tang S, Zhang H, Lee HK (2016) Advances in sample extraction. Anal Chem 88:228–249
10. Dai Y, Rozema E, Verpoorte R, Choi YH (2016) Application of natural deep eutectic solvents to the extraction of anthocyanins from *Catharanthus roseus* with high extractability and stability replacing conventional organic solvents. J Chromatogr A 1434:50–56
11. Dai Y, Witkamp GJ, Verpoorte R, Choi YH (2013) Natural deep eutectic solvents as a new extraction media for phenolic metabolites in *Carthamus tinctorius* L. Anal Chem 85: 6272–6278
12. Xing C, Cui WQ, Zhang Y, Zou XS, Hao JY, Zheng SD et al (2022) Ultrasound-assisted deep eutectic solvents extraction of glabridin and isoliquiritigenin from *Glycyrrhiza glabra*: optimization, extraction mechanism and in vitro bioactivities. Ultrason Sonochem 83: 105946

Chapter 6

Extraction of Essential Oils and Terpene Volatiles from Plants and Identification by GC-MS-Based Techniques

Sandra T. Krause and Maximilian Frey

Abstract

The extraction of plant essential oils (EOs) and analysis by gas chromatography coupled to mass spectrometry (GC-MS) are standard methods when studying aromatic plants and the chemical composition of EOs. Here, two simple methods for the extraction of EO compounds from leaves of *Thymus vulgaris* are described. Organic solvent extraction and solid-phase microextraction (SPME), respectively, are used and the results of the GC-MS analyses are compared. We explain how analyte peaks can be identified and discuss possible rearrangement reactions that can occur during the extraction process or analysis.

Key words Essential oil, Aromatic plants, GC-MS, Organic solvent extraction, SPME, Terpenoids, Monoterpenes, Sesquiterpenes, Mass spectra, Rearrangement reactions

1 Introduction

Plant essential oils (EOs) are of great commercial value as they are used in many pharmaceuticals for their antibacterial, antifungal, and antiviral activities. Because of their pleasant fragrance and taste, EOs are added to body hygiene products and dried herbs containing EOs are commonly applied in food products as aromatic additives [1–4].

From a chemical point of view, EOs consist of mostly lipophilic compounds with a molecular weight lower than 500 Da and can comprise over 100 different compounds that are usually of terpenoid, phenylpropanoid, or fatty acid origin [5, 6]. Among these compounds, terpenoids are predominant and mainly responsible for the health-beneficial properties of EOs [7]. Terpenes form the largest group of plant organic specialized compounds [8]. Generally, they are of lipophilic character as they are built from C5-hydrocarbons, also called isoprene units. The classification of terpenes is determined by the number of isoprene units they are built from. The simplest of all terpenes are the hemiterpenes,

Young Hae Choi et al. (eds.), *Natural Product Isolation and Identification: Methods and Protocols*, Methods in Molecular Biology, vol. 2895, https://doi.org/10.1007/978-1-0716-4350-1_6,

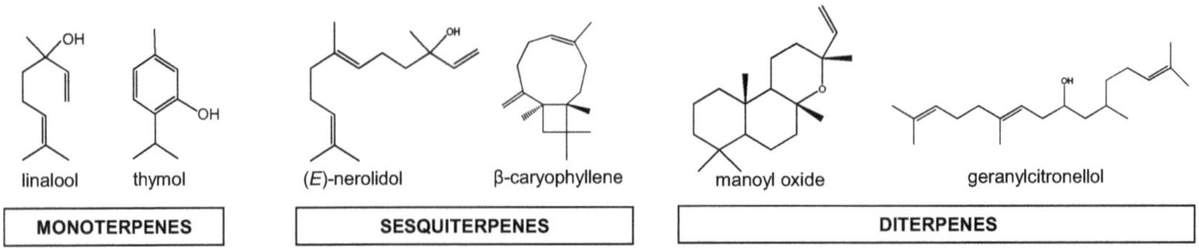

Fig. 1 Terpene classes found in EOs. Two examples for each class are shown. For linalool and *E*-nerolidol R and S isomers can often be found as a racemate that is not resolved on a non-chiral column. Therefore, no R/S stereochemistry is shown

composed of only one C-5 unit. C-10 compounds are called monoterpenes, consisting of two isoprene units. Accordingly, sesquiterpenes (C-15) consist of three isoprene units, and diterpenes (C-20) are built from four isoprene units. High-molecular terpenes consist of C-30 backbones (triterpenes) or C-40 backbones (tetraterpenes) [9]. In EOs, the volatiles of terpenoid origin are monoterpenes, sesquiterpenes, and diterpenes (Fig. 1) [6].

In plants, EOs fulfill various functions such as antiherbivore defense [10] or adaptation to varying climatic conditions [11]. The amount of the respective constituents in the EO can vary due to ecological influences or genetic variation of the EO-producing plants [12]. Because of their cell toxic properties, EOs are often stored in separated plant compartments. This can be oil or resin ducts found in the needles of conifers, or the glandular trichomes typical for Lamiaceae species [6, 13].

EOs are commonly analyzed in order to validate their origin and purity. Gas chromatography coupled to mass spectrometry (GC-MS) is the method of choice and is widely used for the analysis of volatile terpenoids like mono- and sesquiterpenes [14]. It separates compounds in the gas phase of a carrier gas through the passage of a stationary phase fixed on a capillary column. Non-volatile or less volatile compounds, such as many diterpenes, can also be detected, but they often need to be derivatized prior to analysis [15]. Terpene compounds can be identified with the help of mass spectral libraries and by comparison to authentic standards.

There are various methods for the extraction of EOs from plants. The most common are hydrodistillation, organic solvent extraction, and solid-phase microextraction (SPME). The latter is a method using fiber cores coated with stationary phases that are exposed to the headspace of a sample (HS-SPME) or immersed in a liquid sample. Depending on the chemical properties of the analytes and stationary phase, the analytes are absorbed into the fiber and desorbed in the GC injector at high temperatures. The advantages of SPME compared to other methods are solventless extraction and time saving during sample preparation [16].

Here, we present an example of EO analysis of thyme plants (*Thymus vulgaris*). Thyme contains high amounts of EO in the leaves, often dominated by phenolic monoterpenes such as thymol and carvacrol. Due to the antibacterial and spasmolytic activities of these two compounds, thyme is widely used as a medicinal plant, e.g., in plant remedies for infections of the respiratory system [17]. Monoterpene content in individual thyme plants can vary since these plants occur as different chemotypes, each of them dominated by different characteristic monoterpenes [18]. Therefore, thorough analysis is vital to ensure the quality of the EOs used for medicinal purposes. We compare solvent extraction (hexane) with SPME and show differences in extraction results. We also discuss rearrangement reactions of terpene molecules that can occur during the extraction process or analysis by GC-MS.

2 Materials

2.1 Terpene Extraction with Hexane

1. Thyme plants (*Thymus vulgaris*) or frozen plant material.
2. Nonyl acetate (analytical standard quality) in *n*-hexane (10 μg/mL). Dilute 11.52 μL nonyl acetate in 100 ml n-hexane (equals 100 μg/mL). A 1:10 dilution of this solution results in 10 μg/mL nonyl acetate in n-hexane.
3. 0.5 mM thymol (analytical standard quality) in *n*-hexane. Dissolve 75 mg thymol in 10 mL n-hexane (equals 50 mM). A 1:100 dilution of this solution results in 0.5 mM thymol in n-hexane.
4. GC vials 1.5 mL.
5. GC vial lids with silica septum.
6. GC vial glass inserts.
7. Micro scales.
8. Mortar and pestle.
9. Liquid nitrogen.
10. Vortexer.

2.2 Terpene Extraction with SPME

1. 1.5 mL GC vials and lids with silica septum.
2. Micro scales.
3. SPME Fiber Assembly, 100 μm polydimethylsiloxane (PDMS) for the manual holder.
4. Fiber holder for manual sampling.
5. Water bath or mini dry bath incubator.

2.3 GC Analysis

1. Gas chromatograph coupled to a mass spectrometer.
2. Software for running GC-MS and post-run analysis.
3. Mass spectral libraries, e.g.,
   - Wiley Registry of Mass Spectral Data, 2023 edition.
   - NIST 23.
   - Adams, 2007 (focuses on EO components) [19].

3 Methods

3.1 Plant Extraction and Sample Preparation

3.1.1 Extraction with N-hexane (See Note 1)

1. Harvest fresh thyme leaves or use frozen material.
2. Grind to a fine powder with mortar and pestle in liquid nitrogen (*see* **Note 2**).
3. Weigh 25 mg of ground powder into a GC vial (1.5 mL).
4. Add 1 mL of *n*-hexane (supplied with **10** μg/mL nonyl acetate as internal standard) (*see* **Note 3** and **4**).
5. Vortex vigorously for 15 s.
6. Incubate 30 min at RT.
7. Transfer 50–100 μL of the solvent into a new glass vial for GC analysis (*see* **Notes 5** and **6**).
8. Transfer 50 μL of a 0.5 mM solution of thymol standard in *n*-hexane to a GC vial for GC analysis.

3.1.2 Extraction with Solid-Phase Microextraction (SPME)

1. Perform **steps 1–2** in Subheading 3.1.1.
2. Weigh 5 mg of ground powder into a GC vial (1.5 mL) and incubate at 30 °C for 30 min (*see* **Note 7**).
3. Pierce the needle of the SPME fiber holder through the septum of the GC vial lid and expose the fiber to the headspace of the plant powder for 5 min (*see* **Notes 8–10**).

3.2 GC-MS Settings and Conditions

Here, we suggest a general GC protocol applicable for mono- and sesquiterpenes. However, depending on the sample compounds, modification of this protocol (e.g., column and temperature protocol) might be necessary for sufficient separation and detection of compounds of interest (*see* **Note 11**).

3.2.1 Sample Injection

1. Inject 1 μL of hexane extract ***or*** introduce an SPME fiber into the injector (*see* **Note 12**).
2. Injection temperature: 230 °C.
3. Column: FS-Supreme-5 ms Plus (5% Phenylpolysilphenylene, 30 m × 0.25 mm, thickness 0.25 μm) (*see* **Note 13).**
4. Carrier gas: Hydrogen (1 mL/min).

5. Temperature program: 50 °C for 3 min, first ramp 5 °C/min to 170 °C, second ramp 100 °C/min to 300 °C, final 2 min hold.
6. Operation mode: Electron-impact (EI) at 70 eV.
7. Scan range: 46–350 m/z.
8. Transfer line temperature: 250 °C.
9. Ion-source temperature: 220 °C.

3.3 Product Identification by Library Comparison and Authentic Standards

After analysis, obtained GC chromatograms (Fig. 2a) are analyzed with the corresponding GC software. Hexane extraction and SPME of thyme leaves show the same terpene profile, but with different ratios of terpene compounds (*see* **Note 14**). Terpene peaks can be tentatively identified by comparison to mass spectra libraries incorporated in your GC analysis software (*see* Subheading 2.3). The libraries show the best hits compared to the mass spectra of your sample peaks. Figure 2a shows that thyme contains monoterpenes from different classes. Ion masses 69 and 93 are indicators for acyclic monoterpenes (e.g., linalool) and ion masses 93 and 121 are indicators for cyclic monoterpenes (e.g., γ-terpinene). Our thyme sample contains large amounts of the aromatic monoterpenes *p*-cymene and thymol. The indicator ion for *p*-cymene is 119, due to an additional double bond in the aromatic scaffold of the molecule. The indicator ion for thymol, 135, results from an additional OH-group at the aromatic ring. For sesquiterpenes (e.g., β-caryophyllene), typical ion masses are 69 and 93 (Fig. 2b). Authentic standards are used to confirm the identity of compounds by comparison of retention time and mass spectra under the same GC conditions (Fig. 2a) (*see* **Note 15**).

3.4 Rearrangement Reactions

When performing GC-MS analysis of EOs and terpenoids, it is important to keep in mind that not all measured compounds do necessarily reflect the situation in the original sample. Depending on the conditions of the extraction and GC-MS analysis of the volatile metabolites, rearrangement reactions can occur and lead to artifacts, as previously summarized in [20, 21]. Transannular cyclization reactions can occur if an extraction is performed under acidic conditions, for example, if an unbuffered yeast culture is extracted. Another cause could be the use of specific SPME fibers [22]. When analyzing the enzyme product germacrene A acid **1** Nguyen et al. [23] observed acid-induced rearrangement products: α-, β-, and γ-costic acid **2–4** were formed, as well ilicic acid **5** after the addition of a water molecule (Fig. 3a). Heat-induced Cope-rearrangement reactions, the cyclization of 1,5-dienes [24], can be a problem in GC-MS analysis. Germacrene A **6**, for instance, can undergo such a heat-induced rearrangement to β-elemene 7 [25] (Fig. 3b). Most of the described rearrangement products (here: **2–4**, 7) have the same mass as the original product and can

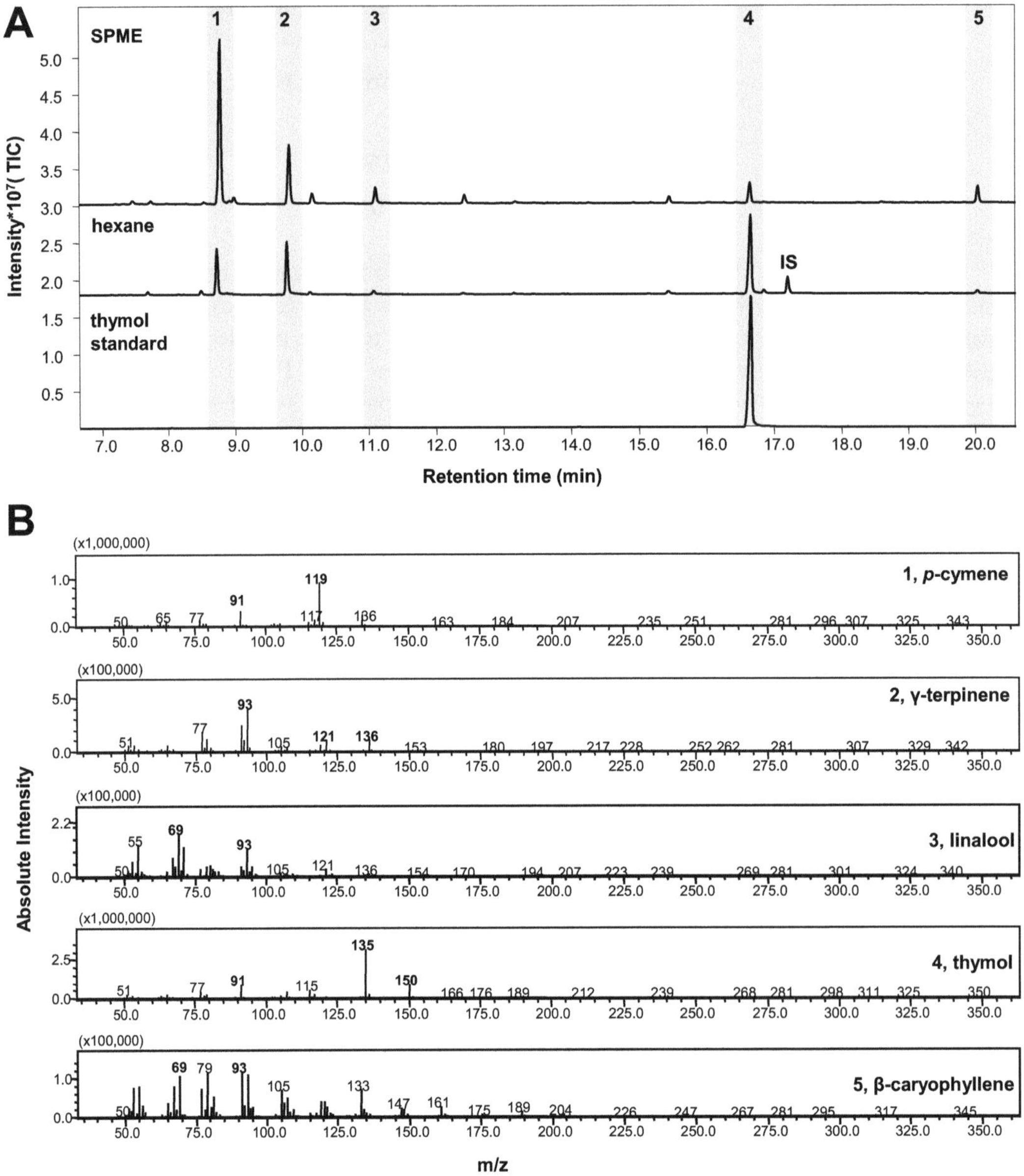

Fig. 2 EO composition of *Thymus vulgaris* leaves. (**a**) GC-MS chromatograms of SPME and hexane extracts show five dominant peaks: (*1*) *p*-cymene; (*2*) γ-terpinene; (*3*), linalool; (*4*), thymol; (*5*), β-caryophyllene. A thymol standard was run under the same GC-MS conditions to validate the identity of the thymol peak detected in the hexane and SPME extracts. (**b**) Mass spectra of dominant terpenes

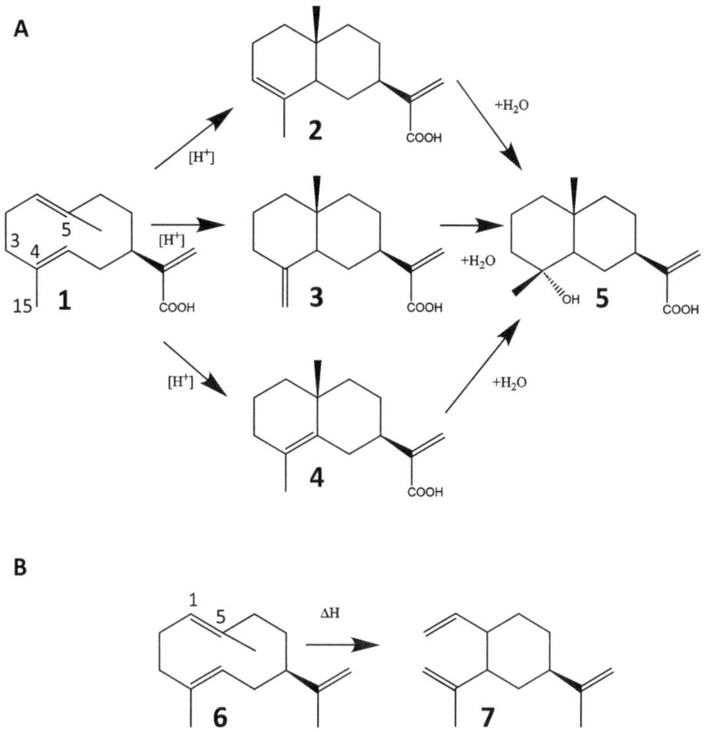

Fig. 3 Possible rearrangement reactions of sesquiterpenoids during extraction and GC-MS. (**a**) Acid-induced rearrangements of germacrene A acid **1** via α-, β-, and γ-costic acids **2–4** to ilicic acid **5**. (**b**) Heat-induced Cope rearrangement of germacrene A **6** to β-elemene **7**

only be differentiated by slightly different retention times and mass spectra of the original product. To avoid acid- and heat-induced rearrangement reactions there are several possibilities: (1) Buffering of the sample that should be extracted, for instance buffering the yeast culture medium [26]. (2) If SPME fibers are used: Comparison of different fibers for optimal conditions and (3) lowering the temperature of the GC-MS inlet [22].

4 Notes

1. Solvents such as *n*-hexane may extract plasticizers from disposable lab equipment such as plastic reaction tubes, culture tubes, and pipet tips that may interfere with the GC-MS analysis. In order to completely avoid such undesired co-extraction switching to glassware is an option. Alternatively, most vendors now give detailed information on the compatibility of their plastic material with solvents and buffers. However, running a solvent control can clarify the profile of plastic extraction and help to identify unspecific co-extracted compounds.

2. While handling liquid nitrogen, make sure to wear protective gloves and safety glasses in order to avoid skin and eye injuries.
3. For analyte quantification, an IS is added at a fixed concentration to all samples. The analyte concentrations in the sample are deduced from the comparison of the peak areas of the analytes to the peak area of the IS. Using an IS helps to avoid quantification errors, e.g., due to partial evaporation of samples or volume errors during sample preparation or injection. IS should resemble the physical and chemical properties of the analytes of interest, but they have to be clearly resolved from the analyte peak in the chromatogram. For monoterpenes, internal standards like p-cymene, camphor, dodecane, or nonyl acetate are widely used. Details of quantification, however, will not be discussed in this protocol.
4. Choice of appropriate solvent is vital for the efficient extraction of specialized plant metabolites. Hexane is a good solvent for apolar lipophilic molecules and is suitable for most terpene extractions from plant material. However, for analysis of polar terpenes, more polar solvents like methanol can lead to better extractions of the compounds of interest.
5. Solvents injected into the GC should be clear and free of any residual plant material. In case there is any plant material floating in the solvent phase, make sure to centrifuge the sample in order to pellet the plant powder to the ground of the vial. Additional filtering of the extract through a 0.45 μm pore size filter is possible.
6. The needle of the syringe of the GC autosampler usually takes the sample from 2 to 3 mm above the bottom of the GC vial. Make sure that the level of the sample is high enough to ensure that the needle reaches into the sample in order to prevent that air is filled into the syringe and injected into the GC. Alternatively, you can use glass vial inserts to elevate the sample level.
7. Screw the lid on GC vial to allow the establishment of equilibrium between the sample and headspace.
8. PDMS-coated fibers are generally suitable for all terpene extractions. However, if you want to extract polar terpenes or other polar compounds, SPME fibers with a more polar stationary phase can be used for efficient extraction, e.g., polyacrylate or PDMS mixed with carboxen (CAR) or divinylbenzene (DVB).
9. Introduce the needle of the SPME fiber holder through the septum of the GC vial lid and expose the fiber to the headspace afterward. Make sure to retract the fiber into the needle before pulling it out of the GC vial.

10. Thyme leaves usually contain high concentrations of EO. However, other plant species might contain less amounts of EO. In that case, you might extend the exposal time up to 45–60 min and increase the incubation temperature.
11. You can use the GC program for both liquid injection and SPME. When using liquid injection, make sure to set the solvent cut time to 4–5 min. It is important that the solvent elutes before the MS turns on, otherwise, the ion source or the detector could get damaged.
12. Leave the SPME fiber for at least 10 min in the GC injector to make sure all residual sample is evaporated from the fiber. The fiber is then conditioned for the next sample.
13. This is a column optimized for very low column bleeding. There are other columns with similar separation properties, which can be used as well. For example, DB-5 (Agilent) or Alltech™ Econo-Cap EC-5 (Grace™).
14. In our example, the SPME fiber (PDMS) shows a strong affinity to *p*-cymene, it is the most abundant peak in the GC chromatogram (Fig. 2a). The hexane extract shows different compound ratios with thymol being the most abundant. This illustrates that the extraction method and chemical properties of the extraction matrix can have a substantial influence on the outcome of the analysis.
15. In case of unclear structure elucidation, lack of standard compounds, or lack of a reliable NIST entry, upscaling and preparative purification for NMR analysis might be necessary, depending on the research context. While in the analysis of known samples comparison of mass spectra and authentic standards is sufficient, the analysis of completely unknown samples and compounds might require NMR analysis for detailed structure information of individual compounds of interest.

References

1. Lang G, Buchbauer G (2012) A review on recent research results (2008–2010) on essential oils as antimicrobials and antifungals. A review. Flavour Fragr J 27:13–39
2. Böhme K, Barros-Velázquez J, Calo-Mata P et al (2014) Antibacterial, antiviral and antifungal activity of essential oils: mechanisms and applications. In: Antimicrobial compounds: current strategies and new alternatives. Springer, Berlin/Heidelberg, pp 51–81. https://doi.org/10.1007/978-3-642-40444-3
3. Mutlu-Ingok A, Devecioglu D, Dikmetas DN et al (2020) Antibacterial, antifungal, antimycotoxigenic, and antioxidant activities of essential oils: an updated review. Molecules 25:4711
4. Hanif MA, Nisar S, Khan GS et al (2019) Essential oils. In: Essential oil research. Springer, Cham, pp 3–17
5. Nakatsu T, Lupo AT Jr, Chinn JW Jr et al (2000) Biological activity of essential oils and their constituents. In: Studies in natural products chemistry. Elsevier, Amsterdam, pp 571–631

6. Baser KHC, Demirci F (2007) Chemistry of essential oils. In: Flavours fragrances: chemistry, bioprocessing and sustainability. Springer, Berlin, p 43
7. Baser KHC, Buchbauer G (2009) Handbook of essential oils: science, technology, and applications. CRC Press, Boca Raton
8. Gershenzon J, Croteau R (1991) Terpenoids. In: Rosenthal GA, Berenbaum MR (eds) Herbivores: their interactions with secondary plant metabolites, The chemical participants, vol I, 2nd edn. Academic Press, San Diego, pp 165–219
9. McGarvey DJ, Croteau R (1995) Terpenoid metabolism. Plant Cell 7:1015–1026
10. Bryant JP, Provenza FD, Pastor J et al (1991) Interactions between woody plants and browsing mammals mediated by secondary metabolites. Annu Rev Ecol Syst 22:431–446
11. Amiot J, Salmon Y, Collin C et al (2005) Differential resistance to freezing and spatial distribution in a chemically polymorphic plant Thymus vulgaris. Ecol Lett 8:370–377
12. Raut JS, Karuppayil SM (2014) A status review on the medicinal properties of essential oils. Ind Crop Prod 62:250–264
13. Stahl-Biskup E (2002) Essential oil chemistry of the genus thymus – a global view. In: Stahl-Biskup E, Saez F (eds) Thyme: the genus thymus. Taylor & Francis, London, pp 75–124
14. Marriott PJ, Shellie R, Cornwell C (2001) Gas chromatographic technologies for the analysis of essential oils. J Chromatogr A 936:1–22
15. Wang Q, Sohrabi R, Tholl D (2014) Analysis of diterpenes and triterpenes from plant foliage and roots. Methods Mol Biol 1153:149–159
16. Vas G, Vékey K (2004) Solid-phase microextraction: a powerful sample preparation tool prior to mass spectrometric analysis. J Mass Spectrom 39:233–254
17. Zarzuelo A, Crespo E (2002) The medicinal and non-medicinal uses of thyme. In: Stahl-Biskup E, Saez F (eds) Thyme: the genus thymus. Taylor & Francis, London, pp 263–292
18. Vernet P, Gouyon RH, Valdeyron G (1986) Genetic control of the oil content in Thymus vulgaris L: a case of polymorphism in a biosynthetic chain. Genetica 69:227–231
19. Adams RP (2007) Identification of essential oil components by gas chromatography/mass spectrometry. Allured Publishing Corporation, Carol Stream
20. Frey M (2020) Traps and pitfalls-unspecific reactions in metabolic engineering of Sesquiterpenoid pathways. Molecules 25:1935
21. Frey M, Vahabi K, Cankar K et al (2024) Sesquiterpene lactones—insights into biosynthesis, regulation and signalling roles. CRC Crit Rev Plant Sci 43:1–27
22. Andersen TB, Cozzi F, Simonsen HT (2015) Optimization of biochemical screening methods for volatile and unstable sesquiterpenoids using HS-SPME-GC-MS. Chromatography 2: 277–292
23. Nguyen DT, Göpfert JC, Ikezawa N et al (2010) Biochemical conservation and evolution of germacrene A oxidase in asteraceae. J Biol Chem 285:16588–16598
24. Cope AC, Hardy EM (1940) The introduction of substituted vinyl groups. V. A rearrangement involving the migration of an allyl group in a three-carbon system. J Am Chem Soc 62:441–444
25. Göpfert JC, Macnevin G, Ro D-K et al (2009) Identification, functional characterization and developmental regulation of sesquiterpene synthases from sunflower capitate glandular trichomes. BMC Plant Biol 9:86
26. Liu Q, Majdi M, Cankar K et al (2011) Reconstitution of the costunolide biosynthetic pathway in yeast and Nicotiana benthamiana. PLoS One 6:e23255

Chapter 7

Isolation of Natural Products from Marine Invertebrates

Yeon-Ju Lee and Chinmayee Bawkar

Abstract

Marine invertebrates such as sponges, cnidarians, bryozoans, and tunicates are considered to be prolific sources of novel biologically active natural products. Specimens from these organisms are composed of soft tissues with a high water and inorganic salt content compared to those from terrestrial counterparts. In addition, some of the secondary metabolites from these organisms tend to be more polar than those of terrestrial. These general characteristics have led to the establishment of differentiated protocols for metabolites derived from marine invertebrates. In this chapter, we describe an available protocol for the isolation of secondary metabolites from marine invertebrates, including extraction, fractionation, and purification procedures. While detailed protocols should be adapted to the characteristics of the specific organism being studied, the methods provided here can serve as an initial guide for designing those.

Key words Marine invertebrate, Natural product, Extraction, Isolation, Purification, Solvent partition, High-performance liquid chromatography

1 Introduction

Marine organisms are relatively untapped sources of natural products that have recently begun to be exploited [1]. Among these, invertebrates such as sponges, cnidarians, bryozoans, and tunicates are receiving increasing attention in natural product research. A recent statistical analysis revealed that a total of 15,442 new natural products were discovered in marine invertebrates from 1990 to 2019, with the last decade being the most productive, with 5630 new compounds discovered [2].

Marine invertebrates share several general characteristics in their ecology, evolution, or development. They are either sessile or motile, with a low level of differentiation in organs and tissues. Some of them, especially those in the phyla Porifera or Cnidaria, which have been one the most prolific sources of natural products, are colonial and filter feeding (Fig. 1). Due to these characteristics, they tend to rely on chemicals to defend themselves against predators or pathogenic microbes, to find the proper habitats, or to

Young Hae Choi et al. (eds.), *Natural Product Isolation and Identification: Methods and Protocols*, Methods in Molecular Biology, vol. 2895, https://doi.org/10.1007/978-1-0716-4350-1_7,

Fig. 1 Photographs of selected *Porifera Hymeniacidon* sp., *Stylissa carteri*, and *Ircinia* sp., Top, in situ; Bottom, specimens after collection for natural product research

track or evaluate foods [3–5]. It is implicated that these compounds are synthesized by holobionts, which are composed of marine invertebrates and microbes associated with them [6, 7].

The isolation and purification of compounds from marine invertebrates require special attention to their characteristics [8–10]; some of these compounds tend to be more polar and thermally unstable than those from their terrestrial counterparts, as expected given their habitat. In this chapter, exemplary protocols for the preparation of extracts from marine invertebrates and for the isolation of natural products from these extracts are provided (Fig. 2). More detailed and appropriate protocols specific to the specimens being studied can be established based on the modification of these protocols.

Recent metabolomics techniques based on ^{1}H nuclear magnetic resonance (NMR) spectroscopy or liquid chromatography-mass spectrometry (LC-MS) can provide a great deal of information about natural products at the extract level. In addition, high-throughput screening (HTS) can efficiently provide the activities of extracts in various assays. It is feasible to select the organism that matches the research objectives based on the characteristics of the extract by utilizing these techniques. In order to streamline the process of isolating natural products, which is usually time-consuming and labor-intensive, it is beneficial to perform targeted compound tracing. Targeted compounds are usually those with specific chemical scaffolds or promising biological activities. The presence of these compounds in a particular fraction or subfraction

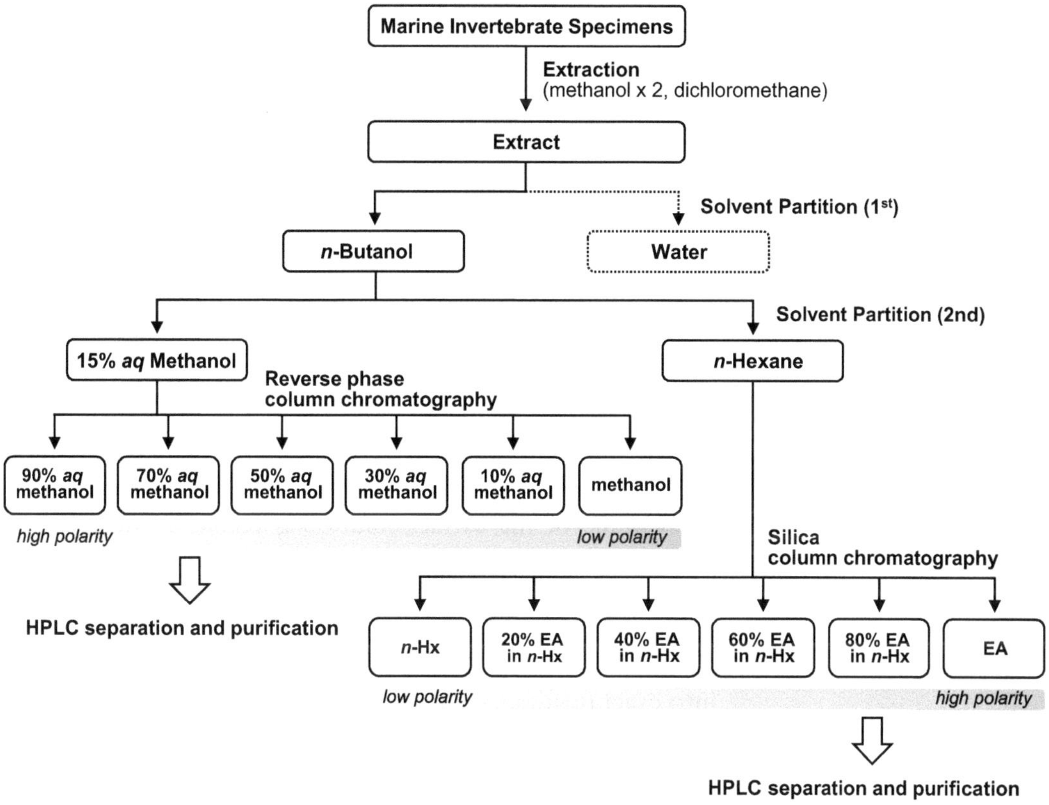

Fig. 2 Overall scheme for the separation of natural products from marine invertebrates

can be validated by the same method used to evaluate the extract. If the study is initiated from the results obtained by ^{1}H NMR analysis of the extract, the entire process can be monitored by measuring the ^{1}H NMR of the fractions and subfractions obtained at each step. Bioactivity-guided separation, which is driven by the activities of the extract and fractions, is also widely recognized and used as an efficient approach.

It should be noted that this chapter does not cover spectroscopic or spectrometric analyses used to evaluate extracts and fractions or to identify isolated compounds. Nor does it cover the measurement of the biological activity of the fractions or compounds. As mentioned above, it is necessary to validate the fractions obtained in each step by ^{1}H NMR, LC-MS, or biological activity measurements, but this chapter focuses only on the isolation procedure.

2 Materials

Collected specimens are immediately frozen and stored at −20 °C until use. The extracts, fractions, and isolated compounds are stored in borosilicate glass vials at −20 °C when not in use. All the glassware such as Erlenmeyer flasks, round-bottomed flasks, test tubes, separatory funnels, and columns are made of borosilicate glass. Separatory funnels and columns are equipped with polytetrafluoroethylene (PTFE) stopcocks. Throughout the process, a rotary evaporator was used for concentration in vacuo.

2.1 Extraction

1. Solvent: MeOH and dichloromethane, either high-performance liquid chromatography (HPLC) grade or purified by distillation, are used.
2. Glassware: Erlenmeyer flasks or beakers with a volume greater than 1.5 times the volume of solvent used are recommended. A glass funnel with a long, narrow stem and an inner diameter greater than 100.0 mm is used for filtration. To collect the filtrate, a round-bottomed flask with a volume more than twice the volume of the extract solution is recommended for safe concentration using a rotary evaporator.
3. Filter paper: Cellulose filter paper of grade 1 (11.0 μm) is folded into a conical shape to facilitate recovery of the solid residue more easily for serial extraction. The paper is folded into exact quadrants and opened up by placing three layers on one side and one layer on the other. The folded filter paper is placed in a glass funnel and moistened with solvent so that it adheres to the inside of the funnel.
4. Instrument: A rotary evaporator equipped with a cooling condenser, vacuum pump, and water bath is used. The vacuum pump should be resistant to organic solvents, and the temperature of the cooling condenser should be kept below −20 °C. A water bath with a thermostat is used to heat the round-bottomed flask containing the solvent. The temperature can be determined by considering the boiling point of the solvent to be removed, but should not exceed 40 °C, as the thermal stability of the compounds to be isolated is not revealed at this point.

2.2 Solvent Partition

1. Solvent: Water is deionized and organic solvents—MeOH, *n*-BuOH, *n*-hexane—are HPLC grade or distilled prior to use.
2. Glassware: A separatory funnel with a PTFE stopcock is used. The funnel volume should be at least 1.5 times the total solvent volume to ensure sufficient mixing of the two layers by shaking. The separatory funnel should be placed stably supported on a ring clamp allowing the layers to separate. Any type of flask, such as an Erlenmeyer flask or round-bottomed flask, can be used as a receiver for each layer; however, the use of round-bottomed flask is recommended when the following concentration method is considered.

3. Instrument: For the use of a rotary evaporator, refer to **item 4**, Subheading 2.1.

2.3 Fractionation by Open Column Chromatography

2.3.1 Reverse Phase Column Chromatography

1. Solvent: Deionized water and HPLC grade MeOH (or distilled) are used.
2. Glassware: Borosilicate glass chromatography column with a fritted disc and a vacuum adapter is used. A Pasteur pipette is used to apply the loading material onto the stationary phase. A round-bottomed flasks are used to collect each effluent.
3. Octadecyl silica (ODS) gel (particle size 75 μm, pore size 12 nm) is used as the stationary phase. The packing slurry is prepared by adding 15% aqueous MeOH (v/v) at a concentration of 30% (w/v).
4. Instrument: For faster elution, a vacuum is applied to the column. A diaphragm pump or aspirator pump is used for the generation of rough vacuum. For the use of a rotary evaporator to concentrate the effluent, refer to **item 4**, Subheading 2.1.

2.3.2 Silica Column Chromatography

1. Solvent: Ethyl acetate (EtOAc) and *n*-hexane of HPLC grade (or distilled) are used.
2. Glassware: The same as those indicated in **item 2**, Subheading 2.3.1.
3. Silica gel (particle size 40-63 μm, pore size 6 nm) is used as a stationary phase. The silica slurry is made by adding the initial mobile phase at a concentration of 30% (w/v).
4. Instrument: Vacuum can be applied as indicated in **item 4**, Subheading 2.3.1, however, the gravity flow is usually enough for silica column chromatography. For the use of a rotary evaporator, refer to **item 4**, Subheading 2.1.

2.4 High-Performance Liquid Chromatography (HPLC)

1. Solvent: Deionized water and HPLC grade MeOH (or AcCN) are used. For maximal separation, acid additives such as formic acid or trifluoroacetic acid of HPLC grade or with purity over 99% are used. Each solvent should be degassed by a rough vacuum with stirring at 25 °C in a solvent reservoir, which will be connected to the HPLC pump until vigorous bubbling stops.
2. Glassware: Erlenmeyer flasks or test tubes are used to collect the effluent corresponding to each chromatographic signal.
3. HPLC columns: Analytical HPLC column (4.6 mm × 250 mm or smaller, ODS (3 or 5 μm)) is used for the screening of various conditions to determine the optimal condition for maximal separation. For preparative HPLC, semi-preparative (10.0 mm × 250 mm, ODS (3 or 5 μm)) or preparative HPLC columns are used.

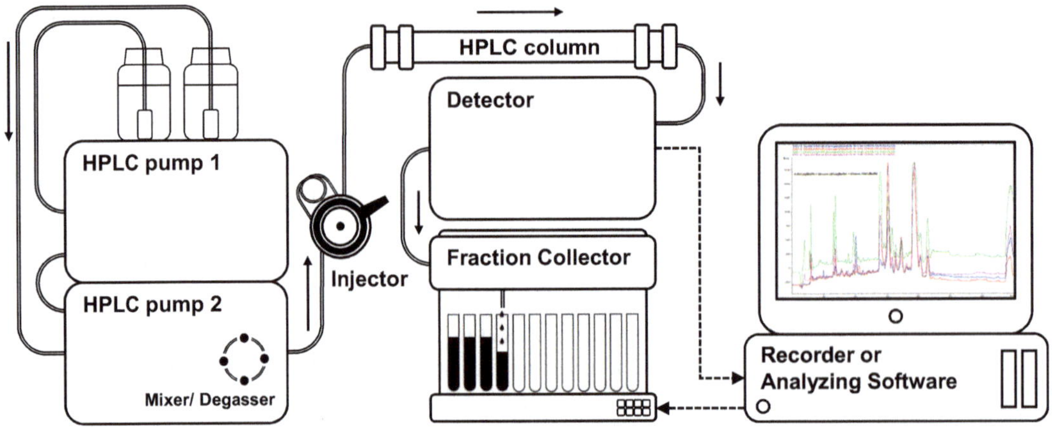

Fig. 3 Schematic diagram of HPLC instrumentation. The automated fraction collector is optional. Without it, each effluent is collected by hand. Recently, analyzing software provided by each manufacturer of HPLC instruments is more commonly used, but classical chromatogram recorder is still in use

4. Instrument: A set of instruments for HPLC consists of an injector, one or more pumps, and a detector, with an optional fraction collector (Fig. 3). Each pump is connected to a solvent reservoir and the solvent flow goes sequentially to the injector, column, detector, and fraction collector. If more than one pump is used, the solvents from each pump are combined and go to the additional pump, then to the mixer (quaternary pump), or directly to the mixer (binary pump). Since detection by UV-vis light is the most commonly used, a photodiode array (PDA) or diode array detector (DAD) is used. The detection method can be modified according to the properties of the compounds in the fraction.

3 Methods

3.1 Extraction

There are several extraction methods for living organisms [11], but the most commonly used method for marine invertebrates is maceration. Specimens are minced, dried, and then extracted using organic solvents such as MeOH and dichloromethane.

1. Frozen specimens are thawed at room temperature for 4–5 h. Once the specimens are soft and at room temperature, the gross specimens are weighed to determine the wet weight.
2. Specimens are diced into small pieces of less than 1.0 cm^3 in size and lyophilized. The dried specimens are weighed, and the dry weight is noted.
3. Dried specimens are placed in an Erlenmeyer flask or beaker. Extraction is initiated by incubating the dried specimens in MeOH for 12 h at 25 °C, maintaining the ratio of 1 L of

MeOH to 100 g of dried specimens (*see* **Note 1**). The extract solution is filtered through filter paper and collected in a round-bottomed flask.

4. The second round of extraction of the remaining residue of the specimens is also performed in MeOH; Subheading 3.1 **step 3** is repeated.
5. The third round of extraction is performed in a less polar solvent—dichloromethane; Subheading 3.1 **step 3** is repeated by simply changing the solvent to dichloromethane.
6. All the extracts obtained in Subheading 3.1 **steps 3–5** are combined in a round-bottomed flask and concentrated in vacuo using a rotary evaporator at below 40 °C to obtain a solid or oily residue. These three cycles are usually sufficient.

3.2 Solvent Partition

Solvent partitioning is the separation of compounds according to their solubility based on phase separation between two immiscible solvents. This process is performed twice, first with *n*-BuOH and water to remove inorganic salts coming from seawater, and then with 15% aqueous MeOH and *n*-hexane for the initial separation of compounds by polarity (Fig. 4). This allows the separation of polar compounds in the 15% aqueous MeOH fraction, while relatively non-polar compounds such as fatty acids, steroids, and terpenoids are obtained in the *n*-hexane fraction.

1. Dissolve the extract in a 5:4 ratio of *n*-BuOH and water (approximately 1.0 L of *n*-BuOH and 0.8 L of water for 20 g of extract) and then transfer it to a separatory funnel. After closing the stopper, shake the funnel vigorously while

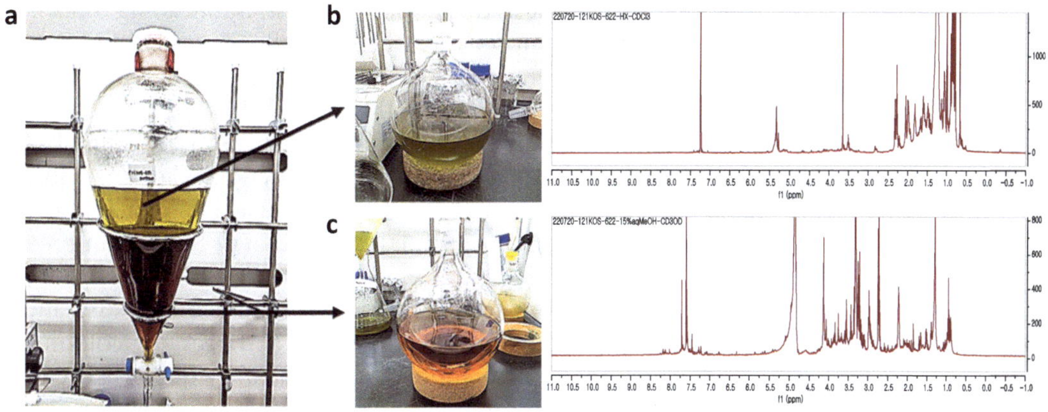

Fig. 4 Photographs of an exemplary solvent partition. (**a**) Phase separation between *n*-hexane and 15% aqueous MeOH. (**b**) The *n*-hexane layer was collected in a round-bottomed flask and the ^{1}H NMR spectra were measured for this fraction. (**c**) The 15% aqueous MeOH layer was collected in a round-bottomed flask and the ^{1}H NMR spectra were measured for this fraction

occasionally opening the stopcock upward to release the vapor pressure inside.

2. After incubation at 25 °C until two immiscible solvents—*n*-BuOH and water—separate into distinct phases, (*n*-BuOH in the upper phase and water fraction in the lower phase), collect each phase separately in a round-bottomed flask. The liquids are separated by drawing off the lower layer and this should be proceeded slowly as the interface approaches the stopcock. The resulting *n*-BuOH solution is concentrated in vacuo using a rotary evaporator (*see* **Note 2** and **3**).
3. The second round of solvent partition is performed using *n*-hexane and 15% aqueous MeOH. The residue obtained by the concentration of the *n*-BuOH phase is dissolved in 15% aqueous MeOH and *n*-hexane (1:1, approximately 0.5 L of each for 10 g of *n*-BuOH fraction), transferred into a separatory funnel, and then the Subheading 3.2 **step 1** is repeated.
4. Let it stand at 25 °C until the compounds are separated into their respective solvents (Fig. 4). When the layers are clearly distinguishable, carefully collect the upper layer of *n*-hexane and the lower layer of 15% aqueous MeOH separately (*see* Subheading 3.2 **step 2**). Each resulting solution is placed in a round-bottomed flask and concentrated in vacuo to give 15% aqueous MeOH and *n*-hexane fractions.

3.3 Fractionation by Open Column Chromatography

Each fraction obtained by solvent partition is further separated on a hand-packed column [12, 13]. For the 15% aqueous MeOH fraction, reversed-phase column chromatography using an ODS resin as the stationary phase and aqueous MeOH as the mobile phase is suitable (*see* Subheading 3.3.1). The *n*-hexane fraction is further fractionated by silica column chromatography using a mixture of *n*-hexane and EtOAc as the mobile phase (*see* Subheading 3.3.2). In both cases, a gradient polarity of the mobile phase should be used for maximum coverage of compounds with varying polarity. In most cases, 90% aqueous MeOH for the 15% aqueous MeOH fraction and *n*-hexane for the *n*-hexane fraction are suitable for starting the elution. The resulting subfractions can be analyzed for their chemical constituents or biological activities, and selected fractions can be further separated by HPLC (*see* Subheading 3.4).

3.3.1 Reverse Phase Column Chromatography

1. In a glass chromatography column with a fritted disc and a PTFE stopcock, add ODS gel slurry and pack the column by flowing the initial mobile phase (90% aqueous MeOH) through the stationary phase (*see* **Note 4**). A solvent volume of 5 to 6 times that of the stationary phase is usually sufficient.
2. Dissolve the 15% aqueous MeOH fraction in a minimal volume of MeOH and apply the solution to the resin using a Pasteur

pipette, distributing it evenly over the surface and allowing it to adsorb onto the stationary phase.

3. Add 90% aqueous MeOH to the column (*see* **Note 5**), make the flow by applying a vacuum, and collect the effluent. Repeat this process by gradually reducing the polarity from 90% aqueous MeOH to 100% MeOH, keeping an interval of 20%, i.e., 90%, 70%, 50%, 30% and 10% aqueous MeOH (v/v), and 100% MeOH.
4. The effluents can be divided into each subfraction based on the eluent used. As a result, six subfractions can be obtained. Each subfraction is collected in a round-bottomed flask and concentrated in vacuo using a rotary evaporator (*see* **Note 6**).
5. The ODS gel is thoroughly washed with MeOH or AcCN, or one of these solvents containing 0.5% formic acid or trifluoroacetic acid, to remove any residues that may be adsorbed on the resin. The washed resin can be stored in approximately 30% aqueous MeOH (or AcCN) and reused.

3.3.2 Silica Column Chromatography

The *n*-hexane fraction obtained by solvent partition is further fractionated by silica column chromatography using a stepped gradient solvent of *n*-hexane and EtOAc. As with reverse phase column chromatography fractionation, six subfractions can be obtained.

1. In a glass chromatography column with a fritted disc and a PTFE stopcock, add silica slurry and pack the column by allowing the initial mobile phase (*n*-hexane) to flow through the stationary phase.
2. Dissolve the *n*-hexane fraction in a minimal volume of EtOAc and apply the solution on top of the resin using a Pasteur pipette, distributing it evenly over the surface. Allow it to adsorb onto the stationary phase.
3. Add *n*-hexane to the column and collect the effluent. For silica column chromatography, gravity elution is usually sufficient. If the flow is too slow, air pressure can be applied (flash column chromatography). Repeat this process by gradually increasing the polarity from *n*-hexane to EtOAc at intervals of 20%, i.e., 20%, 40%, 60%, 80% EtOAc in *n*-hexane (v/v), and 100% EtOAc.
4. The effluents can be divided into each subfraction based on the eluent used. As a result, six subfractions can be obtained. Each subfraction is concentrated in vacuo using a rotary evaporator (*see* **Note 6**).

3.4 High-Performance Liquid Chromatography (HPLC)

HPLC is a liquid chromatography technique of the highest resolution in which the mixture of organic compounds can be separated using a prepacked column specially designed to withstand high pressures of up to 400 bar. The marine invertebrate extracts are usually very complex mixtures of primary and secondary metabolites; it is difficult to obtain single compounds of a certain purity by low to medium-pressure column chromatography. HPLC separation and purification are required at the final stage in the isolation of marine invertebrate natural products. There are different types of columns that have been invented so far, the separation of which depends on their molecular characteristics such as polarity, charge, or size [14, 15]. The most widely used technique for marine natural products is reversed-phase HPLC on an ODS column.

1. The HPLC behavior of the selected subfraction obtained in either Subheading 3.3.1 or 3.3.2 should be monitored using an analytical column. The subfraction of interest is dissolved in the minimal volume of MeOH or AcCN. If the column size is 4.6 mm in diameter and 250 mm in length, approximately 1.0 mg of the subfraction is injected.
2. The HPLC chromatograms obtained under various conditions are examined. The most feasible factors to examine are the column packing material, the composition of the mobile phase, and the flow rate. The most used stationary phase for the separation of marine natural products is ODS with a particle size of 3 or 5 μm (*see* **Note 7**). The mobile phase can be MeOH or AcCN mixed with water. For optimal resolution, acids such as formic acid or trifluoroacetic acid can be added at a concentration of 0.05–0.1%. Either gradient or isocratic conditions can be considered. The flow rate can be adjusted based on the separation observed in the chromatogram, but there is a range of flow rates for each column provided by the manufacturer.
3. The detection method should also be considered. If the chemical scaffold or structure of the compounds in the subfractions can be specified, UV detection using a specific wavelength of light is possible. By referring to the UV absorption pattern (expected or measured) of the compounds, the detection wavelength can be determined. If the UV absorption is too weak or the optimal wavelength is too short, other methods such as refractive index (RI) or evaporative light scattering (ELS) detection can be used as alternatives.
4. Based on the optimal HPLC conditions confirmed at the analytical scale, a preparative scale HPLC separation is performed. If the same type of HPLC column is used, the large-scale separation should be similar to the small-scale separation, but the proper loading amount and flow rate should be verified.

5. Once all the conditions are optimized, inject the sample and collect the effluent separately as each signal is observed. Each collected effluent is concentrated in vacuo and identified by spectroscopic method (*see* **Note 8**).

4 Notes

1. If the color of the solvent extract is too dark or dense, incubating the sample for a longer time after adding more solvent will ensure proper extraction. If the solution is too saturated, extraction may not be effective.
2. The amount of solvent can be varied depending on the amount of extract and its solubility in certain solvents. If precipitate forms, increase the amount of solvent or filter while collecting.
3. It is beneficial to acquire the ^{1}H NMR spectra of the *n*-BuOH fraction in methanol-d_4 or dimethyl sulfoxide-d_6 and the water fraction in D_2O. Most organic compounds are typically found in the *n*-BuOH fraction, while the water fraction contains mostly inorganic salts. If signals from suspected secondary metabolites are observed in the ^{1}H NMR spectra of the water fraction, the extraction of this fraction with *n*-BuOH should be performed again.
4. Column size is variable depending on the amount and the complexity of the sample. Approximately 30.0 cm × 6.0 cm (L x I.D.) stationary phase is used for 5 g.
5. Once the fraction is loaded, allow it to adsorb onto the stationary phase to prevent dissolution in the solvent. Add the solvent from the sides of the column wall or use a glass rod to avoid creating voids in the column that would interfere with proper and linear separation.
6. Each subfraction can be analyzed for its chemical constituents (^{1}H NMR or LC-MS) or biological activity. All or selected subfractions can be subjected to further separation and purification for the isolation of compounds.
7. An ODS HPLC column is usually the first choice for the isolation of marine natural products. However, if it is difficult to obtain the desired separation with an ODS column, a change of the stationary phase can be considered. For the mixtures with predominantly polar compounds, the C-8, C-4, or amide columns can be considered. For the separation of compounds with aromatic functional groups, silica substituted with phenyl groups can be used as the stationary phase. If the retention time of each compound on ODS is too short, hydrophilic interaction liquid chromatography (HILIC) columns are used. If the polarity of the mixture is low, a normal phase column with a

bare silica stationary phase can be used with the non-polar mobile phase such as *n*-hexane and EtOAc.

8. If the purity of the compound does not reach the desired level, additional HPLC purification is required. It is advantageous to use different conditions from the initial HPLC separation, either in the stationary or mobile phase.

Acknowledgments

This work was supported by grants from the Korea Institute of Ocean Science and Technology (PEA0121 and PEA0125).

References

1. Jiménez C (2018) Marine natural products in medicinal chemistry. ACS Med Chem Lett 9(10):959–961. https://doi.org/10.1021/acsmedchemlett.8b00368
2. Calado R, Mamede R, Cruz S, Leal MC (2022) Updated trends on the biodiscovery of new marine natural products from invertebrates. Mar Drugs 20(6):389–402. https://doi.org/10.3390/md20060389
3. Mascuch S, Kubanek J (2019) A marine chemical defense partnership. Science 364(6445):1034–1035. https://doi.org/10.1126/science.aax8964
4. Lauritano C, Ianora A (2020) Chemical defense in marine organisms. Mar Drugs 18(10):518–523. https://doi.org/10.3390/md18100518
5. Kamio M, Derby CD (2017) Finding food: how marine invertebrates use chemical cues to track and select food. Nat Prod Rep 34(5):514–528. https://doi.org/10.1039/C6NP00121A
6. König GM, Kehraus S, Seibert SF, Abdel-Lateff A, Müller D (2006) Natural products from marine organisms and their associated microbes. Chembiochem 7(2):229–238. https://doi.org/10.1002/cbic.200500087
7. Simmons TL, Coates RC, Clark BR, Engene N, Gonzalez D, Esquenazi E et al (2008) Biosynthetic origin of natural products isolated from marine microorganism–invertebrate assemblages. Proc Natl Acad Sci USA 105(12):4587–4594. https://doi.org/10.1073/pnas.0709851105
8. Riguera R (1997) Isolating bioactive compounds from marine organisms. J Mar Biotechnol 5:187–193
9. Wright AE (1998) Isolation of marine natural products. In: Cannell RJP (ed) Natural products isolation, Methods in biotechnology, vol 4. Humana Press, Totowa, pp 365–408. https://doi.org/10.1007/978-1-59259-256-2_13
10. Ebada SS, Edrada RA, Lin W, Proksch P (2008) Methods for isolation, purification and structural elucidation of bioactive secondary metabolites from marine invertebrates. Nat Protoc 3(12):1820–1831. https://doi.org/10.1038/nprot.2008.182
11. Bucar F, Wube A, Schmid M (2013) Natural product isolation—how to get from biological material to pure compounds. Nat Prod Rep 30(4):525–545. https://doi.org/10.1039/C3NP20106F
12. Braithwaite A, Smith FJ (1999) Liquid phase chromatography on open columns. In: Braithwaite A, Smith FJ (eds) Chromatographic methods. Springer, Dordrecht, pp 117–164
13. Blunt JW, Calder VL, Fenwick GD, Lake RJ, McCombs JD, Munro MHG, Perry NB (1987) Reverse phase flash chromatography: a method for the rapid partitioning of natural product extracts. J Nat Prod 50(2):290–292. https://doi.org/10.1021/np50050a039
14. Žuvela P, Skoczylas M, Jay Liu J, Bączek T, Kaliszan R, Wong MW, Buszewski B (2019) Column characterization and selection systems in reversed-phase high-performance liquid chromatography. Chem Rev 119(6):

3674–3729. https://doi.org/10.1021/acs.chemrev.8b00246
15. Latif Z, Sarker SD (2012) Isolation of natural products by preparative high performance liquid chromatography (prep-HPLC). In: Sarker SD, Nahar L (eds) Natural products isolation, Methods in biotechnology, vol 864. Humana Press, Totowa, pp 255–274. https://doi.org/10.1007/978-1-61779-624-1_10

Part II

Identification

Chapter 8

Identification and Profiling of Primary Metabolites Through GC-MS and Associated Data Processing

Minki Shim, Yooseong Jeong, and Dong-Kyu Lee

Abstract

This chapter presents a comprehensive approach to profiling plant-derived primary metabolites using metabolomics, highlighting its critical role in decoding the biosynthesis of bioactive plant compounds. It details the utilization of gas chromatography-mass spectrometry (GC-MS) for the effective analysis and profiling of these metabolites. The process, encompassing extraction methods, chemical derivatization, and data processing, is thoroughly outlined. This methodology outlines comprehensive procedures for each stage of the workflow, encompassing metabolite extraction, GC-MS analysis, and data alignment, to produce a metabolomics dataset.

Key words Metabolomics, Plant primary metabolites, Gas chromatography-mass spectrometry, Metabolite identification, Data processing

1 Introduction

Understanding plant-derived primary metabolites, widely acknowledged as a vital source of bioactive natural products, is at the forefront of drug development [1–3]. The role of primary metabolites in the field of natural product research is to understand the plant-specific interplay between primary metabolites and secondary metabolites [4]. Primary metabolites are fundamental in triggering the synthesis of secondary metabolites, which are often valuable as bioactive compounds in natural products. Metabolites' crucial function enables plants to adapt to external environmental stressors, such as fluctuating cultivation conditions and drought, and they are significant in disease resistance through homeostatic regulation [5, 6]. Their relationship with primary metabolites can be leveraged to enhance the biosynthesis efficiency of active compounds in medicinal plants based on cultivation conditions. Therefore, a comprehensive understanding of the fluctuations in

Young Hae Choi et al. (eds.), *Natural Product Isolation and Identification: Methods and Protocols*, Methods in Molecular Biology, vol. 2895, https://doi.org/10.1007/978-1-0716-4350-1_8,

metabolites, influenced by genetic and environmental factors, is essential [7].

Metabolomics, emerging from the evolution and refinement of diverse chromatographic separation and detecting methods, has gathered extensive attention in natural product research for its pivotal role in identifying novel metabolites [8]. The integration of metabolomics into plant studies offers a comprehensive view of metabolic dynamics, enabling researchers to decipher plant physiology and biochemistry [4]. Especially, gas chromatography-mass spectrometry (GC-MS)-based metabolomics is well suited for the analysis of primary metabolites and low molecular weight polar molecules [9]. GC-MS allows large-scale profiling of metabolites which primarily targets plant-derived molecules including energy metabolism, carbohydrates, amino acids, and organic acids [10].

The profiling of primary metabolites based on GC-MS requires an optimization process from sample preparation to equipment selection. Since the Oliver Fiehn group introduced the method of conferring volatility to polar primary metabolites, various derivatization techniques such as trimethylsilylation have been used for metabolite profiling [11]. Using capillary gas chromatography columns, specifically those with a 95% dimethyl/5% diphenyl polysiloxane composition (indicated by a −5 or −5 ms suffix in their model names), allows for the effective separation of metabolites that have been derivatized by various techniques. This method is notably efficient for conducting simultaneous analyses of a wide range of metabolites. GC-MS-based metabolomics relies on the high reproducibility of electron impact ionization for the efficient identification of metabolites, making it suitable for use with a wide range of mass spectrometers from low-resolution quadrupole mass detector to high-resolution mass spectrometers like time-of-flight (TOF), quadrupole-TOF, and orbitrap. Analyzing the metabolites data obtained by such processes and parameters is notably efficient, owing to the well-established retention index and its associated mass spectral library.

Despite its advantages, GC-MS presents certain limitations when compared to other metabolomics instruments [12]. In contrast with liquid chromatography-mass spectrometry (LC-MS), GC-MS offers less comprehensive molecule coverage and struggles with the detection of larger molecules [13]. Moreover, polar molecules that cannot be made volatile through derivatization pose detection challenges. Results from GC-MS metabolomics can also differ considerably due to variations between laboratories and researchers, leading to discrepancies even when analyzing the same sample—a problem not typically seen with nuclear magnetic resonance (NMR) metabolomics [14]. Therefore, to compensate for these disadvantages, it is necessary to establish a process that is as robust as possible, from sample preparation to GC-MS analysis.

This chapter will describe the whole process of GC-MS-based primary metabolite profiling and identification. Robust and efficient extraction of primary metabolites, considering the structural characteristics of plant samples, is crucial to analyze and identify a maximal number of primary metabolites. The chapter also describes chemical derivatization that allows polar metabolites to be injected and analyzed by GC-MS. After getting the data, data processing including robust identification based on the mass spectral library will be described.

2 Materials

All solutions and reagents used in metabolite extraction (water, MeOH, and chloroform) and chemical derivatization (pyridine, methoxyamine hydrochloride, *N,O*-Bis(trimethylsilyl)-trifluoroacetamide (BSTFA) and 1% trimethylsilyl chloride (TMCS) are recommended to use a GC- or higher grade (MS-grade). Prepare all reagents at room temperature and under a chemical hood if not indicated.

2.1 Metabolite Extraction

1. Ice-cold 50% MeOH (v/v) (*see* **Note 1**).
2. Internal standards mixture: 10 mg/mL of glycine–d_5, succinic acid–d_4, and tryptophan–d_5 dissolved in 50% MeOH (v/v).
3. Vacuum concentrator or nitrogen purge.
4. Probe-type ultrasonicator.
5. Polypropylene disposable syringe without lubricant (NORM-JECT® syringe recommended).
6. 0.45 μm polytetrafluoroethylene (PTFE) syringe filter.
7. 2.0 mL microcentrifuge tube.

2.2 Chemical Derivatization

1. Methoxyamination solution: 20 mg/mL methoxyamine hydrochloride in pyridine by vortexing or ultrasonication over 1 min to be totally dissolved (*see* **Note 2**).
2. Trimethylsilyation solution: BSTFA +1% TMCS (*see* **Note 3**).
3. Heating block or water bath.
4. Autosampler vials (with crimp cap recommended) and vial crimper/decapper.
5. Glass insert.

2.3 GC-MS Analysis

1. Shimadzu GC-MS-QP2010 or other GC-MS (Agilent and Thermo-Scientific, etc.).
2. GC Split/splitless inlet liner and septum.

3. Agilent DB-5 ms (30 m × 0.25 mM, 0.25 μm) (*see* **Note 4**).
4. Kovats index solution (retention index solution).

2.4 Metabolite Identification

1. Vendor-provided data analysis software, e.g., Postrun Analysis for Shimadzu GC-MS.
2. Mass spectral library fit with vendor software, e.g., NIST mass spectral library or Wiley GC-MS spectral database).

3 Methods

3.1 Extraction of Primary Metabolites

1. Dry the samples (freeze-drying recommended). Plant samples without drying can be ground after snap frozen using liquid nitrogen or dry ice.
2. Homogenize plant samples using a mill.
3. Weigh 5–10 mg of the dried plant samples or fresh (snap frozen) ones.
4. Add 500 μL of ice-cold 50% MeOH and 10 μL of internal standards mixture (final concentration of 1 μg/mL after chemical derivatization) (*see* **Note 5**).
5. Extract plant metabolites using a probe-type ultrasonicator for 1 min (depending on the machinery used, the amplitude, pulsation, and time should be adjusted). The sample tube should be put inside the ice bath while ultrasonic-assisted sonication (*see* **Note 6**).
6. Add 250 μL of ice-cold chloroform into the sample tube.
7. Vortex thoroughly (over 20 s).
8. Centrifuge at 13,000 × g for 5 min (at 4 °C recommended).
9. Transfer the supernatant (water layer) into another microtube.
10. Add 500 μL of water and repeat for 6–8 to collect another supernatant.
11. Filter the collected supernatant using a syringe and 0.45 μm PTFE syringe filter.
12. Dry extracts in a vacuum concentrator (nitrogen purging can be a viable option).
13. Dried extract can be stored at −80 °C before chemical derivatization (*see* **Note** 7).

3.2 Multi-Step Chemical Derivatization (See Note 8)

1. Add 100 μL of methoxyamination solution into the dried extract and vortex thoroughly.
2. Reaction at 37 °C for 120 min.
3. Add 100 μL of trimethylsilylation solution into the reactants and react again at 70 °C for 60 min (*see* **Note 9**).

4. Centrifuge 13,000 × g for 5 min.
5. Transfer derivatized samples into an autosampler vial with a glass insert.
6. During the transfer, obtain 10 μL of every aliquot and mix them thoroughly in another vial to prepare a quality control (QC) sample (*see* **Note 10**).
7. Crimp the vial cap using a crimper.

3.3 GC-MS Data Acquisition

1. Use GC-MS conditions as follows: helium as a carrier gas at 1 mL/min flow rate; inlet temperature at 270 °C; interface temperature at 300 °C; electron impact ionization mode at 70 eV; ion source temperature at 200 °C; mass detection range of 40–600 *m/z*; column temperature program initialized at 80 °C for 2 min, an increase to 200 °C at 4 °C/min, a hold for 4 min, an increase to 300 °C at 8 °C/min, and hold for 2 min, for a total of 50.5 min (*see* **Note 11**).
2. Inject 1 μL of each sample aliquot at split/splitless injection (*see* **Note 12**). Inside the whole data acquisition sequence, QC sample data should be obtained for every 5–10 samples.
3. Obtain retention index data at the first and the end of the whole sample sequence.

3.4 Metabolite Identification

1. The data of QC samples should be used for metabolite identification. The mixture of every sample should allow to detect every metabolite that can be analyzed in each sample.
2. Calculate the retention index of metabolite peaks that should be annotated for metabolite identification (*see* **Note 13**).
3. Mass spectrum of each peak should be derived after baseline subtraction or peak deconvolution (*see* **Note 14**).
4. Compare the refined mass spectrum and spectral library. It should include a retention index allowance ±50. Similarity index (Shimadzu) or match factor (MS search from NIST) over 80/100 or 800/1000 can be chosen as putatively identified metabolites (*see* **Note 15**).

3.5 Data Processing

For metabolite profiling among a lot of samples, MetAlign offers an efficient data alignment of the GC-MS metabolomics dataset. Among various data alignment tools, MetAlign is one of the easiest processors that allows data pre-processing (baseline correction and denoising) and alignment of the whole dataset [15]. Detailed manuals about every parameter can be found in the "documentation" folder in the installation directory. Every parameter should be optimized after trying the standard values. To adjust MetAlign parameters in detail, please refer to Lommen et al. [16].

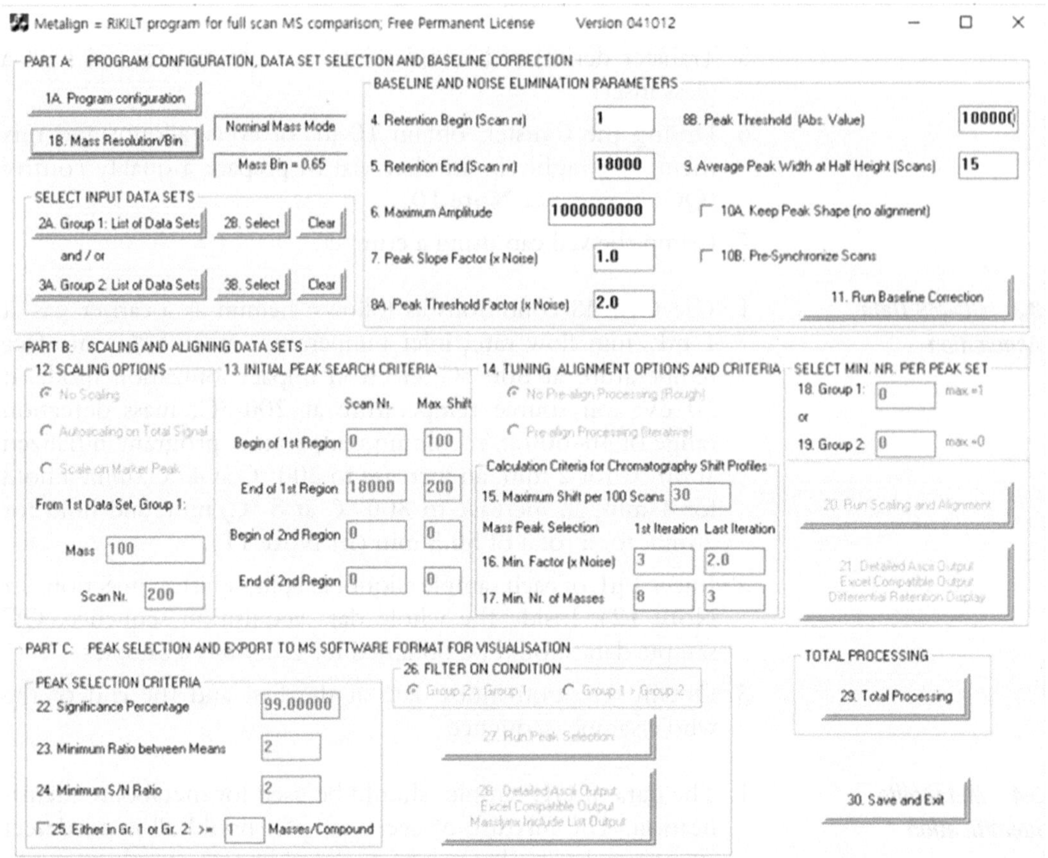

Fig. 1 Main screen of MetAlign

3.5.1 Part A. Program Configuration, Data Set Selection, and Baseline Correction

1. Install MetAlign 3.0 following the instruction.
2. Open MetAlign. The main screen of MetAlign should appear (Fig. 1).
3. In the main screen of MetAlign, set the parameters in 1A. Program configuration and 1B. Mass resolution/bin according to the equipment you have used.
4. Import raw data with proper data format (*.cdf in case using Shimadzu GC-MS) (*see* **Note 16**).
5. In “2A. Group 1: List of Data Sets and 3A. Group 2: List of Data Sets” you can select the imported data into each group. If samples have more than three groups, every sample set can be input to Group 1.
6. Set “Retention Begin” as a scan number of the empty beginning of the chromatogram. “Retention End” should be a scan at the empty end of the chromatogram with maximum baseline noise.

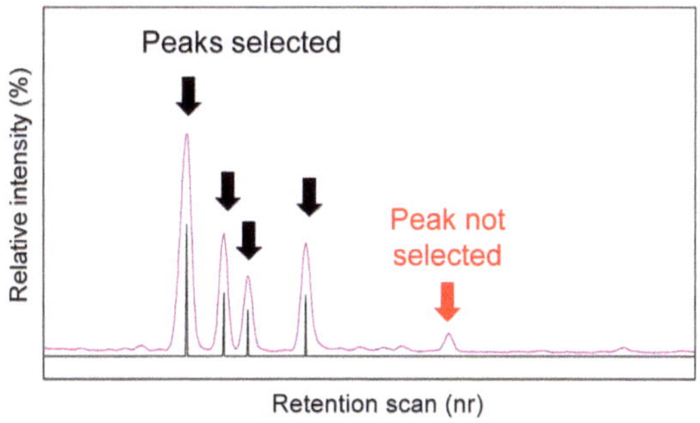

Fig. 2 Overlapped chromatograms before (black) and after (purple) baseline correction. Peaks selected (black arrow) or unselected (red arrow) after baseline correction can be evaluated

7. Set "Maximum amplitude" as 70% of saturated mass peak. In case no saturated peak appears, fill in a value exceeding the maximum peak amplitude in the chromatogram.
8. Set "Peak Slope Factor" as 1 and "Peak Threshold Factor" as 2 (default).
9. Set "Peak threshold (Abs. Value)" that will be lower than the intensity of the lowest mass peak signal of the lowest peak.
10. "Average Peak Width at Haft Height" can be optimized by monitoring the width at half height of unsaturated peaks in the chromatogram. It generally ranges from 8 to 15.
11. Click "Baseline Correction."
12. After the processing, open cdf file of raw and processed data together and check if every metabolite peak is selected in the post-processed data. If the post-processed data missed the metabolite peak, parameters should be re-optimized (Fig. 2).

3.5.2 Part B. Scaling and Aligning Data Sets

1. Set no scaling for "Scaling option" (*see* **Note 17**).
2. In "Initial Peak Search Criteria" choose the same scan number of "Retention Begin" and "Retention End" for "Begin of 1st Region" and "End of 1st Region," respectively. To set the max shift, compare every data and see a "retention shift" of the same peaks at the beginning and the end of the chromatogram.
3. In "Tuning Alignment Options and Criteria" set "No Pre-align Processing (Rough)" (*see* **Note 18**).
4. In "Select Min. Nr. Per Peak Set," fill in the number of QC data in each group.
5. Click "Run Scaling and Alignment."
6. Click "Detailed Ascii Output."
7. In an exported ASCII file, select the peak number that matches the detected metabolite's retention time and representative

mass signal, and choose it as a peak area representing each metabolite. If the original dataset with metabolite features (m/z and retention time without metabolite matching) will be used for non-targeted analysis, omit this step and use the data as is for multivariate statistical analysis.

4 Notes

1. The use of alcohols (MeOH or EtOH) as extraction solvents is suitable for the extraction of metabolites from plant samples. Acetonitrile, a non-aqueous solvent, is not recommended for the extraction of primary metabolites. The extraction solvent should be kept under −20 °C to quench metabolism while immersing samples in the solvent.
2. Methoxyamine hydrochloride, known for its strong hygroscopic properties, should ideally be stored in a desiccator. It's better to prepare the methoxyamine solution just when needed, instead of preparing it in large quantities and storing it in aliquots.
3. In plant metabolomics, N-methyl-N-(trimethylsilyl)trifluoroacetamide (MSTFA) and BSTFA are the predominant reagents employed for trimethylsilylation. A comparison of the derivatization agents for metabolomics, including trimethylsilylation and other classes, is detailed in the study by Jones and coworkers [11].
4. GC column with 5% silica content with adequate temperature program exhibits efficient peak separation. An MS-grade column that exhibits minimal column bleeding at varying temperatures.
5. Measured area of all metabolites (after data alignment) can be normalized using the area of the internal standards. In particular, calculate the relative standard deviation (RSD) of three IS areas in the entire sample, and if it exceeds 30%, it is recommended to adjust procedures including chemical derivatization, and GC-MS conditions, and then reanalyze the whole sample set.
6. If a probe-type sonicator is not available, bath-type sonicator-assisted extraction with optimized time or a bead-type homogenizer can also be used. If plant samples are finely ground, extraction can be achieved through thorough vortexing.
7. When chemical derivatization, the presence of moisture is very critical. If the dried extract is stored in a freezer, it should be left unopened at room temperature for about 30 min, preventing moisture from entering. Additional steps such as using a

vacuum concentrator or freeze-drying can be employed to thoroughly remove moisture.

8. When derivatizing low-concentration metabolite extracts using polypropylene, various interfering substances can be detected. However, the issue can be resolved by correlating the m/z signals originating from specific components through methods such as peak deconvolution.
9. If a large number of incomplete derivatives have a higher intensity than the fully derivatized form (i.e., serine with 1 trimethylsilylated (TMS) vs 3TMS derivative), increase the time and temperature for trimethylsilylation, or increase the amount of derivatizing reagent.
10. QC samples should be analyzed in every 10 samples in the sequence. For smaller sample sets (less than 20 samples), it is advisable to analyze QC samples every 5 samples to ensure a minimum of 4 QC samples are obtained. The area measured for all metabolites in a QC sample should display a RSD of at least 30%.
11. While a range of GC-MS brands such as Agilent Technologies and Thermo-Scientific are applicable, this chapter will primarily focus on terminologies and software associated with Shimadzu.
12. Depending on the concentration of the sample, it is necessary to adjust between split and splitless injection. The initial analysis should start with a split injection to prevent contamination of the column and ion source due to the injection of high-concentration samples.
13. The data analysis software provided by each manufacturer generally includes a feature that allows for the incorporation of retention index data into the main sample data, enabling the calculation of the retention indices for peaks present in the sample.
14. Peak deconvolution can be used for the identification of metabolites. Automated Mass Spectral Deconvolution and Identification System (AMDIS), a free software, facilitates the automatic integration of mass peaks associated with metabolite peaks. This approach of chromatographic peak deconvolution is adept at differentiating metabolite signals from background noise.
15. When utilizing AMDIS connected with NIST MS search software for comparison of the MS spectrum with the spectral library, there is also a reverse search feature. Utilizing both basic and reverse search methods, particularly with scores exceeding 800 points, can significantly enhance the precision of identification. Such putative identification based on mass

spectral library comparison can be validated using metabolite standard compounds.

16. For data formats suitable for MetAlign input, refer to the list under "1A Program Configuration." To import raw data acquired using Shimadzu GC-MS, data can be exported to . cdf file.
17. Data scaling can be performed after the data processing, before conducting multivariate statistical analysis.
18. If you want to select Pre-align Processing (Iterative) in "Tuning Alignment Options and Criteria," please refer to the manual in the MetAlign documentation or Lommen et al., when setting the parameters.

Acknowledgement

This research was supported by the National Research Foundation of Korea (NRF) grant funded by the Korean government (MSIT) (No. RS-2023-00212534).

References

1. Atanasov AG et al (2021) Natural products in drug discovery: advances and opportunities. Nat Rev Drug Discov 20(3):200–216. https://doi.org/10.1038/s41573-020-00114-z
2. Wolfender J-L et al (2019) Accelerating metabolite identification in natural product research: toward an ideal combination of liquid chromatography–high-resolution tandem mass spectrometry and NMR profiling, in silico databases, and chemometrics. Anal Chem 91(1):704–742. https://doi.org/10.1021/acs.analchem.8b05112
3. Wurtzel ET, Kutchan TM (2016) Plant metabolism, the diverse chemistry set of the future. Science 353(6305):1232–1236. https://doi.org/10.1126/science.aad2062
4. Tsugawa H, Rai A, Saito K, Nakabayashi R (2021) Metabolomics and complementary techniques to investigate the plant phytochemical cosmos. Nat Prod Rep 38(10):1729–1759. https://doi.org/10.1039/D1NP00014D
5. Shen S et al (2023) Metabolomics-centered mining of plant metabolic diversity and function: past decade and future perspectives. Mol Plant 16(1):43–63. https://doi.org/10.1016/j.molp.2022.09.007
6. Isah T (2019) Stress and defense responses in plant secondary metabolites production. Biol Res 52(1):39. https://doi.org/10.1186/s40659-019-0246-3
7. Li Y et al (2020) The effect of developmental and environmental factors on secondary metabolites in medicinal plants. Plant Physiol Biochem 148:80–89. https://doi.org/10.1016/j.plaphy.2020.01.006
8. Lee K-M et al (2017) Application of metabolomics to quality control of natural product derived medicines. Biomol Ther 25(6):559–568. https://doi.org/10.4062/biomolther.2016.249
9. Fiehn O (2016) Metabolomics by gas chromatography–mass spectrometry: combined targeted and untargeted profiling. Curr Protoc Mol Biol 114(1):30.34.31–30.34.32. https://doi.org/10.1002/0471142727.mb3004s114
10. Lee D-K et al (2016) Metabolic response induced by parasitic plant-fungus interactions hinder amino sugar and nucleotide sugar metabolism in the host. Sci Rep 6(1):37434. https://doi.org/10.1038/srep37434
11. Jones OAH, Hügel HM (2013) Bridging the gap: basic metabolomics methods for natural product chemistry. In: Roessner U, Dias DA (eds) Metabolomics tools for natural product discovery: methods and protocols. Humana Press, Totowa, pp 245–266

12. Scalbert A et al (2009) Mass-spectrometry-based metabolomics: limitations and recommendations for future progress with particular focus on nutrition research. Metabolomics 5(4):435–458. https://doi.org/10.1007/s11306-009-0168-0
13. Zeki ÖC et al (2020) Integration of GC–MS and LC–MS for untargeted metabolomics profiling. J Pharm Biomed Anal 190:113509. https://doi.org/10.1016/j.jpba.2020.113509
14. Wishart DS et al (2022) NMR and metabolomics—a roadmap for the future. Meta 12(8):678
15. Lommen A (2009) MetAlign: interface-driven, versatile metabolomics tool for hyphenated full-scan mass spectrometry data preprocessing. Anal Chem 81(8):3079–3086. https://doi.org/10.1021/ac900036d
16. Lommen A (2012) Data (pre-)processing of nominal and accurate mass LC-MS or GC-MS data using MetAlign. In: Hardy NW, Hall RD (eds) Plant metabolomics: methods and protocols. Humana Press, Totowa, pp 229–253

Chapter 9

Identification of Flavonoids Using UV-Vis and MS Spectra

Özlem Erol and Sandra Irmisch

Abstract

Flavonoids are a group of specialized metabolites that are ubiquitously found within the plant kingdom. While they fulfill various important functions within the plant, they are also utilized by humans in a variety of different fields such as medicine, food science, and agriculture. Thus, to elucidate the chemical composition of any given plant extract, extraction and identification of flavonoids are of high interest. In this chapter, we outline a general approach for analyzing flavonoids in complex plant extracts. Using blueberry as an example, we start by describing the extraction of plant material, followed by simple spectrophotometric measurements and thin layer chromatography (TLC), to get first insights into flavonoid content and composition. Furthermore, we describe a standard protocol for liquid chromatography-mass spectrometry (LC-MS) for the identification of specific flavonoids. By utilizing and combining the described techniques, general analysis of flavonoid content, composition, as well as identification of specific compounds can be realized within a complex metabolic mixture.

Key words Flavonoids, UV spectroscopy, Thin Layer Chromatography (TLC), Ultra-High-Performance Liquid Chromatography (UHPLC), Mass Spectrometry (MS)

1 Introduction

Flavonoids are a diverse group of polyphenolic specialized metabolites that are ubiquitously found within the plant kingdom. More than 9000 different flavonoids are estimated to exist [1]. They can play crucial roles within the plant, e.g., they can function in plant growth, development, and defense [2]. Flavonoids are most widely known for their function as pigments, which give fruits, vegetables, and flowers their many vibrant colors. Beyond their visual appeal, flavonoids are increasingly recognized for their potential health benefits. They have, for example, been found to possess anti-inflammatory, anti-cancer, anti-diabetes, and cardiovascular benefits [3, 4]. This makes flavonoids subjects of extensive scientific research and a focal point in the field of nutrition and medicine. All flavonoids bear a common structure, which consists of a 15-carbon skeleton with two phenyl rings (A- and B-ring)

Young Hae Choi et al. (eds.), *Natural Product Isolation and Identification: Methods and Protocols*, Methods in Molecular Biology, vol. 2895, https://doi.org/10.1007/978-1-0716-4350-1_9,

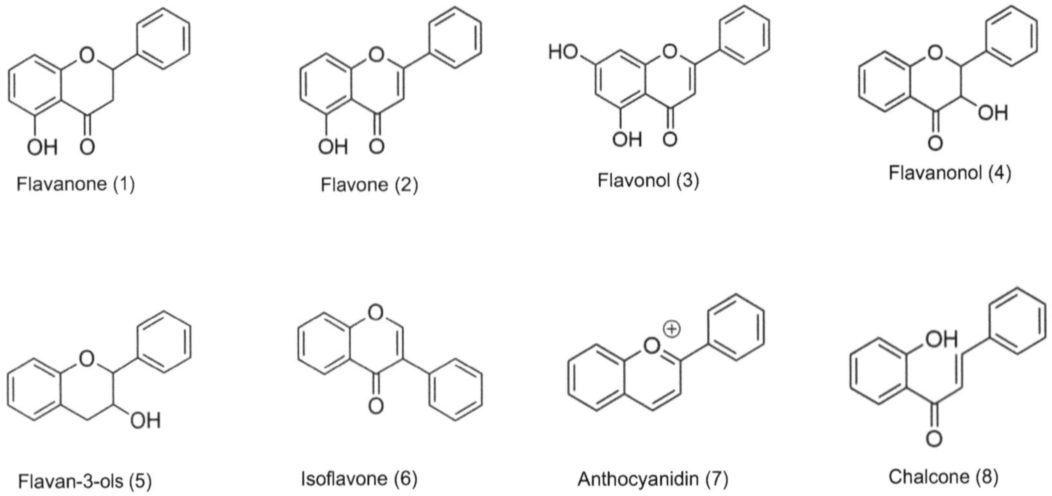

Fig. 1 Basic skeleton structures of flavonoids. Flavonoids can be divided into eight main groups. 1, Flavanones; 2, Flavones; 3, Flavonols; 4, Flavanonols; 5, Flavanols; 6, Isoflavones; 7, Anthocyanidins; 8, Chalcones

connected by a three-carbon chain (C6–C3–C6) which generally forms a heterocycle (C-ring) (Fig. 1). Classification of flavonoids into several subclasses is done based on their chemical structure, mainly regarding the presence of functional groups on the heterocyclic ring (C-ring). There are 8 major subclasses. Flavanones (dihydroflavones) (**1**) possess a C4 ketone group at the heterocycle, flavones (**2**) additionally harbor a C2–C3 double bond, and flavonols (**3**) have both features and a 3-hydroxy group. Flavanonols (dihydroflavonols) (**4**) are similar to flavonols, meaning they harbor a 3-hydroxy group, a C4 keto group but do not possess a C2–C3 double bond, while flavanols (**5**) (i.e., flavan-3-ol) possess the 3-hydroxyl group without ketone at C4. Isoflavones (**6**) are isomers of flavones, with the phenyl ring attached to the C3 position instead of the C2. They are almost exclusively found within members of the Fabaceae. Anthocyanidins (**7**) have two double bonds within the heterocycle (Cl–C2, C3–C4), a positive charge at the oxygen atom, and can be hydroxylated at C3. They represent the major class of water-soluble plant pigments, responsible for colors in many flowers, fruits, and vegetables. Examples are delphinidin, cyanidin, and pelargonidin associated with orange, pink, and blue colors, respectively. The last subclass is (**8**) chalcones which lack the C-ring and can also serve as precursors for other flavonoid classes (Fig. 1) [5, 6].

Apart from their aglycone form, flavonoids within plants often occur as glycosides, meaning they harbor one or more sugar units. Flavonoid glycosides have mainly been reported as *O*-glucosides, whereas flavone glycosides can occur as *O*- or/and *C*-glycosides. Although several different sugars such as glucose, rhamnose, xylose, arabinose, mannose, and galactose can be used for glycosylation, glucosides (glucose attached) are the most prevalent flavonoid glycosides found [7, 8].

The identification and quantification of flavonoids in plant extracts and pharmaceutical formulations are essential for evaluating their therapeutic potential and ensuring product quality. Traditional analytical techniques, such as chromatography-based methods, are effective but often expensive, time-consuming, and labor-intensive. UV spectroscopy offers a rapid alternative for the qualitative and quantitative analysis of flavonoids. This method utilizes the characteristic absorbance of flavonoids within the UV-visible range, yielding distinct absorption spectra due to the presence of conjugated double bonds and functional groups in flavonoids. Specifically, flavonoids possess two absorption maxima (λ_{max}) within 200–400 nM, referred to as Band I (300–380 nM) and Band II (240–280 nM), which allow their selective detection and quantification [9, 10]. Band I is likely derived from B-ring absorption, and Band II is thought to involve the A-ring. While offering a rapid analysis UV spectroscopy has its limitations, e.g., low sensitivity and specificity, especially in complex samples. To overcome this more sensitive methods are available. Thin layer chromatography (TLC) is a powerful technique enabling a quick, initial quantitative inspection of the flavonoid diversity within a plant extract. It is simple, fast, and relatively cheap. The separation of individual flavonoids using TLC is based on their differential migration on a stationary phase [11, 12]. To further gather more detailed information on flavonoid structure and to ultimately identify and quantify specific compounds, liquid chromatography-mass spectrometry (LC-MS) can be used to provide high-resolution separation, accurate mass determination, and thus structural information [13–15].

Blueberries are globally recognized for their vibrant color, unique flavor, and their myriads of health benefits. They are particularly rich in flavonoids including anthocyanins, proanthocyanidins, and flavonols. Anthocyanins not only play a major role in giving blueberries their characteristic blue, purple, and red hues, but they are also major contributors to blueberries' antioxidant capacity. Furthermore, the flavonoids quercetin and catechin are prominent and have been shown to exhibit inflammatory effects and facilitate cardiovascular benefits, respectively [16, 17].

Here we describe the qualitative and quantitative analysis of flavonoids in a blueberry extract using UV spectroscopy, TLC, and LC-MS. A step-by-step protocol including sample preparation, extraction, analysis, and data interpretation is provided, which can serve as a template to explore the flavonoid composition and content of any plant sample of interest.

2 Materials

2.1 Solvents and Chemicals

Solvents used for extraction, separation, and isolation of flavonoids in this protocol should be high-performance liquid chromatography (HPLC) purity grade (MeOH, $CHCl_3$, and EtOAc) and LC-MS grade for LC-MS analysis (AcCN, MeOH, H_2O). $AlCl_3 \times 6H_2O$, NaOH, and $NaNO_2$ (analytic grade) are needed for total flavonoid content (TFC) measurement. Formic acid (American Chemical Society (ACS) reagent ≥96%) and acetic acid are LC-MS (LC-MS grade) are used for TLC and for LC-MS analysis, respectively. LC-MS grade sodium formate solution (10 mM NaOH) is used for the LC-MS calibration. Reference flavonoids are ≥99% pure. Anisaldehyde, 2-aminoethyl diphenylborinate (≥ 97.0%) and sulfuric acid are used for TLC derivatization. Distilled H_2O is used in the extraction process. Liquid nitrogen is used for flash freezing of plant material after harvest (*see* **Note 1**).

2.2 Equipment

Consumables Pipette, pipette tips, 1.5 mL glass vials, 2 mL tubes, 5 mL glass vials, falcon tube, spatula, 0.2 μm syringe filter, pre-coated TLC silica gel 60 plates UV_{254} (20 × 20 cm), high-performance thin layer chromatography (HPTLC) silica gel 60 plates UV_{254} (20 × 20 cm), quartz, and plastic cuvette.

Sample Preparation Freeze-dryer, mortar, and pestle analytical balance; sample extraction: ultrasonicator, Speed-Vac, rotary evaporator, vortex mixer.

TLC TLC Autosampler or glass capillaries/pipette tips; TLC visualization: UV-lamp; derivatizer and hot plate for chemical reaction of TLC plates, glass spray bottle, TLC chamber.

Spectrophotometer UV/VIS-spectrometer.

LC-MS Quadrupole time-of-flight (QTOF) mass spectrometry (MS), ultra-high-performance liquid chromatography (UHPLC) system, C_{18} reversed-phase column (2.1 × 150 mM, 2.6 μm).

2.3 LC-MS Solvent Preparation

1. Prepare Solvent A (H_2O + 0.1% formic acid, v/v) and B (AcCN +0.1% formic acid, v/v).
2. For Solvent A transfer 1 L of HPLC grade H_2O into a 1 L bottle. Remove 1 mL of the H_2O and add 1 mL of formic acid followed by mixing.
3. For Solvent B transfer 1 L of HPLC grade AcCN into a 1 L bottle. Remove 1 mL of the AcCN and add 1 mL of formic acid followed by mixing.
4. Degas Solvent A and B before usage by ultrasonication for 30 min.

2.4 TLC Mobile Phase and Reagent Preparation

1. Prepare 1 mg/mL solution for standard compounds. Weigh 1 mg of the compounds (e.g., rutin, quercetin, or kaempferol) using an analytical balance into a 1.5 mL glass vial. Dissolve the compound in 1 mL of 100% MeOH. Ultrasonicate the solutions for 5 min until dissolved. Stock solutions of flavonoids are stored at −20 °C until use.
2. Prepare 50% MeOH (v/v) for extraction and store at room temperature until use.
3. Preparation of mobile phase EFAW (EtOAc-formic acid-acetic acid-H_2O (100:11:11:27, v/v)).
4. For the preparation of 100 mL of anisaldehyde derivatization reagent, mix the following substances in the following order: 0.5 mL anisaldehyde, 10 mL of acetic acid (mix slowly by hand), 85 mL of MeOH, and 5 mL concentrated sulfuric acid. Store the solution at 4 °C.

2.5 Preparing Solutions for UV Spectroscopy

1. Preparation of 5% $NaNO_2$: Weigh 5.164 g of $NaNO_2$ and dissolve in 100 mL H_2O and store at room temperature.
2. Preparation of 10% $AlCl_3$: For 100 mL dissolve 19.4376 g of $AlCl_3 \times 6H_2O$ (100% purity) in H_2O and store at room temperature.
3. Preparation of 1 M NaOH: For 100 mL dissolve 4 g of NaOH in H_2O and store at room temperature.
4. Preparation of reference flavonoids like rutin or quercetin: Weigh 1 mg of each reference flavonoid and dissolve in 10 mL of MeOH. Ultrasonicate for 5 min until completely dissolved and store at 4 °C.

3 Methods

3.1 UV Measurement

The absorption maxima (λ_{max}) of flavonoids typically fall within the UV range of 200–600 nm. However, the specific λ_{max} can vary depending on the structural characteristics of individual flavonoids and their functional groups (Table 1) [18]. For example, flavones and flavonols often possess a λ_{max} in the range of 330–360 nM, while λ_{max} for flavanones typically falls around 280–310 nM and anthocyanins possess a λ_{max} in the visible range, often around 500–550 nM, giving them their characteristic colors [19, 20]. A digital database that lists the UV absorptions for individual flavonoids can be found online (http://www.photochemcad.com) [21].

3.1.1 Extraction and Preparation of Samples

1. Collect and directly freeze blueberries or other plant material (whole plants or individual organs such as leaves, roots, flowers, fruit, or stems) in liquid nitrogen (*see* **Note 1**).

Table 1
The spectral characteristics of the UV-Vis spectrum of flavonoids

Flavonoid class	Band II (nM)	Band I (nM)
(1) Flavanones	270–295	300–330 (shoulder)
(2) Flavones	250–280	310–350
(3) Flavonols	250–280	350–385
(4) Flavanonols	250–300	300–330 (shoulder)
(5) Flavanols	270–295	–
(6) Isoflavones	240–275	310–330 (shoulder)
(7) Anthocyanidins	240–280	450–560
(8) Chalcones	230–270	340–390

2. Lyophilize frozen materials until dry, usually for 2–3 d, and subsequently grind materials with a mortar and pestle into a fine powder. Store the powder in a closed container (e.g., falcon tubes) at room temperature in the dark (*see* **Note 2**).
3. Weigh 30 mg of dried powder (or 100 mg for frozen) using an analytical balance and transfer it into a 2 mL microcentrifuge tube.
4. Add 1 mL of 50% MeOH to the sample.
5. Vortex for 1 min, then ultrasonicate for 30 min in an ultrasonic bath at room temperature.
6. Centrifuge for 20 min at 20,376 × *g*, to pellet the tissue debris.
7. Weigh an empty 1.5 mL glass vial or 2 mL microcentrifuge tube.
8. Use a pipette to transfer the supernatant (carefully, without touching the pellet) into the vial/tube (*see* **Note 3**).
9. Evaporate the solvent (until dryness) by placing the vial/tube into a Speed-Vac concentrator. Weigh the vial/tube after dryness and determine the weight of the dried blueberry extract (plant extract).
10. Transfer 1 mg of dried material into a fresh tube or 1.5 mL glass vial and dissolve in 1 mL 100% MeOH. Sonicate for 5 min or until completely dissolved. The extract is ready for UV measurement and/or can be stored at −20 °C.

3.1.2 UV Measurement

1. Turn on the UV spectrophotometer and allow it to warm up.
2. Use a quartz cuvette for the measurements. Use one cuvette for the blank solution (1 mL MeOH) and another one for the

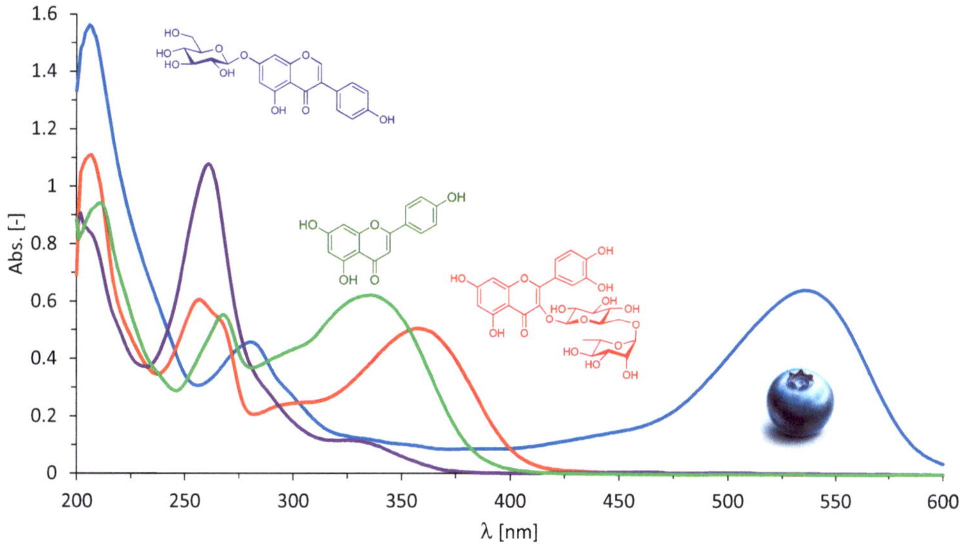

Fig. 2 UV-Vis absorbance spectra of flavonoids. The absorption spectrum of blueberry extracts (blue), rutin (red), apigenin (green), and genistin (purple) measured in methanol are shown. The following maxima were observed: blueberry extract (λ_{max} 280, 536), rutin (λ_{max} 257, 358), apigenin (λ_{max} 268, 536), and genistin (λ_{max} 262, 336 shoulder)

blueberry extract (1 mg /mL) or the reference solutions if available.

3. Set the wavelength to the appropriate UV region, typically around 200 and 600 nM.
4. Perform a baseline correction using a blank solution.
5. Transfer the extract (1 mg/mL) or the reference flavonoid solution (1 mg/mL) into a cuvette (*see* **Note 4**).
6. Insert the filled cuvettes into the UV spectrophotometer and scan the absorbance range.
7. Record the absorbance values.
8. Comparison can be done by using the flavonoid UV database http://www.photochemcad.com [21] or by using reference compounds (*see* **Note 5**).

Data Interpretation of UV-Vis Spectra The blueberry extract shows a λ_{max} at 536 nM due to the hydroxyl cinnamoyl system of the B-ring, clearly indicating the presence of anthocyanins. Another local maximum can be seen at 280 nM pointing toward the benzoyl system of the A-ring (Fig. 2). Both λ_{max} (band I and band II) of the references rutin, apigenin, and genistin can be seen in Fig. 2. Apigenin has a λ_{max} of 336, which falls within the range of general flavone absorption (330–360 nM). Glycosylation of flavones and flavanols causes a bathochromic shift as illustrated by rutin possessing λ_{max} of 358. Genistin shows a λ_{max} at 262 nM in band II and a shoulder at 336 nM. This is in consensus with the literature as in all isoflavonoids, ring B exhibits minimal or no conjugation with the primary chromophore formed by rings A and C. This results in an intense band II and a diminished band I. For isoflavones, band II is characterized by absorption at a λ_{max} of 240–275 nM.

3.2 Measurement of Total Flavonoid Content by UV-Vis Spectra

To determine the TFC in an extract the aluminum chloride colorimetric protocol can be used [22, 23]. Here the absorbance of a solution is measured in a spectrophotometer at 510 nm. For the calibration, either quercetin or rutin can be used.

1. Sample Preparation: Follow **steps 1–9** in Subheading 3.1.1 to prepare the sample.
2. Transfer 1 mg of dried blueberry sample extract into a glass flask and dissolve in 10 mL 100% MeOH.
3. Dilute the quercetin stock solution (0.1 mg/mL) with MeOH to prepare a series of standard solutions with concentrations ranging from 10 to 100 μg/mL.
4. Transfer 500 μL of the blueberry extract and 500 μL of each quercetin standard solution to an individual fresh 5 mL tube.
5. Add 2 mL of H_2O to each tube.
6. Add 0.15 mL of 5% $NaNO_2$ solution to each tube.
7. After 5 min add 0.15 mL of 10% $AlCl_3$ solution to each tube. Let it incubate for 5 min at room temperature.
8. Add 2 mL of 1 M NaOH solution.
9. Mix the contents thoroughly and incubate the reaction mixture at room temperature for 30 min to allow the formation of the aluminum chloride-flavonoid complex.
10. After incubation, transfer 1 mL of the mixture to a cuvette (either plastic or quartz).
11. To prepare the blank, fill another cuvette with 1 mL MeOH. Perform a baseline correction using the blank solution.
12. Measure the absorbance at 510 nM for each sample. Include technical and/or biological replicates as applicable.
13. Prepare a calibration curve by plotting the absorbance values of the quercetin solutions against their respective concentrations. Then calculate the regression equation of the calibration curve (Fig. 3).
14. Use the regression equation to determine the concentration of flavonoids in the blueberry sample based on the sample absorbance. The results can be expressed as milligrams of quercetin equivalents (QE) per gram (or mg) of dry weight (DW) (Fig. 3).

$$TFC_{PlantExtract} = TFC_{Test\ Solution} \times V \times d/m.$$

A

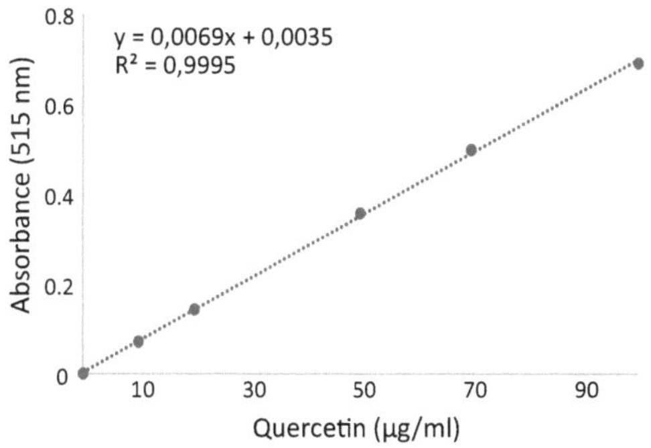

B

Sample	Absorbance	Total flavonoid content [1] (mg * QE * g^{-1})
Blueberry-1	0.252	3,601
Blueberry-2	0,232	3,311
Blueberry-3	0,227	3,239

[1] formula to calculate total flavonoid content:
$TFC_{PE} = TFC_{TS} * V * d * m^{-1}$

Fig. 3 TFC of blueberry extracts. A quercetin calibration curve (**a**) and calculation of the TFC (**b**). Absorption of a quercetin standard curve and blueberry extracts (three technical replicates) was measured at 515 nM. The regression equation can be used to calculate flavonoid concentration as mg × QE per gram of plant DW. TFC_{PE}, quercetin concentration in plant extract in mg QE/g DW; TFC_{TS}, the concentration of quercetin in the test solution obtained using the standard curve; V, volume of extraction solvent; d, dilution factor; m, weight of the dried blueberry extract

$TFC_{test\ solution}$ is the concentration of quercetin in MeOH from the standard curve regression line, V is the volume of the extraction solvent, d is the dilution factor, and m is the weight of the dried blueberry extract (*see* **Note 6**).

Calculation of TFC by UV-Vis Spectra The TFC of the extracts was determined from the regression equation for the calibration curve (in this protocol the question of $y = 0.0069x + 0.0035$; $r^2 = 0.995$ is obtained). This study determined that the TFC in the methanol extract of blueberry is 3.38 mg (equivalent to quercetin content/g).

3.3 Detection of Flavonoids by Thin Layer Chromatography (TLC)

TLC is a useful and robust method that allows rapid screening of plant extracts for specialized metabolites. It is especially useful for flavonoids due to their chromophoric nature, which can be enhanced by the application of anisaldehyde or natural product (NP) reagents. All flavonoids cause fluorescence quenching which can be seen as dark spots at 254 nM. At 366 nM a color can be seen, which depends on the structure of the flavonoids and the derivatization reagent. Observed spots can be yellow, blue, or green.

3.3.1 Sample Preparation

1. Follow **steps 1–9** in Subheading 3.1.1 to prepare the sample.
2. Transfer 5 mg of dried material into a fresh tube or 1.5 mL glass vial and dissolve in 1 mL 100% MeOH to yield a metabolite solution of 5 mg/mL. Ultrasonicate the solutions for 5 min or until completely dissolved and store at 4 °C.
3. Weigh 1 mg of the reference compounds into a fresh tube or 1.5 mL glass vial and dissolve in 1 mL 100% MeOH. Ultrasonicate the solutions for 5 min or until completely dissolved and store at 4 °C.

3.3.2 TLC Conditions

1. The TLC chromatogram in Fig. 4 is generated by using an automated HPTLC system. This automated method provides higher resolution and better-shaped bands compared to manual spotting but does not change the results, especially R_f values. Therefore, as a more general protocol, a conventional TLC protocol is described in this protocol.
2. Fill the mobile phase EFAW (*see* **Note 7**) into the TLC chamber. Use 30 mL for big chambers (TLC plate size 20 × 10 cm) or 10 mL for small chambers (TLC plate size 10 × 10 cm) and let equilibrate for 30 or 10 min, respectively.
3. Carefully with a soft pencil draw a horizontal baseline onto the TLC plate 1 cm from the bottom. Apply 10–15 μL of the extract and 5 μL of the reference compound (*see* **Note 8**), using a microcapillary or a pipette tip, to the baseline (leave 1 cm space from each side and 0.5 cm between samples).
4. Leave the plate to dry for 15 min. Place the plate into the saturated TLC chamber and ensure that the baseline is above the level of the mobile phase. Close the chamber (if no lid is available, covering with aluminum foil suffices).
5. Let the TLC plate develop until the solvent reaches 0.5–1 cm from the top of the plate. Take the plate out, mark the solvent front with a pencil, and allow the plate to air dry (*see* **Note 9**). For faster drying use an additional device like an air gun or a hair drier if available.
6. For visualization of flavonoids place the TLC plate under UV light, specifically 254 nM and 366 nM. Take a picture under the different wavelengths (Fig. 4) (*see* **Note 10**).
7. Record the values for the R_f (retention factor) for each spot by measuring the distance traveled by the spot (from baseline to spot) and the solvent front (from baseline to upper solvent front). This information can be used for compound identification by comparing the R_f values to the R_f-values of reference compounds.
8. Calculate the R_f using the following formula:

$$R_f = \text{distance traveled by the compound/distance traveled by the solvent front.}$$

9. Carry out derivatization with anisaldehyde-sulfuric acid reagent by spraying the plate (use a spray bottle) with 2 mL of the solution and subsequently heating the plate at 100 °C for about 3 min by holding over/placing it onto a hotplate (*see* **Note 11**).
10. Take a picture and report the color changes of the spots.

Data Interpretation of TLC Chromatograms The TLC plate shows the analysis of the blueberry extract and nine flavonoid references (Fig. 4). Detection at 254 nM revealed the presence of

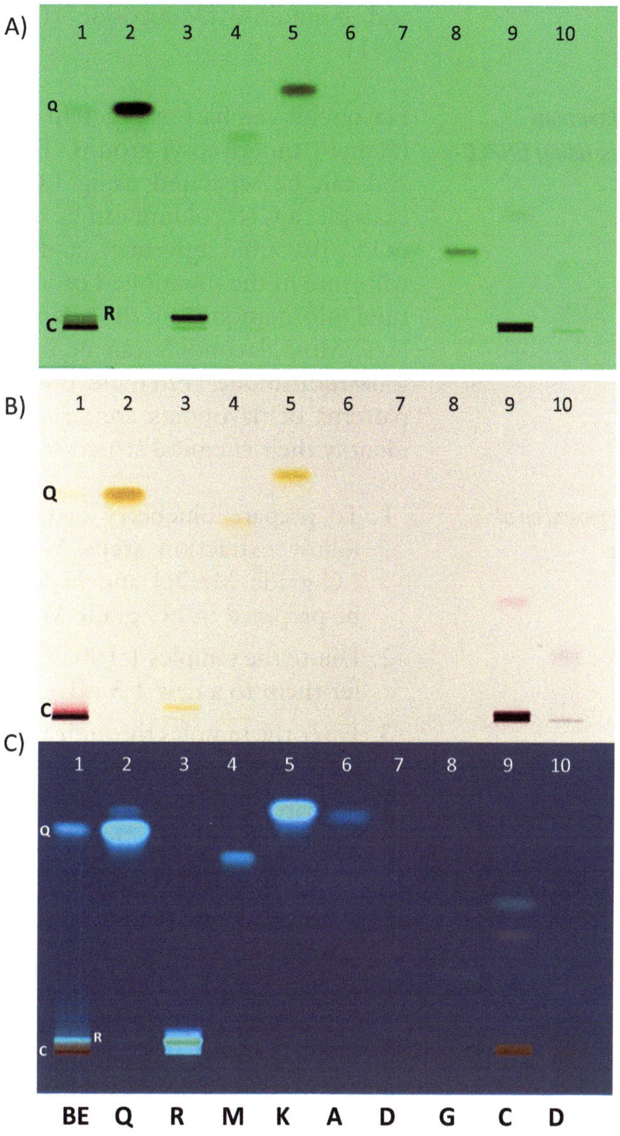

Fig. 4 High-Performance Thin Layer Chromatogram of blueberry extract and flavonoid reference compounds. The blueberry extract and nine flavonoid references were spotted and developed on a silica gel HPTLC plate (20 × 10 cm, F_{254}) using EFAW as a mobile phase. The chromatogram was visualized before derivatization at 254 nM (**a**) and after derivatization with anisaldehyde reagent under visible light (**b**) and at 366 nM (**c**). Sample order: (1) BE, Blueberry extract; (2) Q, quercetin; (3) R, rutin; (4) M, myricetin; (5) K, kaempferol; (6) A, apigenin; (7) D, daidzein; (8) G, genistin; (9) C, cyanidin 3-*O*-glucoside; (10) *D* delphinidin chloride. The identity of the band within the blueberry extract is indicated by a marking next to the band

quercetin (spot 2) in the blueberry extract with an R_f value of 0.71. Using post-derivatization with anisaldehyde and detection at 366 nM and visible light, additional compounds were identified in the blueberry extract, including rutin (spot 3, R_f value: 0.04), and cyanidin-3-*O*-glucoside (orange/purple spot 9, R_f value: 0.01).

3.4 Flavonoid Analysis Using UHPLC-MS/MS

Flavonoids can harbor different functional groups, e.g., hydroxyl-, methyl-, and glycosyl groups. Thus, they show different polarities and can be separated using LC on reverse-phase columns. For example, a C18 column can be used to separate glycosylated flavonoids, flavonoid aglycones, and methylated flavonoids, and these will elute in the mentioned order. This allows to draw some structural information from the retention time of the analyzed metabolites. Most flavonoids can be ionized using negative electrospray ionization mode (ESI) and positive ESI. The MS fragmentation patterns of flavonoids are structure specific and can be used to identify their chemical structures (*see* **Note 9**).

3.4.1 Preparation of Samples

1. To prepare blueberry extracts (or any other plant extract), follow extraction **steps 1–9** in Subheading 3.1.1, but use LC-grade MeOH and H_2O. Standard solutions should also be prepared in LC-grade MeOH (1 mg/mL) (*see* **Note 12**).
2. Dilute the samples 1:100 in 50% MeOH (LC-grade) and transfer them to a new 1.5 mL glass vial (*see* **Note 13**).
3. Filter the samples through a 0.2 μm syringe membrane filter, to clear any particles which could block the column.
4. Prepare a quality control (QC) sample by pooling equal volumes (e.g., 50 μL) of each sample to be analyzed into a new vial (*see* **Note 14**).
5. Prepare a blank sample by using the solvent which was used to dissolve the samples (50% MeOH) (*see* **Note 15**).

3.4.2 LC-MS Setup

1. Ultra-high-performance liquid chromatography (UPLC) system equipped with a C_{18} reversed-phase column (2.1 mM diameter × 150 mM length) (*see* **Note 16**). The chromatography is done using a column temperature of 40 °C, the mobile phase A: 0.1% formic acid in H_2O, B: 0.1% formic acid in AcCN, and a flow rate of 0.3 mL/min (*see* **Note 17**). A linear gradient is applied by increasing mobile phase B from 5% to 95% over 30 min, 95% B for 5 min, then 5% B for 5 min. The injection volume is 1 μl (*see* **Note 18**).
2. Set ESI source positive ionization that can vary depending on instruments: dry gas 12 L/min, dry temperature 250 °C, nebulizer gas: 43.5 psi, and capillary voltage 4 kV.

3. Set MS and auto MS/MS parameters to analyze flavonoids. For Bruker QTOF instruments: funnel 1 radiofrequency (RF) voltage: 300 Vpp funnel 2 RF voltage: 300 Vpp, hexapole RF voltage: 150 Vpp, ion energy: 8 eV, collision energy: 8 eV transfer time 100 μs and pre pulse storage time 5 μs from mass range from 50 to 1650 (*see* **Note 19**).
4. Calibrate the MS by using sodium formate solution (flow rate of 10 μL/min, 30 min run). Accept the calibration when the calibration score is higher than 94%.
5. Set the MS scan range from *m/z* 50 to 1650, spectra rate of 2 Hz/msec or higher, and include the calibrant at the beginning of the gradient for internal mass calibration of the data.
6. Choose the sequence table, the LC method, the MS method, and the specific injection volume (1 μL).
7. Run samples and acquire data.

3.4.3 Data Analysis

LC-MS identification of flavonoids can be performed by comparing retention time, accurate mass, and fragmentation patterns to authentic references, to the literature, and/or to spectral libraries (e.g., Metaboscape, GNPS, MassBank, or HMBD) [24–26]. Some general rules for interpreting flavonoid fragmentation patterns exist. Following this procedure helps to identify different flavonoid classes. Below we describe some steps to gain structural information and identify different flavonoid classes based on their fragmentation patterns.

For flavonoid aglycones, the most important fragments result from the C-ring carbon-carbon bond cleavage [retro-Diels-Alder (RDA) reaction] [27]. This cleavage results in the formation of two fragments: one fragment containing the A-ring and part of the C-ring, and another fragment containing the B-ring and part of the C-ring (Fig. 5). In positive ionization mode, ions are labeled as $^{i,j}A^+$ and $^{i,j}B^+$ (where i and j indicate the C-ring bonds that have been broken). Most fragmentations result in A and B ions requiring cleavage of the C–C bonds at positions 1/3, 0/2, 0/4, or 1/4 of the C-ring. For free aglycones, this results in $^{1,3}A^+$ and $^{1,3}B^+$ ions (Fig. 5). In negative ionization mode, the ions are denoted A^- and B^-, respectively [28].

1. $^{1,3}A^+$-ring and $^{1,3}B^+$-ring ions are common for flavanones, flavones, flavonols and for some isoflavones. Table 2 shows the most characteristic fragment ions for $^{1,3}A^+$, ^{1,3} B^+, $^{0,2}A^+$, and ^{0,2} B^+ obtained for designated flavonoid classes.
2. $^{0,2}B^+$ is a common cleavage for flavones and flavonols, the corresponding $^{0,2}A^+$ can be used to distinguish flavonols as it is not appearing in other flavonoid classes (Table 2).

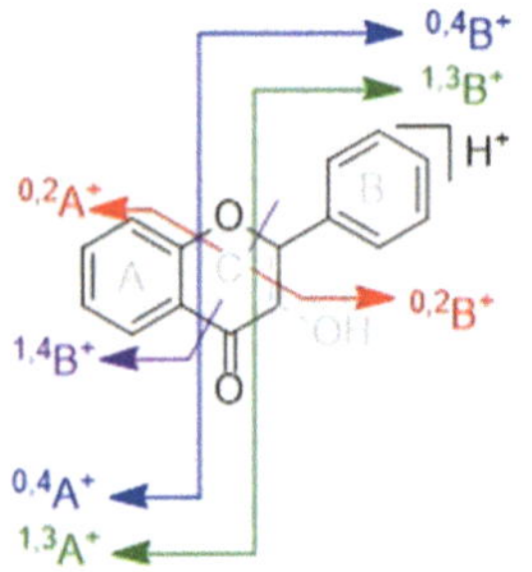

Fig. 5 Identification of flavonoids using fragmentation patterns. The general retrocyclization cleavage of the flavonoid C-ring is indicated [27]

Table 2
Major ESI-MS/MS product ions observed for the $[M-H]^+$ of selected flavonoids

Compound	Fragment ion $[M + H]^+$	ESI-MS/MS product ions $^{0,2}A^+$	$^{0,2}B^+$	$^{1,3}A^+$	$^{1,3}B^+$	$^{1,4}A^+$	$^{0,4}A^+$
Quercetin	303	165	137	153	149		
Kaempferol	287	165	121	153	133		
Rutin	611	303					
Luteolin	287		137	153	135		
Apigenin	271		121	153	119		
Naringenin	273		121	153	119	147	
Genistein	271			153	119		
Catechin	291	123					
Hesperetin	303			153			177

3. Other more general fragments result from the loss of H_2O (M-18), CO (M-28), C_2H_2O (M-42), successively H_2O and CO (M-46).

4. To analyze fragment ions in flavonoid glycosides, Domon and Castello (1988) introduced a specific nomenclature [29]. The ions containing the aglycone are denoted as $^{k,l}Xj$, Yj, and Zj, where j signifies the interglycosidic bond number broken, counted from the aglycone. The superscripts k and l indicate cleavages within the carbohydrate rings. The glycosidic bond linking the glycan part to the aglycone is labeled as Z_0. Z_1 occurs in diglycosides by cleavage between the sugars. Y represents the diglucoside, with fragments containing the aglycone part referred to as Y_1 (indicating the loss of one sugar moiety) and Y_0 (indicating the loss of two sugar moieties).

5. Cleavage at the glycoside *O*-linkages leads to the loss of monosaccharide residues, for example, 162 (hexose), 146 (deoxyhexose), and 132 (pentose).
6. For the identification of anthocyanins, the loss of sugar residues produces specific aglycone fragments: cyanidin (*m/z* 287), malvidin (*m/z* 331), delphinidin (*m/z* 303), petunidin (*m/z* 317), and peonidin (*m/z* 301).
7. For the spectral data analysis, the free software mzmine can be used (https://mzio.io/#mzmine) [30]. For this, convert and centroid the obtained MS data to mzML or mzXML using the software MSConvert (http://proteowizard.sourceforge.net/).
8. Another way to attain structural information is by using the Global Natural Products Social Molecular Networking (GNPS) (https://gnps.ucsd.edu/ProteoSAFe/static/gnps-splash.jsp). Hereby MS/MS data is searched against reference spectral libraries on the GNPS platform. To use GNPS free registration on the website is needed.
9. For GNPS analysis, upload the mzML files to the website and run the Spectral library search with the default parameters given.

Data Interpretation of MS and MS/MS Spectra The identification of flavonoids was initially conducted using MS/MS fragmentation patterns, specifically the RDA reaction, along with bibliographic information on MS/MS spectra. This approach enabled the identification of 11 flavonoids within the blueberry extract based on literature (Fig. 6, Table 3).

Table 3 shows the information on retention times, parent ions, and common fragment ions used for this identification. The flavonoid profile of the blueberry extract includes various glycosylated derivatives of anthocyanins, such as cyanidin, peonidin, malvidin, and petunidin, as well as flavonols like rutin, quercetin, and myricetin. For instance, cyanidin-3-*O*-glucoside was annotated at a retention time of 5.40 min with a parent ion (*m/z*) of 449.0228. This compound exhibited a fragment ion (*m/z*) of 287.007, characteristic of the cyanidin aglycone, with a neutral loss of 162u, corresponding to the glucose moiety. Using quercetin as an example to illustrate RDA fragmentation, we observed a parent ion at *m/z* 303. The fragmentation process specifically yielded the $^{0,2}A^+$ fragment (*m/z* 165), the $^{1,3}A^+$ fragment (*m/z* 153), and the $^{1,3}B^+$ fragment (*m/z* 149). The primary fragments generated from the cleavage of the C-ring of quercetin are illustrated in Fig. 7.

Another possibility as mentioned above is by using the GNPS spectral library search option. The GNPS result of the blueberry extract can be found at https://gnps.ucsd.edu/ProteoSAFe/status.jsp?task=1eb16db88de04c86be710080c58e452a. The

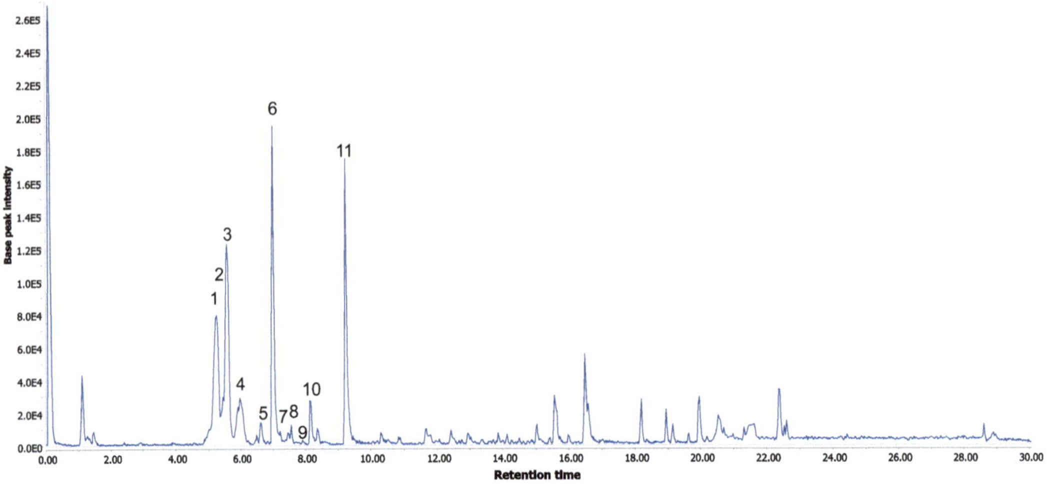

Fig. 6 LC-MS chromatogram of blueberry extract. A methanolic blueberry extract was analyzed on a QTOF instrument in positive ionization mode and data processing was done using mzmine (30). The total ion chromatogram (TIC) is shown. Peaks were tentatively identified 1, rutin; 2, cyanidin 3-*O*-glucoside; 3, cyanidin 3-rutinoside; 4, peonidin 3-rutinoside; 5, cyanidin aglycone; 6) delphinidin 3-rutinoside; 7, unknown; 8, petunidin 3-rutinoside; 9, malvidin 3-rutinoside; 10, myricetin; 11, quercetin

detailed features of the 19 annotated compounds from the GNPS library search are listed in Table 4. This table includes their library class, cosine score, spectral and library *m/z*-values, instrumentation, ion source, and data source. One of the critical features is the cosine score: a score of 1 indicates identical spectra, while a score of 0 denotes no similarity (*see* **Note 20**) [36]. The mirror match between the experimental and library mass spectra can be obtained for all library hits. For example, the mirror match for cyanidin 3-*O*-glucoside shows a close similarity to the experimental data, with a cosine value of 0.98 (Fig. 8). This supports the annotation of the detected *m/z* 449.0251 in the blueberry as cyanidin 3-*O*-glucoside. Another example is the mirror spectral match for quercetin which shows a very close similarity to the blueberry extract data with a cosine value of 0.92. This supports the annotation of the *m/z* 303 peak as quercetin.

4 Notes

1. Flash freezing in liquid nitrogen is a post-harvest method to reduce the degradation of metabolites within the plant material. Alternatively, freeze-drying material can also be ground to a fine powder in liquid nitrogen and stored at −80 °C (frozen material).
2. To avoid degradation of labile metabolites, storage in the dark or wrapping with aluminum foil is recommended.

Table 3
Flavonoids detected within the blueberry extract identified by mass spectra

Compound	RT minutes	Parent ion *m/z*	Common fragment ions *m/z*	Identification or tentatively characterization	References
1	5.20	611.0904	302.9963; 465.0195; 449.0256 303 [^{0,2} A$^+$] 465 [M + H – 146]$^+$ 449 [M + H – 162]$^+$	Rutin	[31]
2	5.40	449.0228	287.0072 287 [M + H – 162]$^+$	Cyanidin 3-*O*-glucoside	[32]
3	5.58	595.0524	449.0228; 287.0041; 1189.1727 449 [M + H – 146]$^+$ 287 [M + H – 146 – 162]$^+$	Cyanidin 3-rutinoside	[33]
4	5.95	609.0716	463.0417; 301.0179; 463 [M + H – 146]$^+$ 301 [M + H – 146 – 162]$^+$	Peonidin 3-rutinoside	[33]
5	6.95	287.0048	287.0037	Cyanidin aglycone	
6	7.00	611.0495	465.0186; 302.9968; 136.9927 465 [M + H – 146]$^+$ 303 [M + H – 146 – 162]$^+$	Delphinidin 3-rutinoside	[33]

(continued)

Table 3 (continued)

Compound	RT minutes	Parent ion *m/z*	Common fragment ions *m/z*	Identification or tentatively characterization	References
7	7.20	478.9970	302.9984 303 [M + H – 176]$^+$	Unknown—a mass difference of 176 is of a glucuronide residue, peonidin monoglucuronide	
8	7.50	625.0593	479.0269; 317.0114 479 [M + H – 146]$^+$ 317 [M + H – 146 – 162]$^+$	Petunidin 3-rutinoside	[34]
9	7.88	639.0463	331.0259 331 [M + H – 146 – 162]$^+$	Malvidin 3-rutinoside	[33]
10	8.12	318.9894	152.9897 153 [^{1,3}A$^+$]	Myricetin	[35]
11	9.22	302.9972	165.0180; 153.0180; 149.0014 165 [^{0,2}A$^+$] 153 [^{1,3}A$^+$] 149 [^{1,3}B$^+$]	Quercetin	[35]

3. After transferring the supernatant into a new tube, verify whether the supernatant is clear. If it is not clear, repeat the centrifugation and transferring step.
4. UV-Vis spectroscopy is a powerful tool for initial characterization, other advanced analytical techniques may be required for a more comprehensive identification and quantification of flavonoid subclasses. Additionally, the presence of multiple flavonoids in a complex mixture may require a combination of techniques for accurate characterization.
5. Employing reference compounds from each flavonoid class can aid in identification. By comparing the absorption spectra of the sample with those of known reference compounds, a preliminary classification can be made.
6. If unusual or unexpected results are obtained, troubleshoot by checking for contamination, ensuring proper dissolution, or

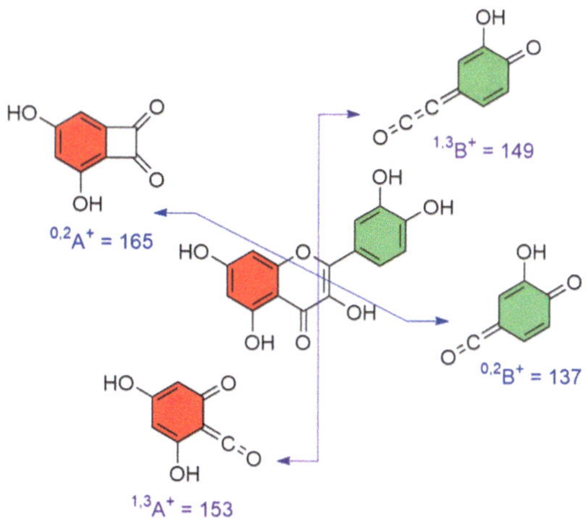

Fig. 7 The retrocyclization cleavage of the C-ring in quercetin in positive ionization mode. Observed fragments include the $^{0,2}A^+$ fragment (*m/z* 165), the $^{1,3}A^+$ fragment (*m/z* 153), the $^{1,3}B^+$ fragment (*m/z* 149), and the $^{0,2}B^+$ fragment (*m/z* 137). The A-ring is depicted in red and the C-ring in green

recalibrating the spectrophotometer if necessary. Work with triplicates.

7. The mobile phase composition can be adjusted based on the specific properties of flavonoids of interest. For example, for flavonoid glycosides EtOAc-formic acid-water [EFW (50:3:3, v/v)] or for chalcones and isoflavones *n*-hexane-EtOAc-acetic acid [HEA (6:4:0.5, v/v)].
8. Include authentic reference compounds on the same TLC plate to verify flavonoid identity.
9. Residual solvent (acids) must be removed from the silica gel layer before further reaction. If necessary, an additional device like a hairdryer, air gun, or waiting for 30 min can be applied to the plates.
10. Not only flavonoids but also other phenolics show absorbance in this range. To confirm further after running the TLC plate, antimony (III) chloride reagent can be used as a derivatization reagent. Flavonoids have a distinct reactivity with antimony (III) chloride compared to other phenolic or polyphenolic compounds. This reaction typically results in visible colors such as yellow, orange, or red, which can be easily observed under visible light. The formation of these colored complexes is specific to the presence of hydroxyl groups on the flavonoid structure.
11. TLC plates with chemical derivatization cannot be used for other reactions. For each reaction, individual plates are required. The most common derivatization reagents for

Table 4
Library hits of compounds within the blueberry extract from a library search in GNPS

Name	Cluster index	Library Class	Cosine	Mass difference	Spectral *m/z* Library *m/z*	Instrument	Data Source	Ion Source
MassBank: FIO00274 quercetin	1080	Bronze	0.92	0.052	303.00 303.05	QTOF	MASSBANK	ESI
Tamarixetin collision energy:102040	1234	Bronze	0.79	0.056	317.01 317.07	Orbitrap	Trent Northen	ESI
Massbank:PR310841 LPC 18:3	1660	Bronze	0.75	0.098	518.23 518.33	QTOF	MASSBANK	ESI
PC (18:3/0:0); [M + H] + C26H49N1O7P1	1662	Gold	0.87	0.095	518.23 518.33	HCD; Velos	Thomas Metz	ESI
PC (18:2/0:0); [M + H] + C26H51N1O7P1	1764	Gold	0.90	0.098	520.24 520.34	HCD; Velos	Thomas Metz	ESI
Spiraeoside	565	Bronze	0.93	0.083	465.02 465.10	QTOF	Stephane GREFF	ESI
Massbank:PR311137 Delphinidin 3-glucoside	570	Bronze	0.93	0.084	465.02 465.10	QTOF	MASSBANK	ESI
Massbank:PR302672 Diosmetin-7-O-neohesperidoside	596	Bronze	0.95	1.865	611.05 609.18	QTOF	MONA	ESI
Massbank:PR302318 Cyanidin-3-*O*-glucoside	626	Bronze	0.98	0.080	449.03 449.11	QTOF	MASSBANK	ESI
Massbank:PR310745 Caffeoyl quinic acid	642	Bronze	0.89	0.060	355.04 355.10	QTOF	MASSBANK	ESI
Massbank:PR303856 Petunidin 3-galactoside	662	Bronze	0.95	0.934	480.05 479.12	QTOF	MASSBANK	ESI

Massbank:PR310523 Cyanidin-3-*O*-alpha-arabinoside	665	Bronze	0.95	0.018	611.13 611.15	QTOF	MASSBANK	ESI
Spectral match to Peonidin 3-*O*-glucoside cation from NIST14	679	Bronze	0.96	0.079	463.04 463.12	QTOF	MASSBANK	ESI
Spectral match to Malvidin 3-*O*-galactoside cation from NIST14	686	Bronze	0.89	0.084	493.05 493.13	QTOF	Data from PC Dorrestein	ESI
7-hydroxy-2-(4-hydroxy-3,5-dimethoxyphenyl)-5-[(2S,3R,4S,5S,6R)-3,4,5-trihydroxy-6-(hydroxymethyl)oxan-2-yl]oxychromen-4-one	714	Bronze	0.82	0.902	494.04 493.13	QTOF	MONA	ESI
NCGC00180851–02 5,7-dihydroxy-2-(4-hydroxy-3-methoxyphenyl)-3-[(2S,3R,4S,5R,6R)-3,4,5-trihydroxy-6-[[(2R,3R,4R,5R,6S)-3,4,5-trihydroxy-6-methyloxan-2-yl]oxymethyl]oxan-2-yl]oxychromen-4-one	759	Gold	0.79	0.095	519.05 519.15	Maxis II HD QTOF Bruker	Jadhav/ Dorrestein	ESI
Spectral match to Kaempferol 3-*O*-rutinoside from NIST14	879	Bronze	0.89	0.130	595.03 595.16	QTOF	Jadhav/ Dorrestein	ESI
Massbank:PR301932 isorhamnetin-3-*O*-rutinoside	887	Bronze	0.88	0.117	625.06 625.18	QTOF	MASSBANK	ESI
Myricetin	962	Bronze	0.80	0.069	318.98 319.05	QTOF	MONA	ESI

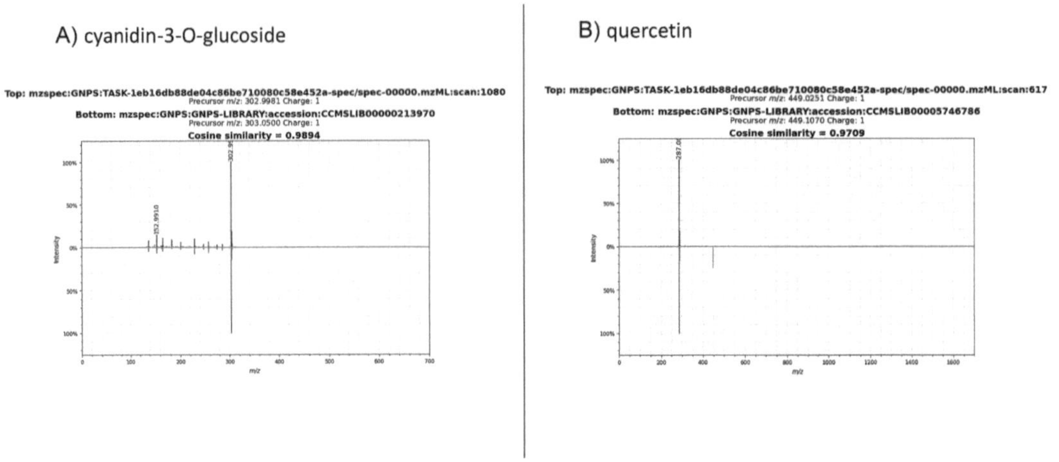

Fig. 8 Identification of flavonoids in blueberry extracts using GNPS mirror match. Cosine similarity measures the degree of overlap between two mass spectra, with high cosine similarity indicating a near-perfect match. (**a**) The mirror match for cyanidin 3-*O*-glucoside shows a close similarity to the experimental data, with a cosine value of 0.98. (**b**) Another example is the mirror spectral match for quercetin which shows a very close similarity to the blueberry extract data with a cosine value of 0.92. LC-MS-QTOF data of methanolic blueberry extracts was processed using mzmine and analyzed using GNPS

flavonoids detection are anisaldehyde-H_2SO_4, NP, and antimony (III) chloride reagent.

12. The fragmentation pattern of metabolites is largely independent of instruments and the used ionization mode (ESI or atmospheric-pressure chemical ionization (APCI)). However, it is important to note that while the overall fragmentation pattern is stable, the intensity of the fragments can vary depending on the ionization technique and the specific instrument settings. These variations in fragment intensity may influence the sensitivity and quantitative aspects of the analysis, but they do not alter the fundamental fragmentation pathways of the metabolites.
13. The detection limits of compounds in LC-MS measurements can vary widely depending on several factors, including the type of instrument, the ionization technique, the nature of the compound, and the sample matrix. Generally, detection limits for LC-MS can range from nanogram per milliliter (ng/mL) to femtogram per milliliter (fg/mL) levels. The average detection limit for flavonoids is 5–10 μg.
14. It is recommended to run QC samples that are prepared by pooling aliquots (e.g., 10 μL) from each sample. For very large sample sets, QCs can be prepared by pooling aliquots from a subset of the samples. QC samples can be measured after every five samples to verify the performance and stability of the employed LC-MS system.

15. A blank sample is recommended to run after QC samples to check for carryover from previous samples.
16. Use LC-MS-grade solvents to minimize background noise, which can obscure or overlap with the signals, reducing the clarity and accuracy of the results. Impurities in solvents can also lead to ion suppression or enhancement affecting the ionization efficiency of the analytes. LC-MS-grade solvents are specifically processed to reduce these effects, ensuring more consistent and reliable ionization. Using high-quality solvents ensures reproducibility in your analyses. Variability in solvent quality can lead to inconsistent results, making it difficult to compare data across different runs or experiments.
17. Choosing a suitable chromatography column (e.g., C18 and C8) is crucial for obtaining optimal results in terms of separation efficiency, resolution, and sensitivity. Several factors must be considered, including column length, internal diameter, and particle size. Longer columns provide better separation efficiency, but they also increase analysis time and solvent consumption. Narrower columns provide higher sensitivity due to reduced dilution and also consume less solvent. Smaller particle sizes provide higher efficiency and better resolution, leading to sharper peaks and better separation. However, smaller particles generate higher back pressure which requires more robust instrumentation capable of handling higher pressure (e.g., UPLC systems). By carefully considering these factors, the most appropriate LC column can be selected to achieve optimal separation.
18. The injection volume is generally 1–5 μL though it depends on the LC-MS system.
19. A collision energy of 30 eV is proposed to give the most informative data for flavonoids as it provides a good balance between fragmentation and signal intensity.
20. Cosine similarity measures the degree of overlap between two mass spectra. A high cosine similarity indicates a near-perfect overlap between the spectra. Only consider annotated compounds with a cosine ≥0.7.

References

1. Buer CS, Imin N, Djordjevic MA (2010) Flavonoids: new roles for old molecules. J Integr Plant Biol 52:98–111. https://doi.org/10.1111/j.1744-7909.2010.00905
2. Petrussa E, Braidot E, Zancani M, Peresson C, Bertolin AI, Patui S, Vianello A (2013) Plant flavonoids—biosynthesis, transport and involvement in stress responses. Int J Mol Sci 14:14950–14973. https://doi.org/10.3390/ijms140714950
3. Dias MC, Pinto DCGA, Silva AMS (2021) Plant flavonoids: chemical characteristics and biological activity. Molecules 26:5377. https://doi.org/10.3390/molecules26175377

4. Kumar S, Pandey KP (2013) Chemistry and biological activities of flavonoids: an overview. Sci World J 29(2013):162750. https://doi.org/10.1155/2013/162750
5. Chen S, Wang X, Cheng Y, Gao H, Chen X (2023) A review of classification, biosynthesis, biological activities and potential applications of flavonoids. Molecules 28(13):4983. https://doi.org/10.3390/molecules28134982
6. Williams CA, Grayer RJ (2004) Anthocyanins and other flavonoids. Nat Prod Rep 21(4): 539–573. https://doi.org/10.1039/b311404j
7. Hofer B (2016) Recent developments in the enzymatic O-glycosylation in flavonoids. Appl Microbiol Biotechnol 100(10):4269–4281. https://doi.org/10.1007/s00253-016-7465-0
8. Du Y, Chu H, Chu IK, Lo C (2010) CYP93G2 is a flavanone 2-hydroxylase required for C-glycosylflavone biosynthesis in rice. Plant Physiol 154:324–333. https://doi.org/10.1104/pp.110.161042
9. Zou L, Li H, Ding X, Liu Z, He D, Kowah JAH, Wang L, Yuan M, Liu X (2022) A review of the application of spectroscopy to flavonoids from medicine and food homology materials. Molecules 27:7766. https://doi.org/10.3390/molecules27227766
10. Mabry TJ, Markham KR, Thomas MB (1970) The ultraviolet spectra of flavones and flavonols. In: The systematic identification of flavonoids. Springer, Berlin/Heidelberg. https://doi.org/10.1007/978-3-642-88458-0_5
11. Gocan S, Cimpan G (2004) Review of the analysis of medicinal plants by TLC: modern approaches. J Liq Chromatogr Relat Technol 27(7–9):1377–1411. https://doi.org/10.1081/JLC-120030607
12. Nile SH, Park SW (2015) HPTLC densitometry method for simultaneous determination of flavonoids in selected medical plants. Front Life Sci 8(1):97–103. https://doi.org/10.1080/21553769.2014.969387
13. Fabre N, Rustan I, de Hoffmann E, Quetin-Leclercq J (2001) Determination of flavone, flavonol, and flavanone Aglycones by negative ion liquid chromatography electrospray ion trap mass spectrometry. J Am Soc Mass Spectrom 12:707–715. https://doi.org/10.1016/S1044-0305(01)00226-4
14. Cuyckens F, Claeys M (2004) Mass spectrometry in the structural analysis of flavonoids. J Mass Spectrom 39:1–15. https://doi.org/10.1002/jms.585
15. De Villiers A, Venter P, Pasch H (2016) Recent advances and trends in the liquid-chromatography–mass spectrometry analysis of flavonoids. J Chromatogr A 1430:16–18. https://doi.org/10.1016/j.chroma.2015.11.077
16. Kalt W, Cassidy A, Howard LR, Krikorian R, Stull AJ, Tremblay F, Zamora-Ros R (2020) Recent research on the health benefits of blueberries and their Anthocyanins. Adv Nutr 11(2):224–236. https://doi.org/10.1093/advances/nmz065
17. Liu J, Hefni ME, Witthöft CM (2020) Characterization of flavonoid compounds in common Swedish berry species. Food Secur 9:358. https://doi.org/10.3390/foods9030358
18. Saha S, Singh J, Paul A, Sarkar R, Khan Z, Banerjee K (2020) Anthocyanin profiling using UV-vis spectroscopy and liquid chromatography mass spectrometry. J AOAC Int 103(1):23–39. https://doi.org/10.5740/jaoacint.19-0201
19. Longo L, Vasapollo G, Rescio L (2005) Identification of Anthocyanins in Rhamnus alaternus L. berries. J Agric Food Chem 53(5): 1723–1727. https://doi.org/10.1021/jf048253p
20. Taniguchi M, Du H, Lindsey JS (2018) PhotochemCAD 3: diverse modules for photophysical calculations with multiple spectral databases. Photochem Photobiol 94:277–289. https://doi.org/10.1111/php.12862
21. Taniguchi M, LaRocca CA, Bernat JD, Lindse JS (2023) Digital database of absorption spectra of diverse flavonoids enables structural comparisons and quantitative evaluations. J Nat Prod 86(4):1087–1119. https://doi.org/10.1021/acs.jnatprod.2c00720
22. Shraim AM, Ahmed TA, Rahman MdM, Hijji YM (2021) Determination of total flavonoid content by aluminum chloride assay: a critical evaluation. 150:111932. https://doi.org/10.1016/j.lwt.2021.111932
23. Julkunen-Tiitto R, Nenadis N, Neugart S et al (2015) Assessing the response of plant flavonoids to UV radiation: an overview of appropriate techniques. Phytochem Rev 14:273–297. https://doi.org/10.1007/s11101-014-9362-4
24. MetaboScape 4.0—untargeted metabolomics solution. Bruker.com Available at: https://www.bruker.com/products/mass-spectrometry-and-separations/ms-software/metaboscape
25. Wang M et al (2016) Sharing and community curation of mass spectrometry data with global natural products social molecular networking.

Nat Biotechnol 34:828–837. https://doi.org/10.1038/nbt.3597

26. Horai H, Arita M, Kanaya S et al (2010) MassBank: a public repository for sharing mass spectral data for life sciences. J Mass Spectrom 45(7):703–714. https://doi.org/10.1002/jms.1777
27. Ma YL, Li QM, Van den Heuvel H, Claeys M (1997) Characterization of flavone and flavonol aglycones by collision-induced dissociation tandem mass spectrometry. Rapid Commun Mass Spectrom 11:1357–1364. https://doi.org/10.1002/(SICI)1097-0231(199708)11:12<1357::AID-RCM983>3.0.CO;2-9
28. Śliwka-Kaszyńska M, Anusiewicz I, Skurski P (2022) The mechanism of a retro-diels–alder fragmentation of Luteolin: theoretical studies supported by electrospray ionization tandem mass spectrometry results. Molecules 27:1032. https://doi.org/10.3390/molecules27031032
29. Domon B, Costello CE (1988) A systematic nomenclature for carbohydrate fragmentations in FAB-MS/MS spectra of glycoconjugates. Glycoconj J 5:397–409. https://doi.org/10.1007/BF01049915
30. Katajamaa M, Miettinen J, Orešič M (2006) MZmine: toolbox for processing and visualization of mass spectrometry based molecular profile data. Bioinformatics 22(5):634–636. https://doi.org/10.1093/bioinformatics/btk039
31. Kazuno S, Yanagida M, Shindo N, Murayama K (2005) Mass spectrometric identification and quantification of glycosyl flavonoids, including dihydrochalcones with neutral loss scan mode. Anal Biochem 347(2):182–192. https://doi.org/10.1016/j.ab.2005.09.020
32. Diaz de Cerio E, Verardo V, Caravaca AMG, Segura CA (2016) Exploratory characterization of phenolic compounds with demonstrated anti-diabetic activity in guava leaves at different oxidation states. Int J Mol Sci 17(699):1–13. https://doi.org/10.3390/ijms17050699
33. Ruiz A, Hermosín-Gutiérrez I, Vergara C, von Baer D, Zapata M, Hitschfeld A, Obando L, Mardones C (2013) Anthocyanin profiles in south Patagonian wild berries by HPLC-DAD-ESI-MS/MS. Food Res Int 51(2):706–713. https://doi.org/10.1016/j.foodres.2013.01.043
34. Nour V, Stampar F, Veberi R, Jakopic J (2013) Anthocyanins profile, total phenolics and antioxidant activity of black currant ethanolic extracts as influenced by genotype and ethanol concentration. Food Chem 141:961–966. https://doi.org/10.1016/j.foodchem.2013.03.105
35. Pedro SI, Fernandes TA, Luís Â, Antunes AMM, Gonçalves JC, Gominho J, Gallardo E, Anjos O (2023) First chemical profile analysis of acacia pods. Plan Theory 12:3486. https://doi.org/10.3390/plants12193486
36. Bittremieux W, Chen C, Dorrestein PC, Schymanski EL, Schulze T, Neumann S, Meier R, Rogers S, Wang M (2020) Universal MS/MS visualization and retrieval with the metabolomics spectrum resolver web service. bioRxiv 2020.05.09.086066; https://doi.org/10.1101/2020.05.09.086066

Chapter 10

Analysis of Salicylic and Phenolic Acids in the Plant by HPLC–Fluorescence Detector

Natali Rianika Mustafa

Abstract

Salicylic acid is a member of benzoic acid derivatives, a group of compounds which have a backbone of C6C1 consisting of one carboxyl group and one (or more) hydroxyl group(s) attached to the aromatic ring. Salicylic acid is a signaling compound in systemic acquired resistance (SAR). An increased level of salicylic acid is found in the plant after a fungi's attack, which further induces the accumulation of phytoalexins, low molecular weight defense compounds. Here, we describe the extraction method and determination of salicylic acid using high-performance liquid chromatography (HPLC) coupled with fluorescence detection.

Key words Phenolic acids, C6C1, Salicylic acid, Acid hydrolysis, Organic solvent extraction, HPLC, Fluorescence detection

1 Introduction

C6C1 phenolic acids are called benzoic acid derivatives because they are biosynthesized from the benzoic acid/phenylalanine pathway (except gallic acid, which is from the shikimate pathway) [1, 2]. Salicylic acid (SA, pKa = 2.97) is an example of C6C1 compounds, also produced by bacteria, but through a shorter pathway, the isochorismate pathway. Further, it has been found that SA [3] and 2,3-dihydroxybenzoic acid (2,3-DHBA or pyrocatechuic acid, pKa = 2.56) [4] can also be synthesized in the plant through the isochorismate pathway. SA is a signaling compound involved in systemic acquired resistance (SAR) as a response of the plant when attacked by certain pathogens (e.g., fungi). Elicitation of *Catharanthus roseus* cell cultures with a fungal extract (*Pythium aphanidermatum*) provided a peak of SA after 24 h [3] followed by an accumulation of a defense compound 2,3-DHBA released in the culture medium [4]. The *Catharanthus* suspended-cells cultures were also able to convert SA spiked in the medium into

Young Hae Choi et al. (eds.), *Natural Product Isolation and Identification: Methods and Protocols*, Methods in Molecular Biology, vol. 2895, https://doi.org/10.1007/978-1-0716-4350-1_10,

2,5-dihydroxybenzoic acid (2,5-DHBA = gentisic acid, pKa = 2.97) and its glucoside [5].

The levels of SA or other C6C1s can vary between individual plants. Analytical methods for determination of C6C1 (or phenolics in general) were developed or reviewed in several studies, including thin-layer chromatography (a common TLC-UV or a more advanced high-performance TLC-MS/MS) [6], gas chromatography (GC-MS) [7–11], high-performance liquid chromatography (HPLC)-UV/PDA [12–17], and LC-MS [18–21]. TLC is an efficient method because many samples can be run at once (on one plate), and C6C1s (phenolics) can be detected by UV and/or colorimetry (derivatization using a color reagent). However, even though a quantitative analysis using TLC-UV (spectrophotodensitometry) is possible, it is less sensitive compared to HPLC-UV. While GC-MS is a very sensitive method (ng-pg level), polar compounds like phenolics need a derivatization step before analysis. LC-MS is often used for plant metabolomics and identification of compound, particularly for compounds which do not have chromophores. The high sensitivity (ng-pg level) and the availability of compound's library are the advantages of LC-MS if compared to a common HPLC-PDA. Nevertheless, HPLC-PDA is mostly preferable for quantitative analysis of a target (known) compound having chromophores like phenolics because the HPLC system and the operation are simpler and cheaper than LC-MS. In general, C6C1s and other phenolics are eluted from a C18 column with the isocratic or gradient mode of mixtures of water and methanol (or acetonitrile) containing formic acid, acetic acid, or trifluoroacetic acid of 0.01–1.0% and can be detected with UV/PDA detection [12–17] depending on the levels of the compounds in the extract. However, SA (a plant hormone) is particularly produced in the plant at a very low level, which is hardly detected by UV/PDA. Hence, a more sensitive detection is necessary. Since SA can fluoresce, an HPLC coupled with a fluorescence detection (more sensitive than UV/PDA) is very suitable. This chapter discusses the extraction method of SA (which is also applicable for other C6C1s and C6C3s) in plants, and analysis of the extract by HPLC–fluorescence [22].

2 Materials

Unless otherwise stated, all solvents employed in this protocol are analytical grade; for example, *n*-hexane, EtOAc, and MeOH are HPLC grade. Reference standard(s) are of 99% purity or higher.

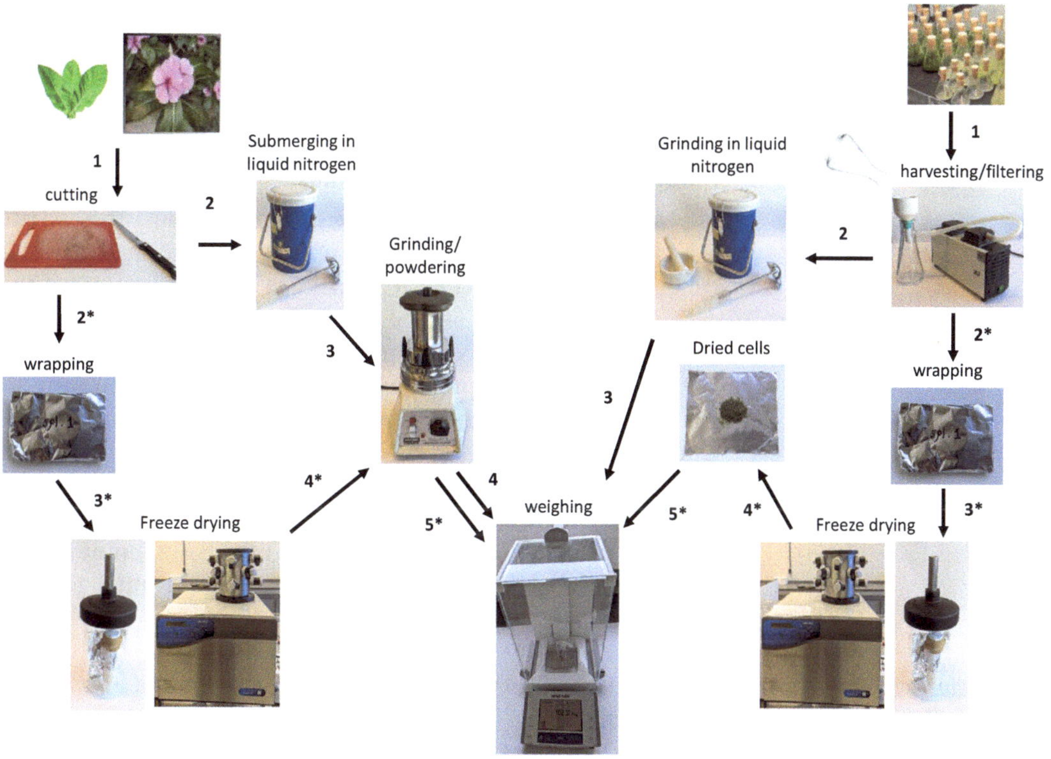

Fig. 1 Preparation of plant material before extraction. *Step of lyophilization

2.1 Preparation of Plant Materials (See Also Fig. 1)

1. Working with fresh plant materials: separate plant materials from different observation groups properly (e.g. different plant organs such as flowers, leaves, stem barks, or roots), cut each group of plant material in small pieces, transfer the cut material immediately on a piece of aluminum foil, wrap, label properly, and plunge it into nitrogen liquid. Transfer the frozen material into a metal blender or a mortar (with a pestle) previously poured with nitrogen liquid, and grind it until powder. Keep the powder material frozen while sampling/weighing it before extraction with organic solvent.
2. Drying plant materials: separate properly plant materials from different observation groups (can be based on different treatments of observation and/or different plant organs such as flowers, leaves, stems/bark, or roots), cut each group of plant material in small pieces, transfer the cut material immediately on a piece of aluminum foil, wrap, label properly, and plunge it into liquid nitrogen for 10–15 min. Transfer the frozen wrapped material to the freeze dryer, and lyophilize it for at least 48 h. When it is dried properly, transfer the material to a blender, blend it until powder, and weigh the powder as a sample for extraction. Store the rest of the powder in a clean

and dry container with a lid/screw cap (e.g., a 50-mL Falcon tube), close the container tightly, and place it at room temperature in the dark and dry storage.

3. *Working with plant cell cultures*: harvest a flask of a "stationary phase" suspended-cells culture by pouring the cell culture on the surface of a round Whatman filter paper placed on a Buchner vacuum-filter system (connecting to a pump) to separate the cells from the culture medium. Collect the medium in a clean 50-mL Falcon tube for analysis (if necessary), rinse the cells on the Buchner filter using 100 mL of de-ionized water (Millipore), do it twice. Transfer some parts of the rinsed cells into a cold mortar (previously poured with nitrogen liquid), pour some liquid nitrogen onto the cells, grind the frozen cells using a cold pestle until powder, transfer (using a cold spoon), and weigh a certain amount of the frozen powder in a clean cold 2-mL microtube. To prepare cells for lyophilization: collect and transfer the rinsed cells on a piece of aluminum foil, wrap and label properly, plunge into liquid nitrogen for 5–10 min, and lyophilize for 48 h.

2.2 Preparation of Reagents and HPLC Mobile Phase

1. To prepare NaOH 0.2 M: weigh 400 mg of NaOH in a 20-mL glass bottle with a screw cap, add 10 mL of de-ionized water (Millipore), close the bottle, and homogenize using a stirring plate or by gentle swirling by hand until all NaOH crystals dissolve.
2. To prepare trichloroacetic acid (TCA) 5% w/v: weigh 5 g TCA in a 100-mL glass bottle with a screw cap, add 100 mL of de-ionized water (Millipore), close the bottle, and homogenize by gentle swirling by hand.
3. To prepare HCl 8 M: transfer 66.2 mL of HCl 37% (= 12.08 M) using a 100-mL volumetric glass cylinder into a 100-mL glass bottle, dilute it by adding 33.8 mL of de-ionized water, close the bottle, and homogenize by gently shaking.
4. To prepare a mixture of EtOAc: *n*-hexane (1:1, v/v): transfer 50 mL of EtOAc into a 100-mL glass bottle (with a screw cap), add 50 mL of n-hexane (or cyclohexane), close the bottle, and mix gently.
5. To prepare HPLC mobile phase (a mixture of 0.1 M ammonium acetate buffer, pH 5.5, and MeOH, 9:1, v/v): weigh 7.7 g of ammonium acetate, transfer to a 1-L glass bottle with a screw cap, add 1 L of de-ionized water, dissolve and homogenize using a magnetic stirring, and add some drops of acetic acid (glacial) (99%) using a glass Pasteur pipette into the solution to adjust the pH to 5.5 (use a pH meter). Transfer 900 mL of the buffer solution into a clean 1-L glass bottle with a screw cap, add 100 mL of MeOH (HPLC grade), close the bottle, and mix gently.

3 Methods

3.1 Extraction of Free SA/Free Phenolic Acids (See Also Fig. 2)

1. Weigh 500 mg fresh weight (FW) of frozen plant powder or 100 mg of freeze-dried plant powder in a 2-mL safelock microcentrifuge tube.
2. Add 1 mL of 100% MeOH (or 90% MeOH in water in case of using dried plant material), and close the microtube.
3. Vortex for a few seconds, and put it in an ultrasonic bath at room temperature (RT) for 30 min.
4. Centrifuge at 20,376 G, 10 min, at RT.
5. Collect the clear supernatant (using a 1-mL micropipette or carefully by decantation) in a clean 2-mL safelock microcentrifuge tube.
6. Repeat extraction on the pellet/plant debris with 0.5 mL of 100% MeOH vortex and sonicate the mixture for 20 min.
7. Centrifuge at 20,376 × g, 10 min, at RT, and collect and combine the supernatant with the previous one in the same microtube (in total 1.5 mL).
8. Add 10 μL of 0.2 M NaOH into the microtube, and homogenize/vortex gently.
9. Evaporate the solvent in the mixture using a SpeedVac concentrator (keeping temperature not above 35 °C) until dryness.
10. Add 250 μL of TCA 5% in water (w/v) into the residue (the microtube), and homogenize/vortex.
11. Add 800 μL of a mixture of EtOAc: *n*-hexane (1:1, v/v), and do partition using a vortex for about 20 s.
12. Centrifuge the sample tube at 20,376 g for 30 s to separate the phases, and pipette and transfer the nonpolar (EtOAc: *n*-hexane) phase into a clean 2-mL safelock microcentrifuge tube.
13. Repeat **steps 11–12**, and combine the nonpolar phase with the previous one at **step 12** (total volume is 1.6 mL).
14. Add 60 μL of HPLC mobile phase (*see* **Note 1**) into EtOAc: *n*-hexane in the microtube. Put a marker on the microtube's wall to mark the height of the polar fraction (60 μL volume) from the bottom of the microtube.
15. Evaporate the nonpolar fraction in the SpeedVac for about 20 min (keeping temperature not above 35 °C) until the 60 μL polar phase remains in the microtube.
16. Add 600 μL of HPLC eluent into the microtube (total volume is 660 μL; *see* **Note 2**), and homogenize/vortex.
17. Centrifuge the microtube at 20,376 × g for 10 min, and transfer the clear supernatant into a 1.5-mL HPLC glass vial. This sample is ready for HPLC analysis.

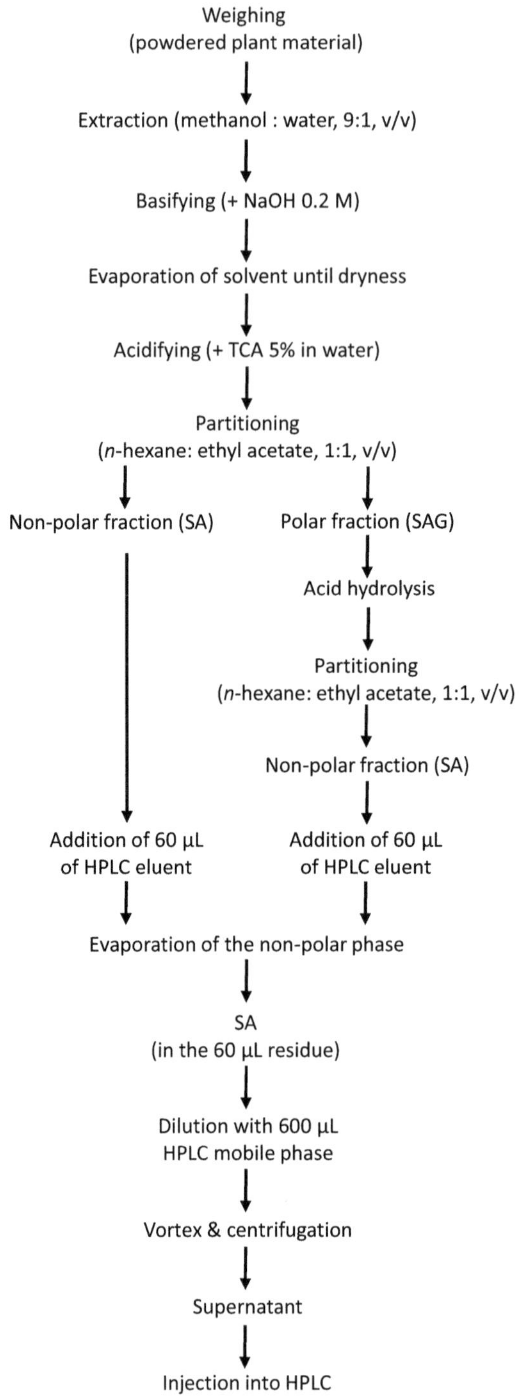

Fig. 2 Scheme of extraction protocol of salicylic acid (phenolic acids)

3.2 Extraction of Ether-Bound SA (Phenolic Acid Glycosides) After Acid Hydrolysis

1. Add 300 μL of 8 M HCl into the remaining water/TCA fraction in the microtube after the partition (**steps 11–13** of Subheading 3.1), and vortex gently.
2. Incubate the acidic mixture at 80 °C for 1 h in a water bath (*see* **Note 3**).
3. Cool down the mixture at RT (or on ice for faster cooling).
4. Add 800 μL of EtOAc: *n*-hexane (1:1, v/v), and do partition/vortex.
5. Centrifuge the sample tube at 20,376 × g for 30 s to separate the phases, and pipette and transfer the nonpolar (EtOAc: *n*-hexane) phase into a clean 2-mL safelock microcentrifuge tube.
6. Repeat partition using another 800 μL of EtOAc: *n*-hexane (1: 1, v/v), and vortex.
7. Centrifuge the sample tube at 20,376 × g for 30 s, and pipette and combine the nonpolar fraction together with the previous one in the same microtube (total volume is 1.6 mL).
8. Add 60 μL of HPLC mobile phase (*see* **Note 1**) into EtOAc: *n*-hexane in the microtube. Put a marker on the microtube's wall to mark the height of the polar fraction (60 μL volume) from the bottom of the microtube.
9. Evaporate the nonpolar fraction in the SpeedVac for about 20 min (keeping temperature not above 35 °C) until the 60 μL polar phase remains in the microtube.
10. Add 600 μL of HPLC eluent into the microtube (total volume is 660 μL), and homogenize by a vortex (*see* **Note 2**).
11. Centrifuge the microtube at 20,376 × g for 10 min, and transfer the clear supernatant into a 1.5-mL HPLC glass vial. This sample is ready for HPLC analysis.

3.3 Sample Preparation for Determination of % Recovery of SA (Target Phenolic Acid)

1. Weigh 500 mg FW of frozen plant powder (or 100 mg of freeze-dried plant powder) in a clean 2-mL safelock microcentrifuge tube. Repeat this step for twelve replicates ($n = 12$), and label as sample no.1 until no.12.
2. Spike SA standard solution into four microtubes (samples 1–4; each of 0.5 μg of SA), and let the other eight samples without spiking.
3. Extract SA in all samples following the protocol of free SA extraction at Subheading 3.1 to provide samples ready for HPLC analysis.
4. Add 300 μL of 8 M HCl into the remaining water/TCA fractions of all microtubes (**step 13** of Subheading 3.1), and homogenize/vortex.

5. Spike SA standard solution into four of the previously non-spiked microtubes (sample no.5–8; each of 0.5 μg of SA), and let the other eight samples (no. 1–4 and 9–12) go without spiking.
6. Do hydrolysis of the SA glycosides (*see* **Note 3**), and extract the free form of SA on all samples following the protocol in Subheading 3.2 to provide samples ready for HPLC analysis.

3.4 Preparation of Standard Stock Solution and Working Solutions for a Calibration Curve of SA/Target Phenolic Acid

1. Weigh 1 mg of SA standard (*see* **Note 2**) in a 1.5-mL safelock microcentrifuge tube, add 1 mL of HPLC eluent, and vortex/homogenize. This is called a stock solution (1 mg/mL).
2. Pipette 50 μL of the stock solution into a clean 1.5-mL safelock microcentrifuge tube, add 950 μL of HPLC eluent, and homogenize using vortex. The concentration of this diluted stock solution is 0.05 mg/mL. Prepare the diluted stock solution in duplicate.
3. Prepare a series of dilution (*see* **Note 2**) from each diluted stock solution by pipetting 20, 50, 100, 200, and 400 μL of it into 1.5-mL HPLC glass vials separately, and add HPLC eluent 980, 950, 900, 800, and 600 μL, respectively, into the microtubes, and vortex/homogenize. The solutions are called working solutions.
4. Analyze the working solutions with HPLC together with the plant extracts and/or samples for determination of % recovery.

3.5 HPLC Analysis

1. Connect an HPLC C18 column (150 × 4.6 mM, 3 μ particle size) properly to the inlet and the outlet of the HPLC system.
2. Put the HPLC mobile phase (HPLC buffer) in a 1-L glass bottle of solvent reservoir with a tubing to allow the mobile phase flowing to the system when pumped (*see* **Notes 4** and **5**).
3. Switch on the HPLC system (*see* **Note 6**) consisting of the pump, the autosampler/autoinjector, the HPLC column, and the lamps/fluorescence detector (*see* **Note 7**) connected to a computer with software for analysis and data acquisition.
4. Create a method (isocratic mode) of mobile phase (*see* **Note 8**) with a flowrate of 0.8 mL/min from the proper solvent reservoir and tubing with a volume of injection of 20 μL (*see* **Note 9**) and an excitation and emission wavelength of 305 nM and 407 nM, respectively (*see* **Note 7**). Save the method.
5. Prepare the sequence table (the order of samples, numbers of injection per sample, the proper/chosen method) and save. Do management of the files properly, always be consistent to save the files in the proper folder to ease finding a saved method, a sequence table, or the HPLC data (chromatograms) afterward.

6. Place the (labeled properly) samples on the rack of the HPLC autosampler following the sequence made in the sequence table.
7. Disconnect the inlet column first, then switch on the pump.
8. Flush the reservoir's tubing with a flowrate of, for example, 3 mL/min for 5 min using the present HPLC mobile phase to remove air bubbles and any other HPLC eluent remained in the tubing previously.
9. Lower the flowrate back to 0.8 mL/min, switch off the pump, and connect again the (inlet) column.
10. Switch on the pump again to allow the present mobile phase flowing through the column until a stable pump pressure and base line are achieved. When the system is ready, click the start button to start the analysis.
11. When the whole analysis is done, flush the column using a mixture of MeOH: water (1:9, v/v) for 30 min, and subsequently a mixture of MeOH: water (1:1, v/v) for 20 min, with a flowrate of 0.8 mL/min (*see* **Note 10**).

3.6 Determination of % Recovery and Calculation of the SA (Phenolic Acid) Level in the Samples

1. Using the data acquisition software, open the chromatograms of the analyzed samples and collect the area under the curves (AUCs) of the target (SA) peak (*see* **Note 8**). An example of a chromatogram of SA analysis is presented in Fig. 3.
2. Create a calibration curve based on the absolute amounts of SA injected (20 μL of the known concentrations of the working solutions; *see* **Note 9**) as the X values, and the corresponding AUCs peak as Y (ordinate) values. Use the average values of X and Y from the two different sets of working solutions to make the calibration curve.
3. Find the regression equation of the calibration curve ($Y = aX \pm b$).
4. Calculate the absolute amount of the target compound in the samples using the regression equation, with which Y = the AUC of the target peak, X = the absolute amount of the target compound injected (can be in the ng or μg unit).
5. Use the average of X values to calculate the % recovery of free SA or bound SA (SAG) as follows: (*see* **Note 11**)

$$\%\text{Recovery} = \left(X_{\text{spiked sample}} - X_{\text{non-spiked sample}}\right) / X_{\text{standard added}}{}^{*}\ 100$$

The SA level in the fresh plant material (μg/g FW): (*see* **Note 9**)

$$= 100/R^{*}\ 660\ \mu L/20\ \mu L^{*}\ X\ \mu g^{*}\ 1000\ mg/500\ mg$$

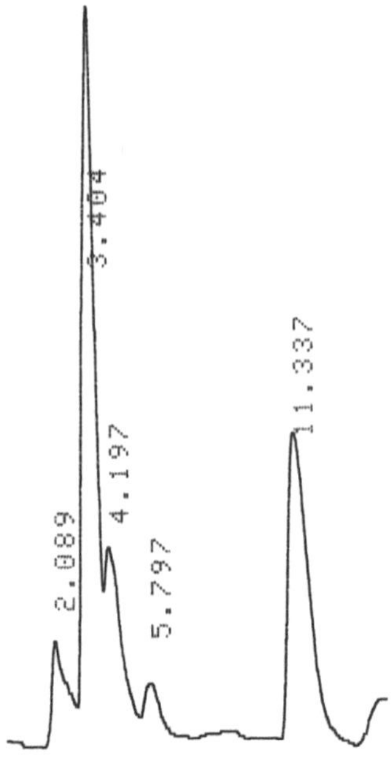

Fig. 3 Analysis of salicylic acid in the plant extract using HPLC–fluorescence. A chromatogram of an extract of *Catharanthus roseus* suspended-cells cultures elicited with a fungal extract. The salicylic acid peak appeared at around 11.3 min. The level of free salicylic acid was about 1.43 ug/g fresh weight of elicited cells

4 Notes

1. The addition of (e.g. 60 μL of) HPLC mobile phase into the nonpolar fraction before evaporation in the SpeedVac is aimed at retaining the SA in the sample. Evaporation until dryness causes loss of parts of SA due to sublimation [22].
2. The final volume of sample ready for HPLC analysis can be adjusted and the concentration of the working solutions as well. The concentration of the target analyte (SA and/or any other C6C1) in the extract/sample should be within the linearity of its calibration curve (Lambert–Beer's regression equation). Standard SA (base form) is used in this protocol; thus, when using a salt form of SA (e.g., sodium SA) as the standard, the absolute amount of SA in the stock solution (for spiking and working solutions) should be calculated properly because of the different molecular weights of SA and sodium SA.
3. During the hydrolysis of the glycosides using heating in a water bath, the gas pressure inside the microtubes containing the samples can open the lids. A (tailor-made) metal microtube rack with a metal plate placed on top of the microtubes (and screwed) can solve the problem.

4. Always calculate and prepare the amount of mobile phase properly for every batch of sample analysis. Put a parameter of the minimum volume of solvent remained in the reservoir to allow the system stop automatically when running out the mobile phase. Avoid lack of mobile phase which can unnecessarily stop the analysis.
5. The use of acetate buffer, pH 5.5, and MeOH (9:1, v/v) as the mobile phase ensures that all SA (pKa of 2.97, or other C6C1 phenolic acids with a pKa below 5.5) are deprotonated/ionized and have an affinity to the polarity of the mobile phase, allowing a reliable quantitative analysis. The pKa(s) of other C6C1 phenolic acids are shown in Table 1.
6. The analytical HPLC system can be of any brand (Waters®, Agilent, Shimadzu etc.), consisting of a pump, an autosampler/autoinjector, and the fluorescence detector connected to a computer with software for analysis and data acquisition. The length of the C18 (3 μ) column used can be of 100 or 150 mM, for example, a LUNA 3 μ C18(2) (150 × 4.60 mM i.d. 3 μm) connected with a Phenomenex SecurityGuard pre-column [22]. The most important thing is that the system should provide a good separation (individual peak of the target compound) within a reasonable observation time (e.g., max 30 min).
7. The level of SA in the plant is very low; therefore, the HPLC is coupled with a fluorescence detector (pg-ng level) because the target compound can fluoresce. In case of phenolic acids which do not fluoresce and/or are produced in plant at a relatively higher concentration, a UV/PDA detector can be used. Table 1 shows some C6C1 information including the excitation/emission wavelengths that can be applied for analysis. The chosen wavelengths are normally the ones that provide maximum (peak) of absorption/emission.
8. The target compounds (SA and/or other fluorescence C6C1) should be separated well in the column, which appear as individual peak(s) in the chromatogram. Any compound which does not fluoresce will not be a problem in the analysis even though it has a similar polarity/retention time with the target peaks. In our HPLC system, SA appeared as an individual peak at around 11.3–11.8 min (Fig. 3). However, a different HPLC, even with a similar C18 column and mobile phase, still can provide different retention time of the target compound. Hence, before running a quantitative analysis, it is important to check the peak separation and the retention time of the target analyte in the chromatogram using its standard compound. Adjustment/modification of the polarity of the mobile phase may be needed to improve peak's separation and retention time. If necessary, an isocratic mode should be replaced by a gradient mode to achieve a better separation.

Table 1
C6C1 phenolic acids, pKa, and the analytical methods

C6C1 Chemical structure	Name	pKa	Examples of the analytical method using fluorescence detection (FLD) and/or other analytical methods
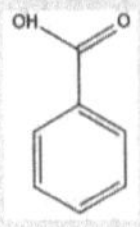	Benzoic acid	4.2	-Fluorescence spectroscopy, excitation–emission matrix (EEM) of benzoic acid/derivatives was provided with an excitation range of 260–330 nM (peak at 280 nM) and emission range of 320–440 nM (peak at 370 nM) [23]. -GC-MS [7, 9, 11] -LC-MS [18]
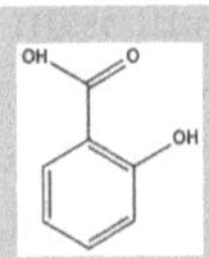	2-Hydroxybenzoic acid = *o*-Hydroxybenzoic acid = salicylic acid	2.97	-HPLC-FLD, Ex/Em = 305/407 nM [3, 22]; HPLC-FLD, Ex/Em = 305/415 nM [24] -GC-MS [7, 9, 11] -NMR [3]
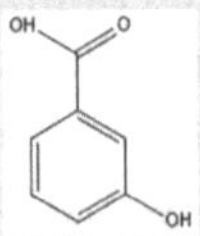	3-Hydroxybenzoic acid	3.84	-GC-MS [9, 11] -HPLC-UV/PDA [25]
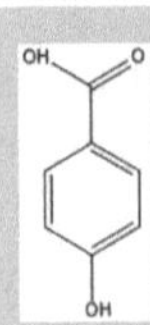	4-Hydroxybenzoic acid = *p*-hydroxybenzoic acid (*p*-HBA)	4.8	-HPLC-MS-FLD, Ex/Em = 285/350 nM [26] -GC-MS [8, 9, 11] -LC-MS [18–21] -HPLC-UV/PDA [14, 16, 17, 25]
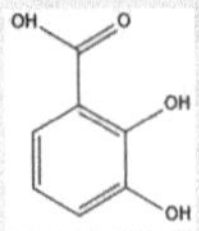	2,3-Dihydroxybenzoic acid = pyrocatechuic acid	2.56	-Spectrofluorometry, Ex/Em = 319/438 nM [27] -GC-MS [7, 9, 11] -HPLC-UV/PDA [12, 24] -NMR [3, 4, 24]
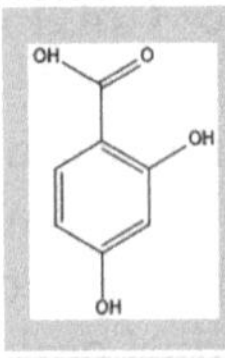	2,4-Dihydroxybenzoic acid = *β*-resorcylic acid	3.11	-GC-MS [9, 11]
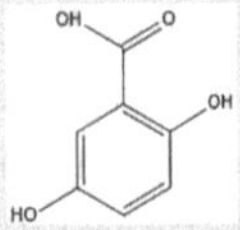	2,5-Dihydroxybenzoic acid = gentisic acid	2.97	-GC-MS [7, 8, 11] -LC-MS [20] -HPLC-UV/PDA [24] -NMR [5]
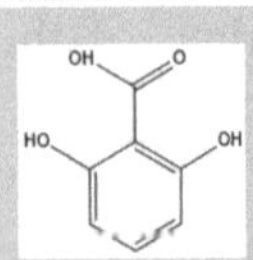	2,6-Dihydroxybenzoic acid = *γ*-resorcylic acid	1.51	-LC-MS [21]

(continued)

**Table 1
(continued)**

C6C1 Chemical structure	Name	pKa	Examples of the analytical method using fluorescence detection (FLD) and/or other analytical methods
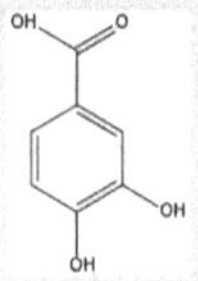	3,4-Dihydroxybenzoic acid = 4,5-dihydroxybenzoic acid = protocatechuic acid	4.16	-HPLC-FLD, Ex/Em = 295/345 nM [22] -HPTLC-UV & HPTLC-MS/MS [6] -GC-MS [11] -LC-MS [18–21] -HPLC-UV/PDA [16, 17]
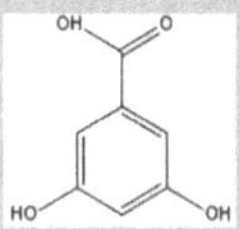	3,5-dihydroxybenzoic acid = α-resorcylic acid	4.04	-LC-MS [21]
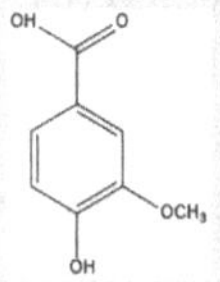	4-Hydroxy-3-methoxybenzoic acid = vanillic acid	4.16	-HPLC-MS-FLD, Ex/Em = 285/350 nM [26]; HPLC-FLD, Ex/Em = 278/360 nM [28]; (synchronous) spectrofluorometry [28] -GC-MS [8, 9, 11] -LC-MS [18–20] -HPLC-UV/PDA [13, 16, 17]
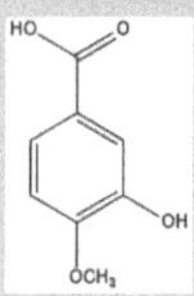	3-Hydroxy-4-methoxybenzoic acid = isovanillic acid	4.15	-HPLC-MS-FLD, Ex/Em = 285/316 nM [26] -GC-MS [11] -LC-MS [21]
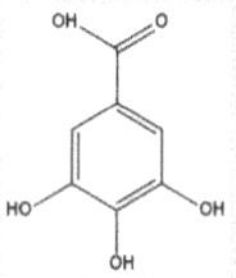	3,4,5-Trihydroxybenzoic acid = gallic acid	4.40	-HPLC-MS-FLD, Ex/Em = 285/350 nM [26]; HPLC-FLD, Ex/Em = 278/360 nM [28]; (synchronous) spectrofluorometry [28] -HPTLC-UV & HPTLC-MS/MS [6] -GC-MS [8, 11] -LC-MS [20, 21] -HPLC-UV/PDA [16]
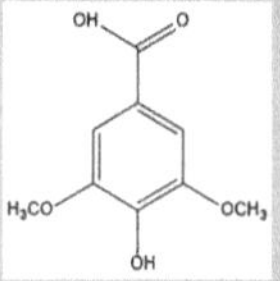	4-Hydroxy-3,5-dimethoxybenzoic acid = syringic acid	3.93	-HPLC-MS-FLD, Ex/Em = 285/350 nM [26]; HPLC-FLD, Ex/Em = 278/360 nM [28]; (synchronous) spectrofluorometry [28] -HPTLC-UV & HPTLC-MS/MS [6] -GC-MS [8, 11] -LC-MS [20, 21] -HPLC-UV/PDA [17]

9. The volume of injection can also be adjusted (e.g., 20, 10, or 5 μL) to avoid a huge peak (overload injection) because the fluorescence detector is very sensitive. Put a blank between two samples to monitor if part of the analyte still remain/contaminate the injection port (appears as peak in the blank chromatogram). In this case, a sample dilution must be done. In case of the absence of the target peak due to a concentration below detection limit, the amount of plant material for extraction should be increased (to increase the final concentration in the HPLC sample). Make a similar injection volume for all samples and the working solutions to reduce deviation caused by technical factor in the analysis. Adjust the injection volume in the formula for the calculation of the level of SA below Subheading 3.6.
10. It is important to flush the tubing/HPLC system with MeOH: water (1:9, v/v) after analysis when using a buffer as mobile phase, for, for example, 30 min with a flowrate of 0.8–1.0 mL/min to avoid possible formation of salts in the HPLC system (for maintenance). The next flushing with MeOH: water (1:1, v/v) is to keep/store the C18 column in this polarity of solvent.
11. In the formula for the calculation of % recovery, the "$X_{standard\ added}$" means the theoretical amount of SA standard in 20 μL of the final volume of the sample (660 μL) spiked at the beginning of the protocol with 0.5 μg of SA. Thus, the "$X_{standard\ added}$" = 20 μL/660 μL * 0.5 μg = 15.15 ng. In our study with *Catharanthus roseus* suspended-cells culture, the average of free SA and bound SA recovery was about 86% and 83%, respectively.

References

1. Mustafa NR, Verpoorte R (2005) Chorismate derived C6C1 compounds in plants. Planta 222:1–5. https://doi.org/10.1007/s00425-005-1554-0
2. Mustafa NR, Verpoorte R (2007) Phenolic compounds in *Catharanthus roseus*. Phytochem Rev 6:243–258. https://doi.org/10.1007/s11101-006-9039-8
3. Mustafa NR, Kim HK, Choi YH, Erkelens C, Lefeber AWM, Spijksma G, van der Heijden R, Verpoorte R (2009) Biosynthesis of salicylic acid in fungal-elicited *Catharanthus roseus* cells. Phytochemistry 70:532–539. https://doi.org/10.1016/j.phytochem.2009.01.009
4. Budi Muljono RA, Scheffer JJC, Verpoorte R (2002) Isochorismate is an intermediate in 2,3-dihydroxybenzoic acid biosynthesis in *Catharanthus roseus* cell cultures. Plant Physiol Biochem 40:231–234. https://doi.org/10.1016/S0981-9428(02)01369-4
5. Mustafa NR, Kim HK, Choi YH, Verpoorte R (2009) Metabolic changes of salicylic acid-elicited *Catharanthus roseus* cell suspension cultures monitored by NMR-based metabolomics. Biotechnol Lett 31:1967–1974. https://doi.org/10.1007/s10529-009-0107-1
6. Jug U, Glavnik V, Kranjc E, Vovk I (2018) High-performance thin-layer chromatography and high-performance thin-layer chromatography–mass spectrometry methods for the analysis of phenolic acids. J Planar Chromatogr 31:13–22. https://doi.org/10.1556/1006.2018.31.1.2

7. Budi Muljono RA, Looman AMG, Verpoorte R, Scheffer JJC (1998) Assay of salicylic acid and related compounds in plant cell cultures by capillary GC. Phytochem Anal 9: 35–38. https://doi.org/10.1002/(SICI)1099-1565(199801/02)9:1<35::AID-PCA379>3.0.CO;2-D
8. Proestos C, Boziaris IS, Nychas G-JE, Komaitis M (2006) Analysis of flavonoids and phenolic acids in Greek aromatic plants: investigation of their antioxidant capacity and antimicrobial activity. Food Chem 95:664–671. https://doi.org/10.1016/j.foodchem.2005.01.049
9. Zuo Y, Wang C, Zhan J (2002) Separation, characterization, and quantitation of benzoic and phenolic antioxidants in American cranberry fruit by GC–MS. J Agric Food Chem 50:3789–3794. https://doi.org/10.1021/jf020055f
10. Zhang K, Zuo Y (2004) GC-MS determination of flavonoids and phenolic and benzoic acids in human plasma after consumption of cranberry juice. J Agric Food Chem 52:222–227. https://doi.org/10.1021/jf035073r
11. Khoddami A, Wilkes MA, Roberts TH (2013) Techniques for analysis plant phenolic compounds. Molecules 18:2328–2375. https://doi.org/10.3390/molecules18022328
12. Budi Muljono RA, Darsono FL, Scheffer JJC, Verpoorte R (2001) Assay of 2,3-dihydroxybenzoic acid and related compounds in plant materials by high-performance liquid chromatography. J Chromatogr A 927: 39–45. https://doi.org/10.1016/S0021-9673(01)01098-6
13. Wen D, Li C, Di H, Liao Y, Liu H (2005) A universal HPLC method for the determination of phenolic acids in compound herbal medicines. J Agric Food Chem 53:6624–6629. https://doi.org/10.1021/jf0511291
14. Ross KA, Beta T, Arntfield SD (2009) A comparative study on the phenolic acids identified and quantified in dry beans using HPLC as affected by different extraction and hydrolysis methods. Food Chem 113:336–344. https://doi.org/10.1016/j.foodchem.2008.07.064
15. Ciulu M, Cádiz-Gurrea M, Segura-Carretero A (2018) Extraction and analysis of phenolic compounds in rice: a review. Molecules 23: 2890. https://doi.org/10.3390/molecules23112890
16. Kelebek H, Selli S (2011) Determination of volatile, phenolic, organic acid and sugar components in a Turkish cv. Dortyol (*Citrus sinensis* L. Osbeck) orange juice. J Sci Food Agric 91:1855–1862. https://doi.org/10.1002/jsfa.4396
17. Heinio R-L, Liukkonena K-H, Myllymaki O, Pihlavac J-M, Adlercreutzd H, Heinonend S-M, Poutanen K (2008) Quantities of phenolic compounds and their impacts on the perceived flavour attributes of rye grain. J Cereal Sci 47:566–575. https://doi.org/10.1016/j.jcs.2007.06.018
18. Barros L, Dueñas M, Ferreira ICFR, Baptista P, Santos-Buelga C (2009) Phenolic acids determination by HPLC–DAD–ESI/MS in sixteen different Portuguese wild mushrooms species. Food Chem Toxicol 47:1076–1079. https://doi.org/10.1016/j.fct.2009.01.039
19. Arráez-Román D, Fu S, Sawalha SMS, Segura-Carretero A, Fernández-Gutiérrez A (2010) HPLC/CE-ESI-TOF-MS methods for the characterization of polyphenols in almond skin extracts. Electrophoresis 31:2289–2296. https://doi.org/10.1002/elps.200900679
20. Zarena AS, Udaya Sankar K (2011) Phenolic acids, flavonoid profile, and antioxidant activity in mangosteen (*Garcina mangostana* L.) pericarp. J Food Biochem 36:627–633. https://doi.org/10.1111/j.1745-4514.2011.00575.x
21. Dziedzic K, Górecka D, Szwengiel A, Sulewska H, Kreft I, Gujska E, Walkowiak J (2018) Plant Foods Hum Nutr 73:82–88. https://doi.org/10.1007/s11130-018-0659-0
22. Verberne MC, Brouwer N, Delbianco F, Linthorst HJM, Bol JF, Verpoorte R (2002) Method for the extraction of the volatile compound salicylic acid from tobacco leaf material. Phytochem Anal 13:45–50. https://doi.org/10.1002/pca.615
23. Santos I-d, Bosman G, Aleixandre-Tudo JL, Toit W-d (2022) Direct quantification of red wine phenolics using fluorescence spectroscopy with chemometrics. Talanta 236:122857. https://doi.org/10.1016/j.talanta.2021.122857
24. Bartsch M, Bednarek P, Vivancos PD, Schneider B, Roepenack-Lahaye E-v, Foyer CH, Kombrink E, Scheel D, Parker JE (2010) Accumulation of isochorismate-derived 2,3-dihydroxybenzoic-3-*O*-β-D-xyloside in *Arabidopsis* resistance to pathogens and ageing of leaves. J Biol Chem 285(33):25654–25665
25. Zhou L, Huang TW, Wang JY, Sun S, Chen G, Poplawsky A, He YW (2013) The rice bacterial pathogen *Xanthomonas oryzae* pv. *Oryzae* produces 3-hydroxybenzoic acid and

4-hydroxybenzoic acid via XanB2 for use in xanthomonadin, ubiquinone, and exopolysaccharide biosynthesis. MPMI 26(10): 1239–1248. https://doi.org/10.1094/MPMI-04-13-0112-R

26. Monasterio RP, Olmo-García L, Bajoub A, Fernández-Gutiérrez A, Carrasco-Pancorbo A (2016) Potential of LC coupled to fluorescence detection in food metabolomics: determination of phenolic compounds in virgin olive oil. Int J Mol Sci 17:1627. https://doi.org/10.3390/ijms17101627

27. Geipel G, Acker M, Vulpius D, Bernhard G, Nitsche H, Fanghänel T (2004) An ultrafast time-resolved fluorescence spectroscopy system for metal ion complexation studies with organic ligands. Spectrochim Acta A 60:417–424. https://doi.org/10.1016/S1386-1425(03)00244-0

28. Žiak Ľ, Sádecká J, Májek P, Hroboňová K (2014) Simultaneous determination of phenolic acids and scopoletin in brandies using synchronous fluorescence spectrometry coupled with partial least squares. Food Anal Methods 7:563–570. https://doi.org/10.1007/s12161-013-9656-y

Chapter 11

Isolation and Identification of Coumarins from *Citrus latifolia* Leaves Using Advanced Chromatographic Techniques

Hocelayne Paulino Fernandes and Luis Francisco Salomé-Abarca

Abstract

Coumarins are a class of natural compounds that exhibit a wide range of biological activities, which include anticoagulant, anti-inflammatory, anticancer, and antimicrobial effects. This class of compounds has drawn considerable interest in recent years. The exploration of such properties contributes to our understanding of the potential of coumarins as valuable agents in various clinical applications. Thus, several scientific articles providing a comprehensive exploration of the isolation of coumarins from various plant species have been published. Therefore, this chapter presents an already tested protocol that ensures the effective extraction and isolation of different coumarins from leaves of *Citrus latifolia*, achieved by liquid and planar chromatography. The identification of coumarins is performed by TLC and ^{1}H NMR analysis.

Key words Coumarins, Isolation, *Citrus latifolia*, Medium-pressure liquid chromatography, Thin-layer chromatography, Structure elucidation, NMR

1 Introduction

Coumarins are a family of benzopyrones that are widely distributed in nature. These organic compounds have a low molecular weight and exhibit intriguing physical, chemical, and pharmacological properties. The fundamental chemical structure of a coumarin comprises a benzene ring fused to an α-pyrone ring. The α-pyrone ring includes a lactone ring containing an oxygen atom [1] (Fig. 1).

Coumarins are generally synthesized through the shikimate pathway, which utilizes aromatic amino acids as precursors. In plants, the synthesis of coumarins from phenylalanine involves a series of enzymatic reactions. One such reaction is the conversion of phenylalanine to trans-cinnamic acid by the action of phenylalanine ammonia-lyase (PAL). This process results in the formation of the central metabolite 4-coumaroyl-S-CoA. Subsequently, this crucial

Young Hae Choi et al. (eds.), *Natural Product Isolation and Identification: Methods and Protocols*, Methods in Molecular Biology, vol. 2895, https://doi.org/10.1007/978-1-0716-4350-1_11,

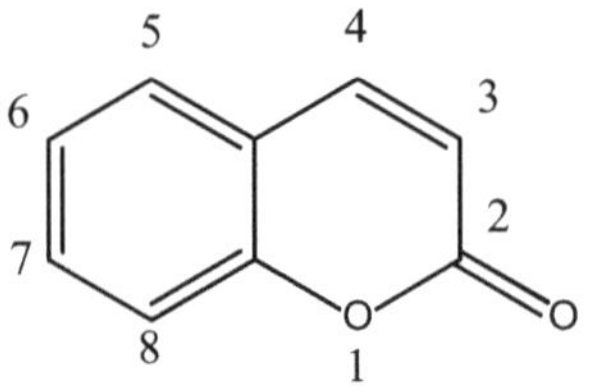

Fig. 1 Basic chemical structures of coumarins

intermediate can be transformed into various classes of phenylpropanoids [2]. Coumarins are also found in fungi and bacteria, exhibiting wide chemical diversity. This diversity may be correlated with their wide range of pharmacological activities, including anticoagulant, anti-inflammatory, anticancer, antidiabetic, and antimicrobial properties [3].

Extensive research has comprehensively examined the medicinal chemistry and pharmacological properties of coumarins. For instance, studies have explored their therapeutic significance, various synthetic methodologies, versatility, and potential applications in drug development [4, 5]. These investigations have enhanced our understanding of the potential of coumarins and their analogues as valuable agents in clinical applications [6, 7]. Additionally, some studies have described the antifungal effects of coumarins in citrus plants [8, 9], demonstrating the antimicrobial potential of citrus coumarins as promising candidates for the development of novel antifungal compounds.

The isolation and identification of these metabolites is crucial for their biological characterization and applications. Numerous methods exist for extracting and isolating coumarins from different plant species, providing valuable guidance on the isolation and identification of these molecules. This chapter presents a previously proven protocol for the effective extraction and isolation of various coumarins from the leaves of *Citrus latifolia*. The isolation is achieved through medium-pressure liquid chromatography (MPLC) and thin-layer chromatography (TLC), and the identification is confirmed by ^{1}H NMR spectroscopy.

2 Materials

1. All solvents used in the separation and isolation of coumarins were high-pressure liquid chromatography (HPLC) purity grade (MeOH) or analytical grade ($CHCl_3$, *n*-hexane, EtOAc, and toluene).
2. Liquid nitrogen is used to grind fresh plant materials to prevent the degradation of chemical constituents during the grinding process.

3. Materials for plant sample preparation: ceramic mortar and pestle (or laboratory mill), 100-mM funnel, sieve (no. 40, < 0.42 mM of particle size), and filter paper (Whatman no. 2), freeze-drier, rotary evaporator.
4. MPLC separation: EcoFlex Silica 50-μm irregular 80 g column, 12 g UltraPure irregular silica gel, 40–63 μm, 60A column.
5. TLC analysis: analytical TLC silica gel 60 F254 aluminum plates, chamber for TLC 10 × 10 cm, 2-aminoethyl diphenylborinate for the derivatization, TLC Visualizer for TLC plate visualization.
6. ^{1}H NMR analysis: 600-MHz NMR, 3-mM NMR tubes, CH_3OH-d_4, and hexamethyldisiloxane (HMDSO).

3 Methods

3.1 Preparation of Plant Material and Sample Extraction

1. Collect citrus leaves from the plant, and flash-freeze them in liquid nitrogen to prevent the degradation of metabolites [10].
2. Grind the plant materials in liquid nitrogen using either a laboratory mill or a ceramic mortar and pestle (*see* **Note 1**).
3. Place the freeze-ground material in a freeze-dryer and dry it for 24–48 h. Once the plant material is completely dried, homogenize its particle size using a sieve (no. 40, < 0.42 mM of particle size).
4. The homogenized materials are stored in sealed conical tubes at room temperature (*see* **Note 2**).
5. Plant extracts are prepared by adding 250 mL of 95% MeOH (v/v) to 25 g of freeze-dried materials followed by ultrasonication for 30 min at room temperature.
6. Filter the supernatant from the extract using a funnel covered with a filter paper. Then, reduce the supernatant using a rotary evaporator at 35–40 °C (*see* **Note 3**).
7. To quantify the overall mass of the dried extract, transfer the reduced extract into preweighed 50-mL conical tubes. Place the tubes in a speed vacuum device, and dry them completely. After drying, record the weight of the tubes containing the dried extract using an analytical scale. Determine the weight of the dried extract by subtracting the weight of the empty conical tube from the weight of the tube with the dried extract.

To isolate coumarins from the citrus leaf extract, MPLC equipment is utilized (*see* **Note 4**). The isolation process is organized into sequential steps to ensure the systematic separation and isolation of compounds.

3.2 Separation and Isolation of Coumarins by Medium-Pressure Liquid Chromatography (MPLC)

1. Weigh 2 g of the dried extract and dissolve it in 2 mL of methanol. Add 2 g of silica gel with a pore size of 60 Å. Subsequently, dry the mixture using a rotary evaporator. Transfer the resulting powder into a sample loading silica precolumn, and then connect it to an EcoFlex Silica 50-μm irregular 80 g column.
2. Prepare the mobile phase for the MPLC separation utilizing a combination of $CHCl_3$, MeOH, *n*-hexane, EtOAc, and toluene. To initiate the separation of coumarins, the elution gradient should commence with a blend of MeOH (A) and $CHCl_3$ (B), with B increasing from 5% to 100% over a duration of 55 min, at a flow rate of 25 mL/min, followed by a 5-min equilibration period under initial conditions. Subsequently, seven fractions are obtained by peak collection utilizing a UV detector with a wavelength range of 254–400 nM, with one of these fractions containing coumarins (*see* **Note 5**). The fractions necessitate TLC monitoring, with coumarin-containing fractions distinguished by intensely fluorescent blue spots (see Subheading 3.3).
3. Dry the coumarin fractions, ascertain their mass, and solubilize them in methanol (1:1, w/v). Subsequently, add silica gel with a pore size of 60 Å in an equal quantity to the weight of the coumarin fraction (yielding approximately 200 mg of coumarin fraction).
4. Adsorb the extract onto silica gel by drying it using a rotary evaporator. Transfer the resulting extract powder into a sample loading silica precolumn, then connect it to a 12 g UltraPure irregular silica gel 40–63 μm, 60A column.
5. To isolate the first coumarin, employ a gradient elution comprising *n*-hexane (A) and EtOAc (B), increasing B from 0% to 100% over 60 min, with a flow rate of 10 mL/min, followed by a 4-minute equilibration period to return to initial conditions. Collect the singular peak using a UV detector with a wavelength range of 254–400 nM. This procedure results in the isolation of a specific coumarin identified as 5,7-dimethoxycoumarin. The residual extract should be retained for the isolation of additional coumarins (*see* **Note 6**).
6. To isolate other coumarins, the residual extract from the preceding step undergoes final purification. Initially, the residual extract is dried, following the same procedure as before. Using the same 12-g silica gel column, a gradient of toluene (A) and EtOAc (B) is applied, with B increasing from 0% to 100% over 50 min, at a flow rate of 10 mL/min, followed by a 5-minute equilibration period to return to initial conditions. This procedure yields the isolation of three additional coumarins identified as 5-geranyloxy-7-methoxycoumarin, 7-geranyloxycoumarin, and furanocoumarin-like structure (*see* **Note 7**).

3.3 Chemical Monitoring Using Thin-Layer Chromatography (TLC)

1. TLC analysis employs TLC silica gel 60 F254 aluminum plates. A nonpolar mobile phase composed of toluene–EtOAc (8:2, v/v) is utilized for sample analysis.
2. Individually dissolve 10 mg of each fraction in 1 mL of a solution of toluene–EtOAc (8:2, v/v).
3. Apply 1 μL of each solution onto a 10 × 10 cm TLC plate using a 10-μL micropipette.
4. Saturate a small TLC chamber with 10 mL of the toluene–EtOAc solution for 10 min.
5. Place the plate into the chamber and allow for compound separation. Once the solvent front reaches 9 cm from the starting line, remove the TLC plate from the chamber and dry it at room temperature under a fume hood.
6. Perform an initial visual inspection using a TLC Visualizer at wavelengths of 254 nM and 366 nM.
7. For derivatization, prepare the natural product (NP) reagent by dissolving 1 g of 2-aminoethyl diphenylborinate in 100 mL of methanol.
8. Apply 2 mL of the solution onto the TLC plate by spraying, and then transfer it to a TLC Visualizer under 366 nM. The observation of a characteristic blue color with high fluorescent emission (Fig. 2) confirms the presence of coumarin-like compounds in the fractions (*see* **Note 8**).

3.4 Sample Preparation and Analysis by NMR

1. To prepare the sample for NMR analysis, dissolve the isolated compounds in 300 μL of CH_3OH-d_4 containing 3.93 mM HMDSO as an internal standard for chemical shift calibration (at 0.06 ppm) and quantitation. Subsequently, transfer the solution into 3-mM NMR tubes (*see* **Note 9**).
2. The ^1H NMR experiments are conducted on a 600-MHz NMR spectrometer operating at a frequency of 600.13 MHz at 298 K (25 °C). CH_3OH-d_4 is employed as a locking solvent. The acquisition parameters are configured as follows: pulse program (PULPROG) zgpr30, acquisition time (AQ) of 2.72 s, recycle delay (D1) of 1.5 s, number of scans (NS) set to 64, pulse width (P1) of 7.5 μsec (90°), mixing time of 0.010 s, receiver gain (RG) of 287, spectral width (SW) of 20 ppm, size of the free induction decay (FID) data points (TD) set to 65.53, and size of the real spectrum (SI) also of 65.53, with exponential line broadening of 0.3 Hz. A presaturation sequence is employed to suppress the residual water signal, utilizing low-power selective irradiation at the H2O frequency during the recycle delay. The raw spectra are Fourier-transformed and processed through manual phasing, baseline correction, and calibration to the HMDSO signal at 0.06 ppm (*see* **Note 10**).

Fig. 2 Typical thin-layer chromatogram of coumarin fraction from *Citrus latifolia* extract obtained from elution gradient MeOH (**a**) and $CHCl_3$ (**b**), using a nonpolar mobile phase, toluene–ethyl acetate (8:2, v/v)

3.5 Structure Elucidation of Coumarins by ^1H NMR Analysis

A typical ^1H NMR spectrum of coumarins exhibits two characteristic resonances corresponding to the pyrone ring. For instance, a doublet around 6.14–6.30 ppm with J = 9.6 Hz is indicative of H-3 of the pyrone ring, while another doublet around 7.60–8.30 ppm with J = 9.6 Hz corresponds to H-4 (*see* **Note 11**). Taking isolated coumarin 1, elucidated as 5,7-dimethoxycoumarin, as an example, in this molecule, the pyrone ring is detected at 6.15 (H-3, d, J = 9.6 Hz, 1H) and δ 8.09 (H-4, d, J = 9.6 Hz, 1H). Additional signals at δ 6.47 (H-6, d, J = 2.2 Hz, 1H) and δ 6.52 (H-8, d, J = 2.2 Hz, 1H) are attributed to protons in the aromatic ring, while signals at δ 3.87 (s, 3H) and δ 3.93 (s, 3H) are assigned to methoxy protons at δ 157.4 (C5) and 164.5 (C7), respectively (Fig. 3). For example, the geraniol moiety from 5-geranyloxy-7-methoxycoumarin typically appears at δ 4.70 (H-1a′, 1b′, d, J = 6.74 Hz, 2H), δ 5.49 (H-2′, t, J = 6.74 Hz, 1H), δ 5.09 (H-6′, t, J = 6.03 Hz, 1H), δ 2.10 (H-4′, m, 2H), and δ 2.15 (H-5′, m, 2H), along with three methyl resonances at δ 1.60 (H-10′, s, 3H), δ 1.64 (H-8′, s, 3H), and δ 1.78 (H-9′, s, 3H)

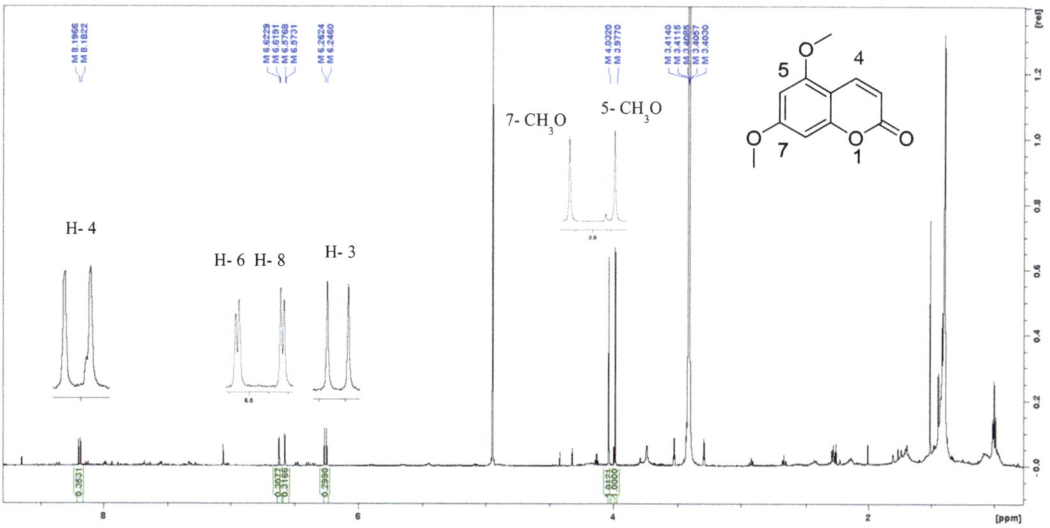

Fig. 3 ^{1}H NMR spectrum of 5,7-dimethoxycoumarin (600 MHz in CH_3OH-d_4)

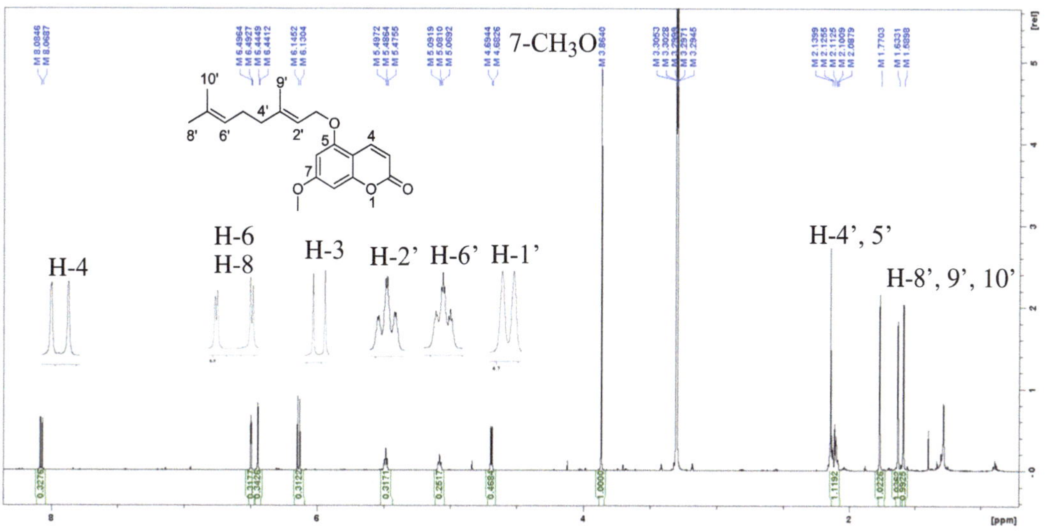

Fig. 4 ^{1}H NMR spectrum of 5-geranyloxy-7-methoxycoumarin (600 MHz in CH_3OH-d_4)

(Fig. 4) (*see* **Note 12**). In most cases, ^{1}H NMR is sufficient for elucidating isolated coumarins. However, in some instances, 2D NMR and mass spectrometry are necessary to confirm the structure (Fig. 5) (*see* **Note 13**).

4 Notes

1. In this procedural stage, either a laboratory mill or an 80-mm ceramic mortar and pestle may be utilized for grinding the plant material. Carefully employ liquid nitrogen, and ensure

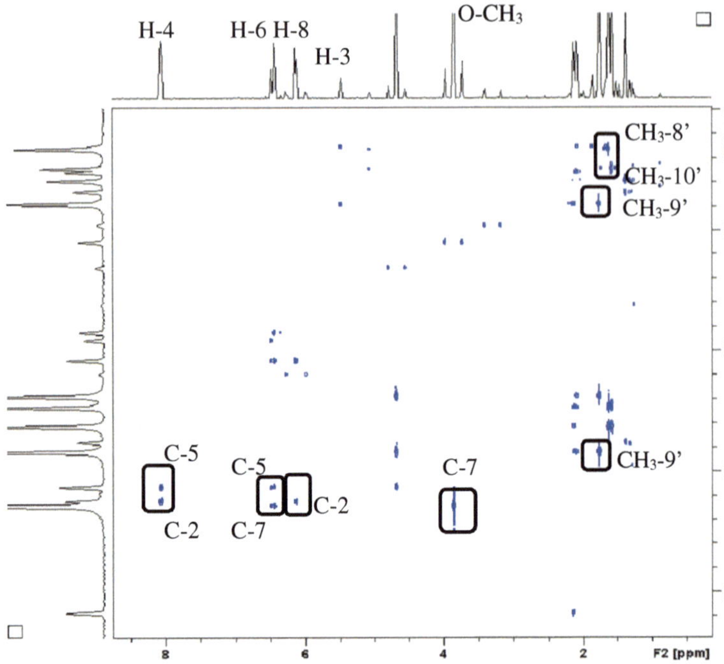

Fig. 5 ^{1}H-^{13}C HMBC spectrum of 5-geranyloxy-7-methoxycoumarin (600 MHz in ^{1}H and 150 MHz in ^{13}C in CH_3OH-d_4)

that the leaves remain frozen throughout the grinding process. Sieving aids in achieving a more uniform and homogeneous sample, enhancing the surface area, and facilitating solvent interaction, thereby improving extraction yields. This homogenization is essential for obtaining consistent and reproducible results in subsequent extraction steps.

2. If it is not possible to grind the material immediately after the harvest, keep the collected leaves at −80 °C until freeze-drying. The lyophilized plant material can be stored at room temperature.

3. If it is not possible to grind the material immediately after the harvest, store the collected leaves at −80 °C until freeze-drying. The lyophilized plant material can then be stored at room temperature.

4. MPLC is a chromatographic technique operating at intermediate pressures (5.17–20.68 bar). It is commonly utilized for isolating and purifying organic compounds, such as coumarins. MPLC is particularly suitable for handling larger sample sizes, in contrast to HPLC, making it applicable for preparative-scale purposes. MPLC offers versatility in column selection, accommodating various sizes and types. This flexibility in column choice is advantageous for optimizing separation conditions,

especially when dealing with complex mixtures, as it allows for tailoring to the specific properties of compounds within the mixture [11].

5. The coumarin fraction represents the second fraction collected using a UV detector. Employing an alternative chromatography system may result in different fractions. Therefore, it is essential to perform TLC analysis for verification purposes.
6. In this step, a single chromatographic peak is detected using UV with a wavelength range between 265 and 400 nM, and isolated with a yield of 1.0 mg from the coumarin fraction. This compound is subsequently identified as 5,7-dimethoxycoumarin. It is imperative to collect the residual extract to isolate additional coumarins.
7. This step resulted in the isolation of three additional coumarins, identified as follows: 5-geranyloxy-7-methoxycoumarin, with a yield of 0.85 mg; 7-geranyloxycoumarin yielding 0.58 mg; and furanocoumarin-like structure, with a yield of 1.75 mg. The isolation of various coumarins involves diverse methodologies. For example, the use of HPLC coupled with mass spectrometry has been employed for the separation and identification of coumarins in *Peucedanum palustre* and *Angelica archangelica* [12]. Furthermore, a high-speed countercurrent chromatography method has been developed for the isolation and purification of coumarins, including xanthotoxin, isopimpinellin, bergapten, imperatorin, osthole, and an unidentified compound from *Cnidium monnieri* [13]. These investigations, among others, have successfully demonstrated different methods and techniques for the extraction and isolation of a wide variety of coumarins.
8. Various methods and techniques utilizing TLC, including experimental procedures, conditions, and mobile phases, have been demonstrated for the identification and quantification of coumarins [14, 15]. TLC proves to be a rapid and effective method for the qualitative analysis of coumarins and other compounds in plants, offering valuable insights into the composition of plant extracts.
9. The addition of HMDSO is optional and is only necessary for quantitation purposes. Alternatively, for calibration of chemical shifts, tetramethylsilane (TMS) can be used. However, the high volatility of TMS may pose challenges for quantitation. The chemical shift reference for HMDSO is at 0.06 ppm, whereas for TMS, it is at 0.00 ppm.
10. For processing ^{1}H NMR data, various software options are available, such as TopSpin (Bruker), Mnova NMR (Mestrelab), and Chenomx NMR Suite (Chenomx).

11. Various NMR instruments may operate at different field strengths. Depending on the strength of the magnetic field, chemical shifts for the same molecule can vary. Similar variations can occur with different solvents, temperatures, concentrations, and other factors. Therefore, slight chemical shift variations are considered acceptable. In ^{1}H NMR analysis, the presence of hydroxyl or alkoxyl groups induces a downfield shift of H-4 protons from δ 7.6–7.8 to δ 8.1–8.3, as observed in compounds such as bergapten [16], isoimperatorin [17], and phellopterin [18].
12. Two-dimensional NMR spectroscopy (2D NMR), such as correlated spectroscopy (COSY), heteronuclear multiple bond correlation (HMBC), and heteronuclear single quantum coherence (HSQC), reveals correlations between protons and carbons. For example, COSY displays correlations between 1H-1H nuclei, whereas HMBC illustrates correlations between 1H-13C nuclei separated by two, three, and sometimes, in conjugated systems, four bonds. HSQC experiments reveal correlations between ^{1}H-^{13}C nuclei connected by single bonds [19]. Moreover, in an MS/MS spectrum acquired in positive mode, the fragments m/z 103.0537 and 91.0543 correspond to the loss of CO_2 from the pyrone ring system and to the protonated benzofuran of coumarins, respectively, and are commonly detected [20].
13. It is essential to consider that the selection of a chromatographic technique depends on several factors, including the specific properties of the target compounds, equipment availability, and the scale of the isolation process. In this protocol, MPLC proved to be a suitable and practical choice for the separation and isolation of coumarins from *Citrus latifolia*.

References

1. Seoane-Rivero R, Ruiz-Bilbao E, Navarro R, Laza JM, Cuevas JM, Artetxe B, Gutiérrez-Zorrilla JM, Vilas-Vilela JL, Marcos-Fernandez Á (2020) Structural characterization of mono and dihydroxylated umbelliferone derivatives. Molecules 25(15):3497. https://doi.org/10.3390/molecules25153497. PMID: 32751979. PMCID: PMC7436329
2. Stefanachi A, Leonetti F, Pisani L, Catto M, Carotti A (2018) Coumarin: a natural, privileged and versatile scaffold for bioactive compounds. Molecules 23(2):250. https://doi.org/10.3390/molecules23020250. PMID: 29382051. PMCID: PMC6017103
3. Sharifi-Rad J, Cruz-Martins N, López-Jornet P, Lopez EP, Harun N, Yeskaliyeva B, Beyatli A, Sytar O, Shaheen S, Sharopov F, Taheri Y, Docea AO, Calina D, Cho WC (2021) Natural Coumarins: exploring the pharmacological complexity and underlying molecular mechanisms. Oxidative Med Cell Longev 2021:6492346. https://doi.org/10.1155/2021/6492346. PMID: 34531939. PMCID: PMC8440074
4. Divya Singh, Pathak DP, Anjali (2016) Coumarins: an overview of medicinal chemistry. Potential for new drug molecules. Int J Pharm Sci Res 7(2):482–504. ref.66
5. Pereira TM, Franco DP, Vitorio F, Kummerle AE (2018) Coumarin compounds in medicinal

chemistry: some important examples from the last years. Curr Top Med Chem 18(2): 124–148., 25. https://doi.org/10.2174/1568026618666180329115523

6. Pangal A, Muiz G, Mane V, Shaikh J (2013) Various pharmacological aspects of Couamrin derivatives: a review. Chem Med
7. Gouda MA, Hussein BHM, El-Demerdash A, Ibrahim ME, Salem MA, Helal MH, Wafaa S (2019) Hamama A review: synthesis and medicinal importance of coumarins and their analogues (part II). Currt Bioact Compd 15(7). https://doi.org/10.2174/1573407215666191111120604
8. Fracarolli L, Rodrigues GB, Pereira AC, Massola Júnior NS, Silva-Junior GJ, Bachmann L, Wainwright M, Bastos JK, Braga GUL (2016) Inactivation of plant-pathogenic fungus Colletotrichum acutatum with natural plant-produced photosensitizers under solar radiation. J Photochem Photobiol B 162:402–411. https://doi.org/10.1016/j.jphotobiol.2016.07.009. Epub 2016 Jul 9. PMID: 27434699
9. Ramírez-Pelayo C, Martínez-Quiñones J, Gil J, Durango D (2019) Coumarins from the peel of citrus grown in Colombia: composition, elicitation and antifungal activity. Heliyon 5(6): e01937. https://doi.org/10.1016/j.heliyon.2019.e01937. PMID: 31245648; PMCID: PMC6582165
10. Kim H, Choi Y, Verpoorte R (2010) NMR-based metabolomic analysis of plants. Nat Protoc 5:536–549. https://doi.org/10.1038/nprot.2009.237
11. Hostettmann K, Terreaux C (2000) Medium pressure liquid chromatography. In: Wilson ID (ed) Encyclopedia of separation science. Academic Press, pp 3296–3303. https://doi.org/10.1016/B0-12-226770-2/01891-3. ISBN 9780122267703. https://www.sciencedirect.com/science/article/pii/B0122267702018913
12. Eeva M, Rauha J-P, Vuorela P, Vuorela H (2004) Computer-assisted, high-performance liquid chromatography with mass spectrometric detection for the analysis of coumarins in *Peucedanum palustre and Angelica archangelica*. Phytochem Anal 15:167–174. https://doi.org/10.1002/pca.764
13. Liu R, Feng L, Sun A, Kong L (2004) Preparative isolation and purification of coumarins from *Cnidium monnieri* (L.) Cusson by high-speed counter-curren chromatography. J Chromatogr A 1055(1–2):71–76., ISSN 0021-9673. https://doi.org/10.1016/j.chroma.2004.09.017
14. Tsvetkova TsM (1977) Tunkosloĭnokhromatografsko opredeliane na niakoi proizvodni na kumarina [Thin-layer chromatographic determination of coumarin derivatives]. Vet Med Nauki 14(1):42–46. Bulgarian. PMID: 898648
15. Kovac-Besović EE, Durić K (2003) Thin layer chromatography-application in qualitative analysis on presence of coumarins and flavonoids in plant material. Bosn J Basic Med Sci 3(3):19–26. https://doi.org/10.17305/bjbms.2003.3523. PMID: 16232145
16. Kawaii S, Tomono Y, Katase E, Ogawa K, Yano M (1999) Isolation of furocoumarins from bergamot fruits as HL-60 differentiation-inducing compounds. J Agric Food Chem 47(10): 4073–4078. https://doi.org/10.1021/jf990155u. PMID: 10552768
17. Fujioka T, Furumi K, Fujii H, Okabe H, Mihashi K, Nakano Y, Matsunaga H, Katano M, Mori M (1999) Antiproliferative constituents from umbelliferae plants. V. A new furanocoumarin and falcarindiol furanocoumarin ethers from the root of Angelica japonica. Chem Pharm Bull (Tokyo) 47(1): 96–100. https://doi.org/10.1248/cpb.47.96. PMID: 9987830
18. Bergendorff O, Dekermendjian K, Nielsen M, Shan R, Witt R, Ai J, Sterner O (1997) Furanocoumarins with affinity to brain benzodiazepine receptors in vitro. Phytochemistry 44(6): 1121–1124. https://doi.org/10.1016/s0031-9422(96)00703-0. PMID: 9055449
19. Field LD, Li HL, Magill AM (2015) Organic structures from 2D NMR spectra. ISBN: 978-1118-86894-2, 328 pages
20. Ledesma-Escobar CA, Priego-Capote F, Luque de Castro MD (2015) Characterization of lemon (Citrus limon) polar extract by liquid chromatography-tandem mass spectrometry in high resolution mode. J Mass Spectrom 50(11):1196–1205. https://doi.org/10.1002/jms.3637. PMID: 26505764

Chapter 12

Identification of Lignans

Hye Mi Kim and Chul Young Kim

Abstract

Lignans have garnered significant interest in various fields of pharmaceuticals, nutrition, and pesticides due to their diverse biological activities. This chapter primarily focuses on the extraction and purification methods of lignans from *Schisandra*, followed by an extensive examination of qualitative and quantitative analytical techniques, including thin-layer chromatography (TLC) and high-performance liquid chromatography (HPLC). Additionally, methods for structural analysis using nuclear magnetic resonance (NMR) are briefly summarized. The chapter also details approaches for determining the relative and absolute composition of lignan compounds, with illustrative examples provided.

Key words Lignans, Chromatography, TLC, HPLC, NMR, *Schisandra chinensis*, Dibenzocyclooctadiene lignans

1 Introduction

Lignans, a group of compounds with a long history, are widely distributed in nature. They have been identified in various plant parts, including wood, roots, bark, leaves, flowers, fruits, and seeds. Lignans are present in species belonging to over 70 families, encompassing pteridophytes, gymnosperms, and angiosperms. They typically occur either in their free form (Fig. 1) or as glycosides linked to various carbohydrates [1].

Lignans exhibit a range of biological properties, including anticancer, anti-inflammatory, immunosuppressive, cardiovascular, antioxidant, antimicrobial, and antiviral effects, which have garnered significant interest in pharmacy and nutrition fields [2]. Podophyllotoxin, a well-known lignan, is concentrated in the roots and rhizomes of *Podophyllum* species (Berberidaceae). Derivatives of podophyllotoxin, such as etoposide and teniposide, have been widely used in clinical treatments for various cancers, including lymphocytic leukemia [3], specific brain tumors, and lung tumors, for nearly two decades due to their potent pharmacological activities [4]. The exploration of cytotoxic metabolites among

Young Hae Choi et al. (eds.), *Natural Product Isolation and Identification: Methods and Protocols*, Methods in Molecular Biology, vol. 2895, https://doi.org/10.1007/978-1-0716-4350-1_12,

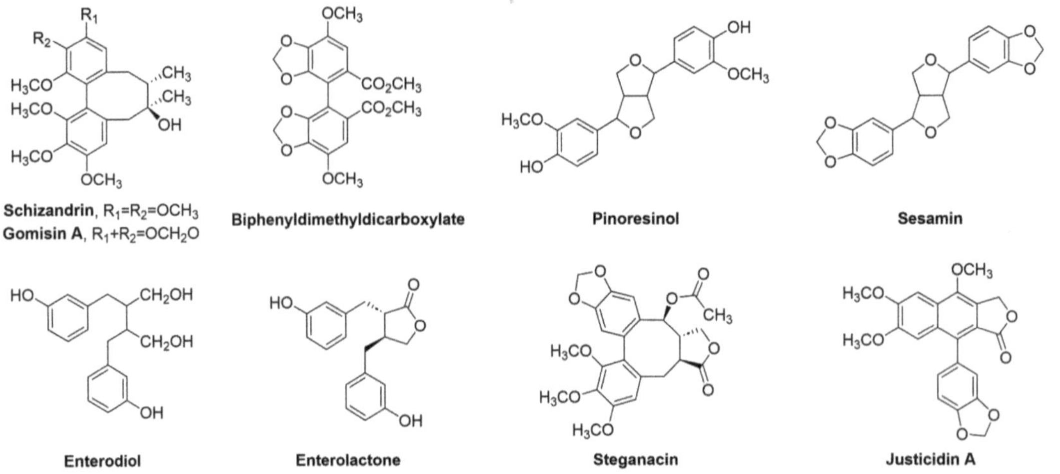

Fig. 1 Chemical structures of typical lignan compounds

polyphenols, including lignans and neolignans, across various structural subtypes has led to intriguing findings, further contributing to the growing interest in these compounds.

Research on active ingredients such as lignans requires the use of high-purity compounds with elevated content and well-defined standardized extracts. In developing lignan utilization, critical steps include optimizing extraction methods to enhance lignan content and ensuring high purity through effective separation and identification processes [5–7]. This comprehensive approach guarantees the reliability and quality of lignan compounds for various research and medical applications.

Therefore, this chapter strategically focuses on the methods for the separation, purification, and structural elucidation of lignan components from *Schisandra chinensis* ("Omija" in Korean, "Wuweizi" in Chinese, or "Gomisi" in Japanese), which play a significant role in liver protection [8]. Additionally, a detailed explanation of quantitative analysis methods for lignan components in this plant is provided.

2 Materials

1. The solvents utilized for extraction, solvent partitioning, and column chromatography, such as MeOH, *n*-hexane, EtOAc, and *n*-BuOH, are of analytical grade.
2. For high-performance liquid chromatography (HPLC), HPLC-grade water, AcCN, and trifluoroacetic acid (TFA) are used as solvents.
3. Chloroform ($CDCl_3$) is employed as the solvent for nuclear magnetic resonance (NMR) analysis.

4. Various materials including Erlenmeyer flask, separatory funnel, clamp, filter paper, round-bottom flask, capillary, TLC chamber, tweezer, dryer, heat gun, open column, long glass rod, beaker, sand, 3 mm NMR tube, pipette, LC vial, 1 mL syringe, and syringe filter (0.45 μm) are employed for the preparation of extraction, separation, isolation, and sample preparation for NMR analysis.
5. Ultrasonicator and rotary evaporator are utilized for the preparation of extracts.
6. The stationary phase for column chromatography consisted of silica gel with particle sizes of 70–230 mesh and 230–400 mesh, while silica gel 60 Å F_{254} plates are used for TLC.
7. TLC plates are visualized either by UV detectors at 254 and 365 nm, or by 10% methanolic sulfuric acid.
8. Quantitative analysis is performed using an HPLC system equipped with a C18 column (4.6 × 250 mm, 5 μm) and a photodiode array (PDA) detector.

3 Methods

3.1 Extraction and Liquid–Liquid Partitioning (Fractionation)

1. To isolate lignans from *S. chinensis* fruit, extraction is first conducted using a suitable solvent (*see* **Note 1**). For *S. chinensis* fruits, 10 g of dried samples is ground and placed into a flask, followed by the addition of 100 mL of MeOH. Ultrasonication is carried out for 1 hour, and this process is repeated three times.
2. Filter the extracts using filter paper, and concentrate the filtrate under reduced pressure using a rotary evaporator.
3. After adding 100 mL of distilled water, transfer the mixture to a separatory funnel for phase separation.
4. Add 100 mL of *n*-hexane, shake well (*see* **Note 2**), and let it stand for 1 hour (repeat three times).
5. If the mixture separates into two layers, collect the *n*-hexane layer, transfer it to a weighed concentration flask, concentrate it under reduced pressure, and measure the final weight.
6. Reintroduce the aqueous layer into a separatory funnel, sequentially add 100 mL of EtOAc and 100 mL of *n*-BuOH, and perform fractionation using the same method described above (*see* **Note 3**).
7. Concentrate each solvent fraction, including *n*-hexane, EtOAc, *n*-BuOH, and the aqueous layer, using a rotary evaporator, and subsequently store them at −20°C until required. Due to the

presence of numerous components in both the extract and solvent fractions, thin-layer chromatography (TLC) analysis is useful for the qualitative and quantitative analysis of the lignans to be separated [7].

3.2 Chemical Monitoring Using TLC

1. Draw a line approximately 0.5 cm from the bottom of the TLC plate (5 × 5 cm) using a pencil and mark the spots for spotting at 1 cm intervals.
2. Using a capillary tube, apply small spots of both the *S. chinensis* extract and the isolated lignans (schisandrol A, gomisin A, and gomisin N) at each marked point.
3. Prepare the TLC solvent system by mixing 7.5 mL of petroleum ether, 2.5 mL of ethyl formate, and 0.5 mL of formic acid. Pour 10 mL of the developing solvent into the TLC developing chamber to a height of approximately 0.5 cm and cover with the lid.
4. Place the silica plate in the TLC chamber for development and remove it when the solvent front is about 1–2 cm from the top edge of the plate. Thoroughly dry the plate using a dryer.
5. Place the developed TLC plate inside a UV lamp box, and press the buttons corresponding to each wavelength (short wave: 254 nm; long wave: 365 nm). Examine the plate and mark the spots that appear brightly or darkly using a pencil (*see* **Note 4**).
6. To prepare the TLC visualizing reagent, carefully add 1 mL of sulfuric acid to 9 mL of MeOH (*see* **Note 5**).
7. Uniformly spray the entire plate with 10% methanolic sulfuric acid (*see* **Note 6**), then heat it for 5–10 min at 105°C. Evaluate the plate under UV (365 nm) or visible light (Fig. 2).

3.3 Isolation of Lignans Using Open Column Chromatography

TLC is also beneficial for subsequent processes involved in lignan separation and purification. Its correlation with column chromatography makes it suitable for preliminary experiments in selecting adsorbents and elution solvents [6]. In this study, open column chromatography is employed to separate lignans from the concentrated *n*-hexane fraction.

1. Preparing the column: use non-fritted columns, and securely plug them with cotton at the narrowest part to prevent the loss of the stationary phase (*see* **Note 7**).
2. Sample preparation: there are two methods for loading the column—wet and dry.
   (a) Wet method: dissolve the *n*-hexane fraction for purification in a minimal quantity of solvent (e.g., hexane, acetone), and load this solution onto the column for further processing (*see* **Note 8**).

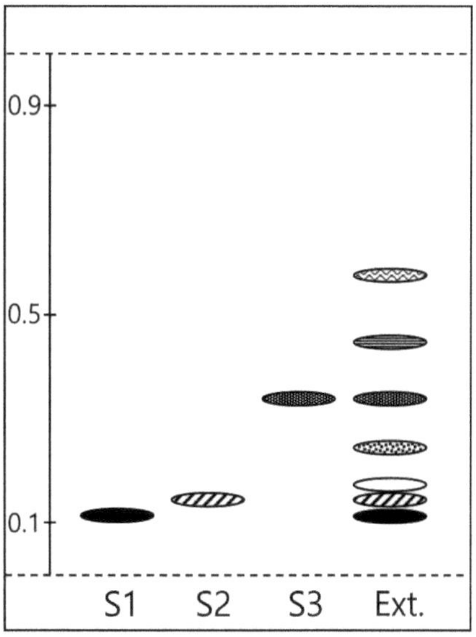

Fig. 2 Thin-layer chromatogram of dibenzocyclooctadiene lignans (S1: schisandrol A; S2: gomisin A; and S3: gomisin N) and *Schisandra chinensis* extract

(b) Dry method: dissolve the sample in an appropriate solvent and transfer it to a round-bottom flask if necessary. Add 1 to 1.5 times the weight of the concentrated sample in silica gel to the dissolved sample. Once mixed, concentrate the mixture and adsorb the sample onto the silica gel. Evaporate the solvent, leaving a dry powder. Transfer this dry powder to the top of the prepared column to initiate the packing process.

3. Packing the column: mix an amount of silica gel (20–100 times the sample quantity) with the elution solvent (mobile phase) (*see* **Note 9**) to create a 25% mixture (*see* **Note 10**). Pack the column with this silica gel mixture (*see* **Note 11**).
4. Sample application: carefully apply the sample to the top of the column and add a minimal amount of elution solvent to ensure proper sample adsorption.
5. Elution: introduce the initial elution solvent (*n*-hexane-EtOAc, 9:1 *v/v*), adjust the flow rate to initiate elution (*see* **Note 12**), and collect the effluent into an Erlenmeyer flask in consistent amounts.
6. TLC analysis: analyze the collected fractions using the TLC method described above to confirm the eluted compounds. The TLC solvent used in this step should be applied under the same conditions as the column elution solvent.
7. Solvent change: if the TLC analysis indicates no further separation with the current eluent, increase the solvent polarity to *n*-

hexane–EtOAc (8:2, *v/v*), and continue eluting the column until no more product is observed in the TLC analysis.

8. Fraction collection: at regular intervals, collect the eluted solution showing identical UV patterns on TLC, concentrate, and set aside the fractions corresponding to the same spot.
9. Emptying the column: once the compounds are isolated, elute all the solvent using compressed air and allow air to flow through the column for approximately 15 min. This will yield dry, free-flowing silica that is easy to pour into the silica waste container (*see* **Note 13**). Clean the column by rinsing with water and acetone.

3.4 NMR Analysis for Structural Elucidation of Lignans

To ascertain the identity of the lignan compound isolated via chromatography, it is imperative to conduct an examination of its chemical structure. Consequently, we utilized NMR spectroscopic analysis, a fundamental technique in structural elucidation [7]. Our objective is to delineate the structure of the elucidated lignan based on the findings obtained from the analysis.

1. Sample preparation: place 3–4 mg of the isolated lignan (schisandrol A) into a vial.
2. Dissolve it in 600 μL of NMR solvent ($CDCl_3$), and transfer the solution to an NMR tube, ensuring that the solution's height in the tube is 4 cm or greater.
3. The prepared sample undergoes analysis using an NMR instrument operating at over 400 MHz.
4. Data Analysis:

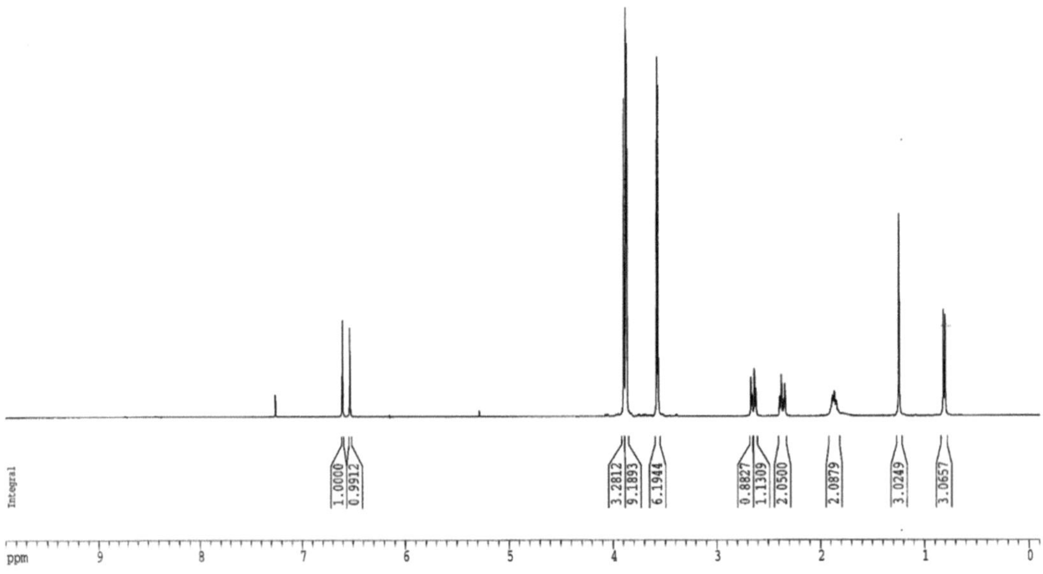

Fig. 3 ^{1}H NMR spectrum of schisandrol A (400 MHz, $CDCl_3$)

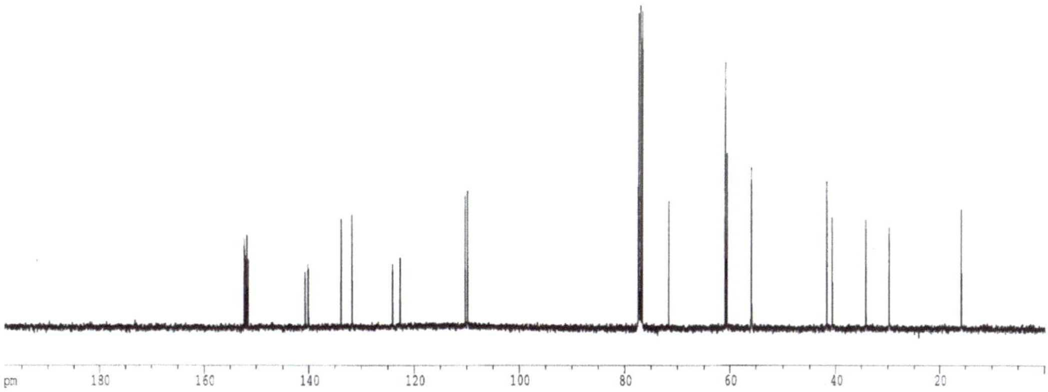

Fig. 4 ^{13}C NMR spectrum of schisandrol A (100 MHz, $CDCl_3$)

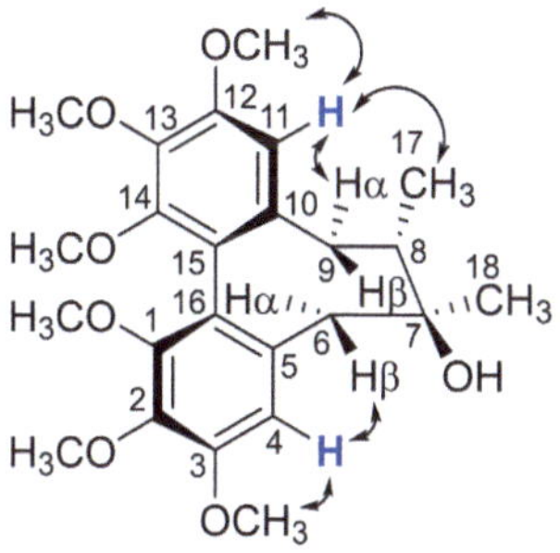

Fig. 5 Key NOESY correlations of schisandrol A

- The ^{1}H NMR spectrum (Fig. 3) of the isolated lignan reveals two aromatic singlet signals (H-4 and H-11) with chemical shifts between δ 6.53 and δ 6.61 ppm (Fig. 3).
- The ^{1}H NMR spectrum (Fig. 3) also indicates the presence of six methoxy groups (δ_H 3.58, 3.59, 3.87, 3.88, 3.89, and 3.90) on its aromatic rings and two methyl groups, one attached to a quaternary carbon (δ_H 1.26, s, CH_3-18) and the other to a tertiary carbon (δ_H 0.82, d, CH_3-17, $J = 6.0$ Hz).
- The ^{13}C NMR spectra (Fig. 4) of the isolated lignan reveal a total of 24 carbons, which include six methoxy carbon substitutions (δ_C 56.2, 56.2, 61.2, 61.2, 60.8, and 60.9) (*see* **Note 14**), two methyl carbons (δ_C 16.1 and 30.0), two benzylic methylenes (δ_C 34.5 and 41.1), a methine (δ_C 42.1), a quaternary carbon (δ_C 72.2), and 12 biphenyl ring carbons (δ_C 110.2, 110.7, 123.0, 124.5, 132.0, 134.1, 140.5, 141.1, 151.8, 152.1, 152.3, and 152.7).
- Nuclear Overhauser Effect Spectroscopy (NOESY) correlations (Fig. 5) between H-11/CH_3-17 and H-9 indicate an

α-orientation for CH_3-17. Additionally, NOESY correlations between H-4 and H-6β and the absence of correlation between H-4 and CH_3-18 suggest an α-orientation for CH_3-18 as well. These observations indicate that the isolated lignan possesses an absolute configuration of C-7 (*R*) and C-8 (*S*).

- The circular dichroism (CD) spectrum of the isolated lignan exhibits two positive Cotton effects at approximately 236 and 250 nm, along with a negative Cotton effect around 213 nm [10], suggesting that it possesses an *R*-biphenyl configuration (*see* **Note 15**).

3.5 Quantitation of Lignans by HPLC-PDA

Consequently, the chemical structure of the isolate is identified as schisandrol A based on its spectroscopic data measurement and comparison with published values. To further analyze and quantify the marker components of the *S. chinensis* extract utilized in the experiment, marker component quantification analysis was performed using HPLC-PDA. The employed method is outlined as follows:

1. Prepare the mobile phase (AcCN: 0.1% TFA in water = 50:50, *v/v*), and equilibrate the HPLC system under the specified solvent conditions (*see* **Note 16**).
2. Sample preparation: accurately weigh 10 mg of *S. chinensis* extract (sample), and dissolve it in 1 mL of MeOH. After filtration using a syringe filter, transfer the filtered sample into an analysis vial (*see* **Note 17**).
3. Standard solution preparation: precisely weigh 1 mg of schisandrol A (standard), and dissolve it in MeOH. After filtration using a syringe filter, transfer the filtered sample into an analysis vial (*see* **Note 17**).
4. Analyze the blank (MeOH) at an absorbance of 254 nm using HPLC-PDA to detect any contamination in the analysis system.
5. Perform a preliminary analysis of the *S. chinensis* extract (10 mg/mL) and schisandrol A (1 mg/mL) solutions at an absorbance of 254 nm using HPLC-PDA to establish the calibration curve concentration.
6. Dilute the calibration curve standard solution into four different concentrations (15.625, 31.25, 62.5, and 125 μg/mL), considering the concentration of the sample (*see* **Note 18**).
7. Analyze the calibration curve standard solutions and the sample at an absorbance of 254 nm under the mobile phase conditions (AcCN: 0.1% TFA in water = 50:50, *v/v*) using HPLC-PDA.

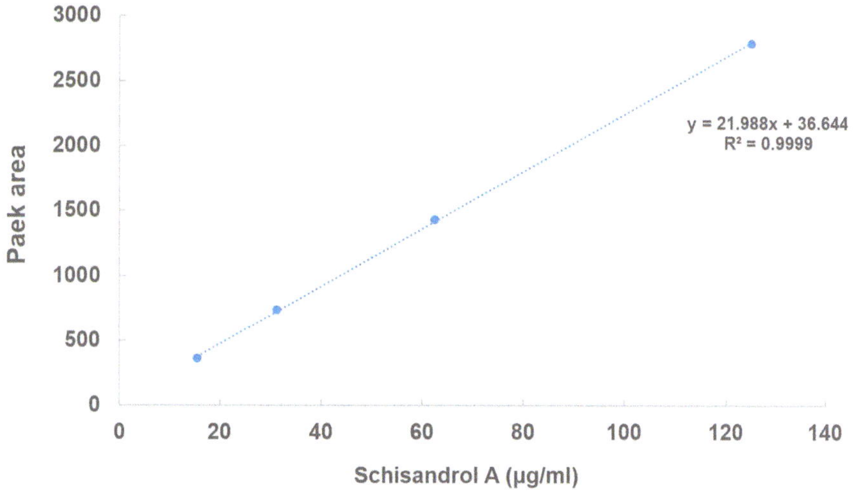

Fig. 6 HPLC calibration curve of schisandrol A

8. Quantify by substituting the absorbance of the sample into the regression line (calibration curve) obtained from the calibration curve standard solutions (*see* **Note 19**). The calibration curve of schisandrol A is represented by y = 21.988x + 36.644, and the correlation coefficient (r^2) is 0.9999 (Fig. 6).

4 Notes

1. When selecting an extraction solvent for lignans, consideration should be given to the physicochemical properties of the target compounds, including polarity, solubility, and lipophilicity. Typically, polar solvents such as ethanol and methanol are employed for extracting lignans due to their relatively lipophilic nature. However, when dealing with a mixture of lignans and lignan glycosides, it is advisable to incorporate approximately 5–10% water into the solvent to enhance extraction efficiency. Although certain lignans can be extracted using nonpolar solvents like dichloromethane, chloroform, or *n*-hexane, these are typically utilized in subsequent stages such as solvent partitioning, separation, and isolation (*see* Ref. [5]).
2. Hold the separating funnel by supporting its rounded portion with the palm of your hand while gripping the stopcock with your thumb and forefinger to manipulate it. Vigorously shake the funnel; however, be cautious as excessive pressure from solvent gas may occasionally build up, necessitating frequent opening of the stopcock.
3. Solvent partitioning involves leveraging the distribution between immiscible solvents, with common examples

including water and hexane, water and ethyl acetate, and water and butanol. These solvents possess differing solubility properties, facilitating the effective separation of compounds within the mixture, hence their widespread use in solvent partitioning.

4. Lignans possess the ability to absorb UV light, enabling visualization of plates without chemical treatment under a 254 nm wavelength UV light. Alternatively, another detection method involves spraying plates with various reagents, such as a 5% sulfuric acid solution in ethanol (*see* Ref. [7]).
5. Sulfuric acid is a potent, corrosive, and highly reactive chemical. Handle it with care, wearing gloves and goggles to protect your eyes and skin. Additionally, add the sulfuric acid to the methanol, not vice versa, as adding methanol to concentrated sulfuric acid can trigger a dangerous exothermic reaction.
6. Wear gloves when handling and spraying. Evenly spray the dried TLC plate with a 10% (*v/v*) solution of sulfuric acid in methanol within an operating hood.
7. The cotton ball should be large enough to seal the bottom of the column without excessively restricting solvent flow. A piece of cotton or glass wool that is about the size of the tip of your little finger is typically suitable for most columns.
8. Occasionally, the chosen solvent may be more polar than the eluting solvents. In such cases, if utilizing the wet method of column loading, it is crucial to only use a few drops of solvent to load the sample. Excessive loading solvent may interfere with elution and subsequent mixture separation. Therefore, in such instances, the dry method of column loading is recommended.
9. TLC often serves as an initial method to guide mobile phase selection. Select several candidate solvents or solvent mixtures, spot a small amount of extract on the baseline of the TLC plate, develop the plate using one of the selected mobile phases, and observe compound separation on the TLC plate (the Rf value of the target spot typically falls below 0.3).
10. Gradually add silica to the solvent while swirling to create a slurry, using a glass rod or spoon-spatula to mix it. Ensure solvent drainage to prevent overflow.
11. Gently tap the column to encourage bubble release and silica settling. Rinse the column interior by pipetting solvent down the inner edge, and drain it until the solvent level aligns with the stationary phase surface.
12. Adjust the flow rate by manipulating the stopcock or Teflon tube position. The flow rate in open column chromatography typically depends on the solvent type, sample characteristics, and desired separation degree. Adjustments are made to enable

a flow rate of 1–2 column volumes per hour, with specifics varying according to experimental conditions.

13. Inhalation of silica dust poses hazards. Handle silica within a fume hood, and wear appropriate protective clothing to minimize dust generation.
14. Methoxy carbons adjacent to the biphenyl bond exhibit a chemical shift approximately 5 ppm upfield compared to other methoxy carbons. For instance, 3-OCH_3 and 12-OCH_3 have a chemical shift at 55.0 ppm, while other methoxy carbons generally appear around 60.0 ppm.
15. CD spectroscopy aids in determining the absolute configuration of the biphenyl chromophore. In a lignan derivative's CD spectrum, the presence of both a positive Cotton effect at 220 nm and a negative Cotton effect at 254 nm typically indicates an *S* configuration for the biphenyl unit. Conversely, an *R*-configured lignan exhibits a CD spectrum featuring a negative Cotton effect at 220 nm and a positive Cotton effect at 254 nm (*see* Ref. [10]).
16. Prepared HPLC solvents must undergo degassing to remove dissolved gases, particularly air, ensuring chromatography system stability. To degas HPLC solvents, pour them into a solvent storage bottle and sonicate with the lid slightly open to allow the release of dissolved gases.
17. Conduct filtration using a syringe and filter to remove insoluble substances or particles from the sample intended for analysis, maintaining the sample in a clean, transparent state.
18. Dilute the standard stock solution to an initial concentration of 125 μg/mL, then prepare a double dilution series (15.625, 31.25, and 62.5 μg/mL) using methanol-filled tubes. Thoroughly mix each tube before transfer.
19. Record chromatograms for standards at various concentrations to determine peak areas or heights. Construct a calibration curve plotting component amount versus peak area or height. Under established conditions, introduce the sample for analysis, quantifying each component's absolute amount based on the calibration curve.

References

1. Robbers, J. E., Speedie, M. K., Tyler, V. E. (1996). Pharmacognosy and pharmacobiotechnology
2. Badal S, Delgoda R (eds) (2017) Pharmacognosy. Fundamentals, applications and strategy. Academic Press, Amsterdam
3. Yu X, Che X, Xu H (2017) Recent advances in the chemistry and biology of podophyllotoxins. Chem Eur J 23(19): 4467–4526. https://doi.org/10.1002/chem.201602472
4. Kamal A, Reddy TS, Polepalli S, Shalini N, Reddy VG, Rao AS, Jain N, Shankaraiah N (2014) Synthesis and biological evaluation of podophyllotoxin congeners as tubulin polymerization inhibitors. Bioorg Med Chem

22(19):5466–5475. https://doi.org/10.1016/j.bmc.2014.07.031

5. Willför SM, Smeds AI, Holmbom BR (2006) Chromatographic analysis of lignans. J Chromatogr A 1112(1–2):64–77. https://doi.org/10.1016/j.chroma.2005.11.054
6. Slanina J, Glatz Z (2004) Separation procedures applicable to lignan analysis. J Chromatogr B 812(1–2):215–229. https://doi.org/10.1016/j.jchromb.2004.06.047
7. Patyra A, Kołtun-Jasion M, Jakubiak O, Kiss AK (2022) Extraction techniques and analytical methods for isolation and characterization of lignans. Plan Theory 11(17):2323. https://doi.org/10.3390/plants11172323
8. Liu S, Yang Y, Hussain N, Jian Y, Li B, Qiu Y, Yu H, Wang H, Wang W (2023) Dibenzocyclooctadiene lignans from the family Schisandraceae: a review of phytochemistry, structure-activity relationship, and hepatoprotective effects. Pharmacol Res 195:106872. https://doi.org/10.1016/j.phrs.2023.106872
9. Hu D, Yang Z, Yao X, Wang H, Han N, Liu Z, Wang Y, Yin J (2014) Dibenzocyclooctadiene lignans from *Schisandra chinensis* and their inhibitory activity on NO production in lipopolysaccharide-activated microglia cells. Phytochemistry 104:72–78. https://doi.org/10.1016/j.phytochem.2014.04.014
10. Ikeya Y, Taguchi H, Yosioka I, Kobayashi H (1978) The constituents of *Schizandra chinensis* Baill. I. Isolation and structure determination of five new lignans, gomisin A, B, C, F and G, and the absolute structure of schizandrin. Chem Pharm Bull 27(6):1383–1394. https://doi.org/10.1248/cpb.27.1383

Chapter 13

Extraction, Annotation, and Purification of Resin Glycosides from the Morning Glory Family (Convolvulaceae)

Lars H. Kruse, Alexandra A. Bennett, Vishwa J. Baruah, Mohammad Irfan, and Gaurav D. Moghe

Abstract

The metabolic diversity of plants, comprising over a million different metabolites across the plant kingdom, harbors enormous potential for pharmaceutical and biotechnological applications. Resin glycoside (RG) acylsugars from the Convolvulaceae are of interest due to their medicinal and agricultural potential. However, understanding the biological relevance of RGs is challenging as they exhibit a high lineage-specific structural diversity. Liquid chromatography–tandem mass spectrometry (LC-MS/MS) coupled with computational peak annotation can provide insights into this diversity. Here, we present a comprehensive protocol for the characterization of RG diversity using a sensitive LC-MS/MS instrument, a knowledge-based computational pipeline, and a web tool for peak annotation. The described experimental approach provides a step-by-step guide for RG sampling, extraction, purification for downstream analyses such as bioassays, and structural annotation using LC-MS/MS and computational metabolomics. The protocol focuses on qualitative analysis for putative annotation (Annotation Level 2 as defined by the Metabolomics Standards Initiative) of RGs and can serve as a valuable template for researchers exploring plant metabolic diversity beyond RGs and acylsugars.

Key words Plant metabolic diversity, Resin glycosides, Acylsugars, Computational metabolomics, Annotation Level 2 prediction, Tandem mass spectrometry, LC-MS, Morning glories, Convolvulaceae

1 Introduction

Metabolic diversity is one of the most remarkable features of plants, and about 1 million different metabolites are estimated to exist across the plant kingdom [1]. About 40% of today's pharmaceuticals are derived or inspired from plant metabolites, suggesting a huge potential for medicine and biotechnology in the undiscovered metabolite space of plants [2, 3]. As opposed to the rather uniform chemistry and characteristics of DNA and proteins enabling standardized workflows for capturing the genome, transcriptome, or proteome of a plant, metabolites exhibit high structural diversity. Therefore, capturing the metabolome requires specialized

Young Hae Choi et al. (eds.), *Natural Product Isolation and Identification: Methods and Protocols*, Methods in Molecular Biology, vol. 2895, https://doi.org/10.1007/978-1-0716-4350-1_13,

analytical approaches for different metabolite classes and to map the entirety of the metabolome. While current advances in LC-MS/MS and bio- or chemoinformatic approaches allow more systematic exploration of metabolic diversity, capturing the entirety of plant metabolic diversity and identifying new metabolites useful for pharmaceutical or biotechnological application remain a fundamental challenge [3]. Recently, advances have been made in coupling LC-MS/MS with computational peak annotation, some of which applied machine learning (ML) algorithms [reviewed in 4]. Overall, these approaches have been demonstrated to be powerful in mapping and annotating metabolic diversity of plants, for example, in capturing the metabolic response to abiotic stress in *Brachypodium distachyon* [5], to shed light on tissue-specific metabolite diversity [6, 7], and to access metabolic diversity in *Euphorbia* [8].

Recently, we presented a strategy that is suited to perform LC-MS/MS-based peak annotation of more challenging metabolite classes such as resin glycoside (RG) acylsugars in Convolvulaceae [9]. RGs in Convolvulaceae demonstrate an extensive lineage- and tissue-specific diversity and structural complexity [9–11], posing a challenge for structural annotation (Fig. 1a). RGs are typically found in all tissues; however, their occurrence is restricted to the morning glory family (Convolvulaceae). Over 300 different RG structures are known to date [12, 13]. RGs can inhibit bacterial and fungal growth, are believed to provide protection against herbivores, participate in allelopathic interactions, and are used in traditional medicine in both Mexico and India [12–16]. RGs typically comprise a 2- to 7-sugar-long oligosaccharide, a 14- to 18-carbon-long hydroxy or dihydroxy fatty acid that can form a macrolactone ring, and several aromatic or aliphatic acyl side chains (Fig. 1a). A high-throughput characterization of RG diversity would enable the discovery of the specific RGs or RG mixtures responsible for their beneficial activities. This would enable the identification of RGs that can be used for medicinal and biotechnological applications and help decipher roles of RGs within ecological systems. The high-recall workflow described previously explained the fragmentation patterns of a high percentage of detected RGs [9].

Here, we describe our experimental approach in detail and provide a step-by-step guide for RG sampling, extraction, purification, and structural annotation of RGs using LC-MS/MS and computational metabolomics. The provided protocol is suited for qualitative analysis and prediction of RGs to Annotation Level 2 as defined by the Metabolomics Standards Initiative [17, 18] and does not consider variable growth stages of the sample species. Purified RGs are suitable for bioassays or subsequent structural elucidation using nuclear magnetic resonance (NMR) spectroscopy analysis (not described in this chapter). All scripts required for the computational analysis are provided on our GitHub page.

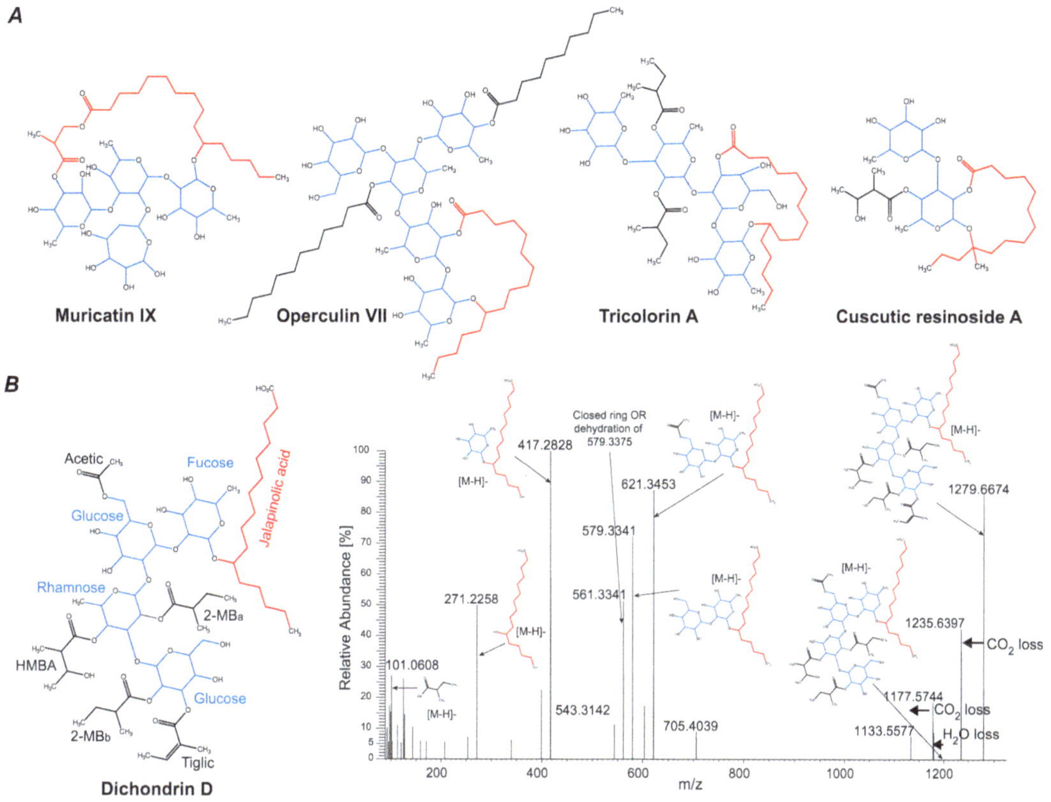

Fig. 1 Resin glycoside structural diversity and typical fragmentation. (**a**) Four different resin glycoside (RG) structures are shown that were isolated from different Convolvulaceae species. Muricatin IX, operculin VII, tricolorin A, and cuscutic resinoside A were previously isolated from *Ipomoea muricata* (L.), *Ipomoea batatas* (L.) (sweet potato), *Ipomoea tricolor* (L.), and *Cuscuta chinensis* (Lam.), respectively. (**b**) Structure and observed mass spectra of dichondrin D, isolated from *Dichondra argentea* and recorded in negative ionization mode. The mass spectrum is annotated with the structures of the corresponding fragment $(M–H)^{-}$. CO_2 and H_2O losses are indicated. The structural components of the different RGs are annotated. Sugars, smaller ester decorations, and hydroxyacyl chains are highlighted in blue, black, and red, respectively. HMBA, hydroxymethylbutyric acid; 2-MB, 2-methylbutyric acid. The figure was modified from Kruse et al. [9]

2 Materials

If not indicated otherwise, prepare all solutions using Milli-Q-purified H_2O (18 MΩ·cm at 25 °C) and analytical grade reagents. Reagents can be stored at room temperature unless indicated otherwise.

2.1 Plant Growth and Sampling

1. 15% (v/v) sulfuric acid.
2. Glass beaker.
3. 10% (w/v) trisodium phosphate solution.
4. Sterile water.
5. Student grade filter paper.

6. Sterile petri dish.
7. Parafilm.
8. 4″ plant pots (or larger pots for larger morning glory species needing extended cultivation).
9. Standard potting soil.
10. Growth chamber or grow room with a constant light/dark (16 h/8 h) regime at 24 ° C.

2.2 Extraction of RGs for LC-MS/MS Analysis

1. Plant materials (e.g., entire plants or plant tissue samples from the morning glory family).
2. Extraction solvent: HPLC grade MeOH with 10 μM telmisartan (*see* **Note 1**).
3. 50-mL falcon tubes (cold-resistant, must withstand liquid nitrogen).
4. Liquid nitrogen.
5. Lyophilizer (freeze-drier).
6. Spatula (metal or disposable plastic).
7. 2-mL round-bottom centrifuge tubes which fit snugly within a mixing mill.
8. 2-mL centrifuge tube with O-ring and screw top (*see* **Note 2**).
9. Precision scale.
10. Precision pipette (500 μL range).
11. Vortex shaker.
12. 2.3-mM metal beads.
13. Sample rocking platform shaker.
14. Mixer mill/homogenizer.
15. OPTIONAL: anti-static ionizer.
16. Centrifuge capable of reaching 14,000 *g*.
17. 2-mL amber HPLC vials.

2.3 Extraction and Purification for Downstream Bioassays or Structural Elucidation

1. Plant materials (fresh or freeze-dried).
2. Metal grain mill (or any other type of temperature-resistant bulk homogenizer).
3. Liquid nitrogen.
4. Liquid nitrogen-resistant container for sample collection (polypropylene, but preferably polycarbonate).
5. 50-mL Erlenmeyer flask.
6. HPLC grade MeOH.
7. Orbital shaker.
8. Whatman grade 1 qualitative filter paper.

9. Büchner funnel.
10. 50-mL vacuum Erlenmeyer flask with rubber gasket.
11. Vacuum pump and rubber tubing.
12. Rotary evaporator.
13. 2-mL amber LC vial.
14. Preparative C18 column: Agilent Eclipse XDB-C18 Semi-Prep (5 μm, 9.4 × 250 mM; *see* **Note 3**).
15. Mobile phase A: H_2O with 0.1% (v/v) HPLC grade formic acid.
16. Mobile phase B: AcCN with 0.1% (v/v) formic acid, both HPLC grade.
17. LC system suitable for preparative work (e.g., larger column size, fraction collector).
18. Mass spectrometer (*see* **Note 4**).
19. Fraction collector with collection vials.

2.4 LC-MS/MS Analysis of RGs

1. Mobile phase A: H_2O with 0.1% (v/v) LC-MS grade formic acid.
2. Mobile phase B: acetonitrile with 0.1% (v/v) LC-MS grade formic acid.
3. LC column: Phenomenex Kinetex 2.6 μm C18 100 Å, LC Column 150 × 2.1 mM (00F-4462-AN) (*see* **Note 5**).
4. Thermo Scientific Q Exactive Hybrid Quadrupole–Orbitrap Mass Spectrometer coupled with Thermo Scientific Dionex UltiMate 3000 HPLC or other liquid chromatography–mass spectrometry system capable of acquiring data-dependent MS/MS data (*see* **Note 6**).

2.5 LC-MS/MS Data Processing and Knowledge-Based MS/MS Peak Processing and Annotation

1. Personal computer with a Windows working environment.
2. MS-DIAL version 4.9 (https://systemsomicslab.github.io/compms/msdial/main.html) (*see* **Note** 7 and **8**).
3. Unix working environment with Python3 with pandas and NumPy package installed.
4. Python scripts downloaded from the Moghe Lab GitHub page (https://github.com/moghelab/ResinGlycoside_Pipeline/tree/main).

3 Methods

3.1 Plant Growth and Sampling

1. Depending on the species and types of seeds, two different germination protocols can be used (*see* **Note 9**):
   (a) For seeds with thick seed coat (e.g., *Ipomoea tricolor*): soak for 15 min in 15% (v/v) sulfuric acid.
   (b) For seeds with thin seed coat (e.g., *Dichondra argentea*): wash for 10 min in 10% (w/v) trisodium phosphate solution.
2. After sterilization or acid scarification, wash seeds 5 times with sterile, Milli-Q-purified H_2O.
3. For germination, put seeds on moist student grade filter in a petri dish and carefully enclose with a parafilm to prevent water evaporation.
4. Keep seeds in the dark at room temperature until the radicle is clearly visible. This usually takes 2–10 days, depending on the species (*see* **Note 10**).
5. Transfer germinated seeds into 4″ pots with nutrient-rich potting soil, and grow in the growth chamber for about 4 weeks. Larger species will need to be repotted and transferred to the greenhouse until sampling.

3.2 Extraction for LC-MS Analysis

1. Harvest roots (leaves optional) from morning glory plants (*see* **Note 11**) into 50-mL falcon tubes.
2. Snap-freeze tissue by immersing tubes into liquid nitrogen (*see* **Note 12**).
3. Keep samples in liquid nitrogen, and use a pre-chilled metal spatula to crush the tissue roughly and scrape any that may be frozen to the sides of the tube. At this stage, the plant material can be stored at −80 °C until further processing.
4. If the sampled tissues are ready for lyophilization, start by pre-chilling the lyophilizer.
5. Loosely unscrew the screw caps on falcon tubes such that air can get in and out but the cap cannot easily be pulled off (*see* **Note 13**).
6. Place the frozen sample into a pre-chilled lyophilizer.
7. Seal lyophilizer and begin pumping down the vacuum chamber.
8. Cover vacuum chamber or keep lights off to prevent light from shining on the samples (*see* **Note 14**).
9. Lyophilize samples for 1 to 2 days, depending on the amount of tissue harvested, until dry.

10. Remove samples from lyophilizer. Keep tubes completely sealed when not in use to prevent samples from getting moist again. The sample can be stored in the −80 °C freezer at this point for subsequent processing.
11. Using a spatula, transfer approximately 100 mg of tissue from the 5-mL falcon tube to a 1.5-mL centrifuge tube which is also suitable for the mixer mill being used. Store the unused sample at −80 °C.
12. Add four metal beads to the tube containing 100 mg of the sample.
13. Place the sample in a mixer mill. Homogenize the samples at 30 bpm for 10 min.
14. OPTIONAL: remove possible static charge from samples using an anti-static ionizer (*see* **Note 15**).
15. Centrifuge the sample at 14,000 *g*.
16. Using a spatula, weigh out 50 mg of powdered tissue from centrifuge tube and transfer to a fresh 1.5-mL centrifuge tube which is suitable for organic solvents. Write down the exact mass of tissue collected.
17. Add 500 μL of extraction solvent per 50 mg of tissue to the centrifuge tube (*see* **Note 16**). Write down the exact amount of solvent used. Seal tube.
18. Vortex the sample for 30 s. Place the sample on a rocker and secure them in place with tape. Set the rocker to maximum speed, and extract samples for 15 min. Vortex the sample for another 30 s.
19. Centrifuge the sample for 30 s at 14,000 *g*. Check that particulate has fully pelleted.
20. Transfer 90% of the supernatant to a fresh, clean labeled microcentrifuge tube. Check to ensure that no particulate matter is floating in the transferred supernatant. If particulate matter is seen, repeat **steps 18** and **19** until there is no visible particulate floating in the sample.
21. Transfer the supernatant to a labeled LC vial, and store at −80 °C until ready for analysis (*see* **Note 17**).

3.3 Extraction and Purification of RGs for Downstream Analysis

Individual RGs may need to be purified to enable their use as standards, in bioassays or for structural elucidation in NMR spectroscopy. The following procedure can be used for purification of individual RGs:

1. Harvest all mass of the plant containing the RG of interest. Snap-freeze with liquid nitrogen, or barring availability of that much liquid nitrogen, freeze at −80 °C (*see* **Note 18**).

Table 1
Gradient profile for preparative LC

Time	% mobile phase A	% mobile phase B
0.0	60	40
3.0	60	40
20.0	0	100
25.0	0	100
25.1	60	40
32.0	60	40

The flow rate and injection volume depend on the instrument used, and the below gradient profile should be used as a template for developing an instrument-specific method. Mobile phase A is H_2O with 0.1% (v/v) formic acid (HPLC grade), and mobile phase B is acetonitrile with 0.1% (v/v) formic acid (HPLC grade)

2. Set up a metal grain mill with a clean liquid nitrogen-resistant container below the mill to collect the ground plant tissue as it comes out. Always maintain a small amount of liquid nitrogen in the container.
3. Chill metal grain mill with liquid nitrogen (*see* **Note 19**).
4. Place plant material in mill and grind plant tissue. Continue to add liquid nitrogen sparingly but as much as needed to chill both mill and container until all tissue is ground.
5. After ensuring that all liquid nitrogen has evaporated but tissue has not thawed, seal the lid of the container and then store plant tissue at −80 °C until extraction (*see* **Note 20**).
6. Pour 9 mL of MeOH in a 50-mL Erlenmeyer flask. While keeping the sample frozen, use a pre-chilled metal spatula to weigh and add approximately 3 g of ground tissue to flask.
7. Extract RGs on an orbital shaker for 15 min at 25 °C and 180 rpm.
8. Prepare the filtration step by placing a Whatman grade 1 qualitative filter paper in a Büchner funnel. Place the Büchner funnel in a 50-mL vacuum Erlenmeyer flask with a rubber gasket connector. Attach the vacuum flask to a vacuum pump with rubber tubing.
9. Wet the filter paper slightly with MeOH. Pull off MeOH with vacuum, and discard this MeOH.
10. Place the sample in the Whatman filter-lined Büchner funnel. Attach to an Erlenmeyer vacuum flask attached to a vacuum pump. Apply vacuum at the highest possible setting. Vacuum the sample until the solid tissue in the funnel is almost dry (1–5 min).

11. Take the filtrate and place it in a round evaporation flask. Attach round-bottom flask to a rotary evaporator, and dry the sample to completion with a water bath temperature set to 25 °C, cooling water temperature set to 4 °C, a vacuum of 50 mbar, and the rotation set to 250–280 rpm. This should take approximately 1 h (180 rpm; *see* **Note 21**).
12. Reconstitute the solid sample in 350 μL MeOH in the evaporation flask. Place on a shaker in a holder, and shake for 15 min at 25 °C and 180 rpm.
13. Transfer the sample to an LC vial and keep at −80 °C until purification (can be kept at 4 °C if purification will be done on the same day).
14. Attach the preparative C18 column to the preparative LC system.
15. Attach adequate amounts of mobile phase A (H_2O with 0.1% formic acid) and mobile phase B (acetonitrile with 0.1% formic acid) to the system (*see* **Note 22**).
16. Generate preparative LC method with an injection volume of 30 μL for the collection method, a flow rate of 3.5 mL/min, and no column temperature regulations. The 32-min gradient should be set according to Table 1.
17. OPTIONAL: determine the elution of your compound(s) of interest on your preparative method.
    (a) Generate a negative ionization mode mass spectrometric method (*see* **Note 23**).
    (b) Load your LC-MS method, and inject one sample.
    (c) Assess the elution time(s) of your SIM peak(s).
18. Generate method for the fraction collector. If you have assessed the elution of your compound of interest, collect only that peak and send everything else to waste (*see* **Note 24**).
19. Load both LC and fraction collection methods. Place the sample in an autosampler. Place collection vials in a fraction collector.
20. For a total of 10 times, inject and collect sample fractions with the RG of interest.
21. Assess your fractions for compounds and purity on a suitable LC-MS system with high sensitivity and mass resolution. Samples should be at approximately 90% purity; that is, all other peaks added together should not be greater than 10% of the total sum of all peaks.

3.4 LC-MS/MS Run Procedure

LC-MS/MS analysis should be performed on a Thermo Scientific Q Exactive Hybrid Quadrupole—Orbitrap Mass Spectrometer coupled with Thermo Scientific Dionex UltiMate 3000 HPLC

Table 2
LC method for LC-MS analysis of resin glycosides

Time	% mobile phase A	% mobile phase B
0.0	75	25
0.1	75	25
5.60	0	100
6.60	0	100
6.61	75	25
7.51	75	25

This method is for Thermo Fisher Dionex UltiMate 3000 UHPLC. The column oven is set to 40 °C. Mobile phase A is H_2O with 0.1% (v/v) formic acid (LC-MS grade), and mobile phase B is acetonitrile (LC-MS grade). The flow rate is 0.6 mL/min, and the injection volume is set to 2 μL

equipped with an autosampler and a 150-mM C18 column as described above (2.4); otherwise, the provided parameters for the run conditions and MS-DIAL analysis should be adjusted. The control of LC-MS/MS systems is performed using Thermo Scientific Xcalibur software, version 4.4.

1. Check the mobile phase reservoirs' liquid levels to make sure there is enough mobile phase for the analysis.
2. Turn on instrument and install the C18 column in the column compartment of the LC system, and ensure the doors of all compartments are closed.
3. Create an LC-MS/MS method using the detailed parameters listed in Tables 2 and 3. Save and load the method.
4. Using Xcalibur, operate the LC-MS/MS and set the flow rate initially to 0.3 mL/min and observe the pump pressure. If the pressure remains below 400 bar, start to slowly increase the flow rate to 0.6 mL/min and equilibrate column and system for 5–10 min with 95% mobile phase A and 5% mobile phase B. Monitor the pressure which should be stable (*see* **Note 25**).
5. After equilibration, inject 2 μL of the blank containing internal standard into the system via the autosampler using the described method.
6. Check if the peak of the internal standard looks good and is at the right retention time. If correct, continue to inject samples (2 μL per sample) and record the data.
7. Once the run is complete, wash the column with alternating high concentrations (95%) of mobile phase A and B for 10 min. Store column in 50% acetonitrile.

Table 3
MS/MS method parameters for LC-MS analysis of resin glycosides

Parameter	Value
Full MS	
Resolution	70,000
AGC target	5e6
Maximum IT	30 ms
Scan range	m/z: 500 to 2000
dd-MS^2	
Resolution	17,500
AGC target	5e5
Maximum IT	110 ms
Loop count	10
TopN	10
Isolation window	m/z: 1.5
Scan range	m/z: 200 to 2000
NCE	20, 40
Minimum AGC target	5.00e3
Intensity threshold	4.5e4
Dynamic exclusion	7.0 s
Source	
Spray voltage (−)	2500.00
Capillary temperature (−)	380.00
Sheath gas (−)	70.00
Aux gas (−)	20.00
Max spray current (−)	100.00
Probe heater temp. (−)	400.00
S-lens RF level	75.00
Ion source	HESI

These parameters are for Thermo Fisher Q Exactive Hybrid Quadrupole–Orbitrap Mass Spectrometer in negative ionization mode. Parameters might need instrument-specific adjustments

Table 4
Detailed MS-DIAL parameters

Parameter	Value
Project	
MS1 data type	Profile
MS2 data type	Profile
Ion mode	Negative
Target	Metabolomics
Mode	ddMSMS
Data collection parameters	
MS1 tolerance	0.005 Da
MS2 tolerance	0.01 Da
Retention time begin	0 min
Retention time end	100 min
Mass range begin	0 Da
Mass range end	2000 Da
MS/MS mass range begin	0 Da
MS.MS mass range end	2000 Da
Maximum charged number	2
Consider Cl and Br elements	Do not check
Number of threads	1 to 5 threads (*see* **Note 31**)
Execute retention time corrections	Do not check
Peak detection parameters	
Minimum peak height	10,000 amplitude (*see* **Note 32**)
Mass slice width	0.1 Da
Smoothing method	Linear weighted moving average
Smoothing level	4 scan (*see* **Note 33**)
Minimum peak width	5 scan
Exclusion mass list (mass & tolerance)	None (*see* **Note 34**)
MS2Dec	
Sigma window value	0.5
MS2Dec amplitude cut off	0
Exclude after precursor	Check
Keep isotope until	0.5

(continued)

Table 4
(continued)

Parameter	Value
Keep original precursor isotopes	Do not check
Identification	Not used
Adduct ion setting	
$[M-H]^-$	Check
$[M-H_2O-H]^-$	Check
$[M + FA-H]^-$	Check
Alignment parameters setting	
Reference file	QC.abf (*see* **Note 35**)
Retention time tolerance	0.05 (*see* **Note 36**)
MS1 tolerance	0.005 (*see* **Note 37**)
Retention time factor	0.5
MS1 factor	0.5
Peak count filter	0
N% detected in at least one group	0
Remove feature based on blank information	Check
Sample average / blank average	5
Keep 'referenced matched' metabolite features	Do not check
Keep 'suggested (w/o MS2)' metabolite features	Do not check
Keep removable features and assign the tag for checking	Check
Gap filling by compulsion	Check
#Tracking of isotope labels	
Tracking of isotopic labels	FALSE
Ion mobility	
Ion mobility data	FALSE
Project	
MS1 data type	Profile
MS2 data type	Profile
Ion mode	Negative
Target	Metabolomics
Mode	ddMSMS

(continued)

Table 4
(continued)

Parameter	Value
Data collection parameters	
Retention time begin	0
Retention time end	7 (*see* **Note 38**)
Mass range begin	0
Mass range end	2000
MS2 mass range begin	0
MS2 mass range end	2000

Detailed information for the selected MS-DIAL parameters. These parameters are specific for the used instrument and can be used as a starting template when a different instrument is being used

3.5 LC-MS Data Analysis and Manual Identification of RGs

1. Convert Orbitrap *.RAW files to *.abf files using Abf Converter (*see* **Note 26**).
   (a) This step is not needed in MS-DIAL 4.9+.
2. Create a new MS-DIAL project (File >> New project).
3. Import LC-MS/MS data abf files using appropriate MS-DIAL settings. For the demo data, use the following settings:
   (a) **Ionization type:** soft ionization.
   (b) **Separation type:** chromatography.
   (c) **MS method type:** conventional.
   (d) **Data type (MS1):** profile, **(MS/MS):** profile.
   (e) **Ion mode:** negative ion mode.
   (f) *Click Next.*
   (g) **Analysis file paths:** click Browse, and select all abf/raw files obtained from a *single species/variety*, as well as Blank. Specify analytical order and type of sample.
   (h) *Click Next.*
   (i) **Analysis parameter settings:** choose all default settings except:
      (i) **Ion mode**: negative.
      (ii) **MS1 tolerance:** 0.005 Da (both in data collection and in alignment).
      (iii) **MS2 tolerance:** 0.01 Da.
      (iv) **Minimum peak height:** 10000 amplitude.
      (v) **Smoothing level:** 4 scans.
      (vi) **Adducts:** $[M\text{-}H]^-$ (m/z: -1.007276), $[M\text{-}H_2O\text{-}H]^-$ (m/z: −19.01839), and $[M + FA\text{-}H]^-$ (m/z: 44.998201). You may specify other

Table 5
Typical resin glycoside MS/MS fragments

Fragment type	Name(s)	Formula	Exact mass	m/z $[M-H]^-$
Longacyl	$C_{14}OH$: hydroxytetradecanoic	$C_{14}H_{28}O_3$	244.2038	243.1966
Longacyl	$C_{15}OH$: hydroxypentadecanoic acid	$C_{15}H_{30}O_3$	258.2195	257.2122
Longacyl	$C_{14}OHOH$: dihydroxytetradecanoic acid	$C_{14}H_{28}O_4$	260.1988	259.1915
Longacyl	$C_{16}OH$: hydroxyhexadecanoic acid Jalapinolic acid	$C_{16}H_{32}O_3$	272.2351	271.2279
Longacyl	$C_{15}OHOH$: dihydroxypentadecanoic acid	$C_{15}H_{30}O_4$	274.2144	273.2071
Longacyl	$C_{17}OH$: hydroxyheptadecanoic acid	$C_{17}H_{34}O_3$	286.2508	285.2435
Longacyl	$C_{16}OHOH$: dihydroxyhexadecanoic acid	$C_{16}H_{32}O_4$	288.2301	287.2228
Longacyl	$C_{18}OH$: hydroxyoctadecanoic acid	$C_{18}H_{36}O_3$	300.2664	299.2592
Longacyl	$C_{17}OHOH$: dihydroxyheptadecanoic acid	$C_{17}H_{34}O_4$	302.2457	301.2384
Longacyl	$C_{18}OHOH$: dihydroxyoctadecanoic acid	$C_{18}H_{36}O_4$	316.2614	315.2541
Sugar	Pentose	$C_5H_{10}O_5$	150.0528	149.0455
Sugar	Deoxyhexose	$C_6H_{12}O_5$	164.0685	163.0612
Sugar	Hexose	$C_6H_{12}O_6$	180.0634	179.0561
Smallacyls	Acetic	$C_2H_4O_2$	60.02113	59.01313
Smallacyls	Acrylic_or_hydroxypropinol(sugar_degradation)	$C_3H_4O_2$	72.02113	71.01313
Smallacyls	Propionic	$C_3H_6O_2$	74.03678	73.02878
Smallacyls	Crotonic	$C_4H_6O_2$	86.03678	85.02878
Smallacyls	Butyric	$C_4H_8O_2$	88.05243	87.04443
Smallacyls	Tiglic	$C_5H_8O_2$	100.0524	99.0444
Smallacyls	Acetoacetic_or_2-oxobutanoate (sugar_degradation)	$C_4H_6O_3$	102.0317	101.0237
Smallacyls	Pentanoic	$C_5H_{10}O_2$	102.0681	101.0601
Smallacyls	Hexanoic	$C_6H_{12}O_2$	116.0837	115.0757
Smallacyls	HMBA	$C_5H_{10}O_3$	118.0630	117.0550
Smallacyls	Benzoic	$C_7H_6O_2$	122.0368	121.0288
Smallacyls	Octanoic	$C_8H_{16}O_2$	144.1150	143.1070
Smallacyls	Cinnamic	$C_9H_8O_2$	148.0520	147.0440
Smallacyls	Coumaric	$C_9H_8O_3$	164.0473	163.0393
Smallacyls	Gallic	$C_7H_6O_5$	170.0215	169.0135
Smallacyls	Decanoic	$C_{10}H_20O_2$	172.1463	171.1383

(continued)

Table 5 (continued)

Fragment type	Name(s)	Formula	Exact mass	m/z [M–H]$^-$
Smallacyls	Caffeic	$C_9H_8O_4$	180.0423	179.0343
Smallacyls	Hydroxydecanoic	$C_{10}H_20O_3$	188.1412	187.1332
Smallacyls	Ferulic	$C_{10}H_{10}O_4$	194.0579	193.0499
Smallacyls	Dodecanoic	$C_{12}H_{24}O_2$	200.1776	199.1696
Smallacyls	Hexadecanoic	$C_{16}H_{32}O_2$	256.2402	255.2322
Smallacyls	Octadecanoic	$C_{18}H_{36}O_2$	284.2715	283.2635
Smallacyls	Eicosanoic	$C_{20}H_{40}O_2$	312.3028	311.2948

Resin glycoside fragments observed in negative ionization mode on Thermo Fisher Q Exactive Hybrid Quadrupole–Orbitrap Mass Spectrometer

adducts depending on your instrument and run conditions.

(vii) Identification: *see* **Note 27**.

(viii) **Alignment:** specify any sample as the reference file except for the Blank (*see* **Note 28**).

(ix) See Table 4 for additional details.

(x) *Click Finish.*

4. Once MS-DIAL aligns all chromatograms, manually confirm that the alignment is good by checking the internal standard (Telmisartan m/z: 513.2296) and, if possible, 3–4 other peaks. If the alignment is not good, realign by setting different retention time tolerances and reprocessing the alignment (*see* **Note 29**).
5. Perform exploratory analysis of the data by checking for the presence of RGs using the hydroxyacyl fragment masses as described in Kruse et al., and shown in Fig. 1b [9]. For example, most species use 11-hydroxyhexadecanoic acid (m/z: 271.2279) as a long acyl chain. Other fragments can also be used for exploratory analyses (see Table 5). Use the MS/MS fragment search option to visualize peaks in the alignment containing this MS/MS fragment.
   (a) Enter tolerance (Da) of 0.05 (or your MS2 mass accuracy value).
   (b) Set ion abundance to 5%.
   (c) Ensure search type is for Product Ion.
   (d) Click "Search."

   (e) Check "Unique Ions" in the Peak spot navigator, which will highlight only the peaks having the specified MS/MS fragment or their combinations. If it is already checked, you may need to uncheck and recheck it.

6. Confirm that the observed MS/MS fragmentation pattern is as per the expected fragmentation patterns of RGs.
   (a) Typically, a well-fragmented peak will show the following fragment masses:
      (i) Hydroxyacyl chain (HAC).
      (ii) HAC + 1 sugar, HAC + 2 sugars, HAC + 3 sugars, ...
      (iii) Acyl chain fragments <m/z: 200 as described earlier [9].
      (iv) A typical fragmentation pattern of RGs is shown in Fig. 1b. A comprehensive list of observed MS/MS peaks can be found in Table 5.
   (b) If you do not see enough MS/MS peaks, you may need to re-adjust your LC or MS data gathering parameters (*see* **Note 30**).
7. Export the data using the following settings:
   (a) Go to Export > > Alignment results.
   (b) Select export type of interest:
      (i) Peak height or area (for quantitation). Do not blank filter, because you could lose RGs which carry over in the analytical method.
      (ii) No need to replace zero values as gap filling was performed, and there will be no zero values.
      (iii) Representative spectra (for in-depth assessment of fragmentation patterns). Select MGF or MSP based on our preferred format.
      (iv) Parameter (this exports the preprocessing parameters and is always good to export for tracking purposes).
      (v) No need to filter by isotope labeling as no isotope labeling was performed.
      (vi) Click "export."
8. At the end of these steps, there should be at least two files exported: Area_**.txt (Peak Area file) and *.mgf (MGF file).

3.6 Automatic Extraction of RG Peaks

1. Ensure all Python scripts downloaded from the Moghe Lab GitHub link are in the same folder as the exported data files. Demo files to run the following steps are provided in the linked GitHub repository.

2. Make (mkdir) four subfolders titled 0_RgFrags, 1_PeakArea, 2_MGF, and 3_predictMotifs.
3. Move (mv) the fragment pair combinations into 0_RgFrags. These combinations are used to extract any peak containing a given pair. Three files with fragment lists are provided:
   (a) **frags.list.mz132.pairs:** lists all the fragment pairs that would occur if a hydroxyacyl fragment was bonded to a pentose.
   (b) **frags.list.mz146.pairs:** lists all the fragment pairs that would occur if a hydroxyacyl fragment was bonded to a deoxyhexose.
   (c) **frags.list.mz162.pairs:** lists all the fragment pairs that would occur if a hydroxyacyl fragment was bonded to a hexose.
4. Move all Area_*.txt files to the 1_PeakArea folder and all *.mgf files to 2_MGF folder.
5. Move the motifs.groups file obtained from the Moghe Lab GitHub link into the 3_predictMotifs folder.
6. Create a new tabular file containing the following information in individual columns. This file may contain multiple rows, each with information about different species/samples.
   (a) Name of the Peak Area file in the 1_PeakArea folder.
   (b) Species name (replace space with underscore or hyphen).
   (c) Indexes in the Peak Area file that house the peak areas of the samples (comma separated).
   (d) Indexes in the Peak Area file that house the peak areas of the blanks (comma separated).
   (e) Index in the Peak Area file that contains Post-curation information.
   (f) Index in the Peak Area file that contains Fill-Percentage information.
   (g) Index in the Peak Area file that contains the Signal:Noise (S/N) information.
   (h) Index in the Peak Area file that contains the MS/MS fragment data.
   (i) Name of the MGF file in the 2_MGF folder corresponding to the species and Area file (*see* **Note 39**).
7. Run parse_MSDIAL_wrapper.py. This script runs all other parse_*.py files:
   (a) Command line: *python parse_MSDIAL_wrapper.py –filelist < tabfile created above >*.

(b) Example command line: *python parse_MSDIAL_wrapper.py -filelist fileList.tab.*

8. This script produces multiple outputs, of which three are the most important:

   (a) ***.combined.stats:** enumerates the number of peaks identified for each sample.

   (b) ***.uniq.tab:** all extracted RG-like peaks containing the searched fragment pair.

   (c) ***.nid1.id2.tab:** final output of the pipeline containing each peak annotated with the likely identities of the different MS/MS fragments.

9. The above pipeline can also be run interactively from the webserver https://tools.moghelab.org/webapp/RGscout/. For each file described in the fileList.tab, one mgf file (containing MS/MS fragmentation patterns) and one Peak Area file should be uploaded. Additional details of how to use the interactive interface are described on the above website.

4 Notes

1. Other solvents such as acetonitrile, isopropanol, or mixtures thereof can be used. The choice depends on the solubility of the analytes of interest. In addition, other internal standards can be used as long as they ionize well under the MS method chosen and do not occur naturally in the morning glory family or plant of interest.

2. Any tube that fits into the used mixer mill and does not leak MeOH can be used. We prefer screw cap tubes due to their low susceptibility to leaking for this step.

3. Other preparative LC columns with similar chemistries and dimensions can be used.

4. Any mass spectrometer that is capable of acquiring MS signals of the RG of interest can be used. Often, in these contexts, a basic single quad MS is sufficient.

5. Other C18-based columns with a length between 100 and 150 mm are sufficient for efficient chromatographic separation of most RGs.

6. Any other LC and high-performance mass spectrometer can be used; however, the parameters described need to be carefully adjusted to the specific instrument used. The provided protocol can be used as an initial template. Data-dependent ionization where parent and fragment masses can be associated is essential.

7. The described workflow has been extensively tested with MS-DIAL v4.9. The protocol is likely suitable for newer versions of MS-DIAL; however, it has not been tested sufficiently.
8. While the newest versions of MS-DIAL can read *.RAW format files, older versions require a file converter such as Abf Converter (https://www.reifycs.com/abfconverter/) or ProteoWizard version 3.9.22155 or higher (https://proteowizard.sourceforge.io/) to convert data to abf format before importing into MS-DIAL.
9. Both sulfuric acid and trisodium phosphate are highly corrosive and should be handled in a fume hood with proper personal protective equipment (PPE).
10. Continue checking the seeds for possible fungal growth, which is more likely as the time for germination grows longer. Manually remove contaminated seeds.
11. Roots may have excess soil or substrate attached. Try to gently shake substrate from roots without disrupting the root system too much. If this does not work, gentle washing of roots in a shallow dish of deionized water can be done. Gently dry roots before the next step. Not drying roots will increase lyophilization time.
12. Be careful not to get liquid nitrogen into the tubes, this can increase pressure and cause tubes to pop open and cause tissue to be lost.
13. This will slow the drying process, but it will prevent accidental sudden changes in vacuum pressure, causing the sample to be drawn out of the tube. Often, the tissue becomes statically charged during the drying process and can also be drawn out of the tube this way. Keeping the cap loosely attached prevents samples from leaving the tube and contaminating one another.
14. Since the stability of RGs under light is not well studied, it is the best practice to keep samples away from light for extended periods of time as a cautionary step.
15. This step is not necessary but highly beneficial for accurate weighing.
16. The ratio of extraction solvent to plant material can be varied, based on the expected concentration of the metabolites of interest. Ratios such as 1:4 or 1:5 have been successfully tested. Generally speaking, using more extraction solvent results in a more complete extraction and improves quantification results.
17. It is not recommended to store samples for extended periods of time to prevent metabolite breakdown and formation of artifacts.

18. The amount of tissue needed for successful purification of enough amounts of RGs for downstream applications depends on the concentration of RG of interest and the downstream application. Comparing the peak's size to the internal standard can be used as a rough measure for estimating how much plant tissue needs to be extracted for the specific downstream application.
19. Alternatively, a liquid nitrogen-chilled mortar and pestle can be used to pulverize plant material for extraction and can be more efficient if relatively small amounts of tissue are to be extracted.
20. Liquid nitrogen should never be stored in a tightly sealed container such as a tube, box, or enclosed freezer to avoid explosions. Make sure all liquid nitrogen has evaporated before sealing tubes or placing samples into storage. Always handle liquid nitrogen with the required PPE including cryogenic gloves and safety goggles.
21. Be careful to watch for "bumping" (when a superheated liquid nucleates, resulting in a sudden rapid boil that expels liquid from the evaporation flask, causing contamination of the system). At 25 °C, MeOH begins to boil at a pressure of 127.05 mbar. The rotation on the evaporator prevents boiling from creating volatile enough bubbles that could escape the evaporation flask. Because there is water in the plant tissue, this will also increase the boiling temperature and lower the boiling pressure. But it is still good to watch for bumping as cleaning a contaminated system would then be necessary.
22. Other mobile phases such as MeOH can be used and will depend on the analyte of interest, the chromatographic system, and the availability.
23. This method is dependent on the mass spectrometer used. You should know your compound(s) of interest beforehand and be able to search for it/them in single ion monitoring (SIM) mode. This type of data acquisition is possible for most mass spectrometers, including single quad.
24. You can also collect multiple fractions along the whole gradient if unsure about exact elution time. Half-a-minute-long fractions are recommended.
25. Column pressure is dependent on the column and LC system used. Therefore, for this setup, and the Phenomenex Kinetex 150 mm C18 column, the pressure should not exceed 600 bar at any point.
26. Centroiding data if obtained in profile mode will drastically speed up data analysis time; however, it needs to be performed with proprietary software from the machine vendor.

27. No library is required for the identification of RGs. However, if you want to annotate other peaks using spectral information, you may use the MS-DIAL negative-mode spectral library available on the MS-DIAL website.
28. Ideally, for peak alignment, you can use a quality control (QC) sample (a mixture of all samples so that it contains all metabolites present across all the analyzed samples).
29. If peak alignment fails or produces unsatisfactory results, the retention time tolerance can be increased, especially if the retention time of known peaks such as the internal standard shows variation between different injections. Furthermore, alignment using a different reference file can improve peak alignment. Try settings until known peaks are properly aligned.
30. If you do not see the expected MS/MS fragmentation patterns, several parameters can be optimized, including the LC method, adding an exclusion list of contaminating ions, reducing the MS1 scan range, increasing/decreasing the fragmentation voltage, or reducing the automatic gain control (AGC) for fragmentation.
31. The number of threads is based on computing power.
32. Peak height/amplitude is highly influenced by instrument/detector type. Peak height filters should be adjusted in a way so that it filters out noise reliably. Consider the smallest known peak of interest as a reference point. Alternatively, this value can be obtained experimentally by using a calibration curve.
33. Smoothing can be increased if the peak is oversampled by the instrument.
34. You can add your calibrant ion in case one was used.
35. Here, you can select whichever file has the most comprehensive number of metabolites. A quality control sample that contains all peaks of interest can be generated by mixing equal parts of the samples analyzed. For a large number of samples, several quality control (QC) samples should be prepared to not dilute lower abundant peaks below the detection limit.
36. Retention time shifting and thus tolerance depend heavily on the chromatography method and column used. Check internal standard peaks to adjust accordingly for proper alignment.
37. Mass accuracy and thus MS1 tolerance are strongly instrument-specific and need to be adjusted for instruments with lower or higher mass accuracy.
38. The retention time range needs to be adjusted for the specific method and should include all eluting RGs of interest.

39. The index corresponds to the column number in the Peak Area file minus 1 (e.g., for columns #1, #12, and #24, indexes will be 0, 11, and 23, respectively).

Acknowledgements

This work was supported in part by USDA-NIFA award #2023-67013-39630 and USDA-NIFA-Hatch award #7002995 to GM and by Deutsche Forschungsgemeinschaft award #411255989 to LK.

References

1. Afendi FM, Okada T, Yamazaki M, Hirai-Morita A, Nakamura Y, Nakamura K, Ikeda S, Takahashi H, Altaf-Ul-Amin M, Darusman LK, Saito K, Kanaya S (2012) KNApSAcK Family Databases: Integrated Metabolite–Plant Species Databases for Multifaceted Plant Research. Plant Cell Physiol 53:e1–e1. https://doi.org/10.1093/pcp/pcr165
2. Medicinal Botany. https://www.fs.usda.gov/wildflowers/ethnobotany/medicinal/index.shtml. Accessed 29 May 2023
3. Moghe GD, Kruse LH (2018) The study of plant specialized metabolism: challenges and prospects in the genomics era. Am J Bot 105: 959–962. https://doi.org/10.1002/ajb2.1101
4. Mahood E, Kruse L, Moghe G (2020) Machine learning: a powerful tool for gene function prediction in plants. Appl Plant Sci 8:e11376–e11376. https://doi.org/10.1002/aps3.11376
5. Mahood EH, Bennett AA, Komatsu K, Kruse LH, Lau V, Rahmati Ishka M, Jiang Y, Bravo A, Louie K, Bowen BP, Harrison MJ, Provart NJ, Vatamaniuk OK, Moghe GD (2023) Information theory and machine learning illuminate large-scale metabolomic responses of Brachypodium distachyon to environmental change. Plant J 114:463–481. https://doi.org/10.1111/tpj.16160
6. Li D, Heiling S, Baldwin IT, Gaquerel E (2016) Illuminating a plant's tissue-specific metabolic diversity using computational metabolomics and information theory. Proc Natl Acad Sci 113:E7610–E7618. https://doi.org/https://doi.org/10.1073/pnas.1610218113
7. Li D, Halitschke R, Baldwin IT, Gaquerel E (2020) Information theory tests critical predictions of plant defense theory for specialized metabolism. Sci Adv 6:eaaz0381. https://doi.org/10.1126/sciadv.aaz0381
8. Ernst M, Nothias L-F, van der Hooft JJJ, Silva RR, Saslis-Lagoudakis CH, Grace OM, Martinez-Swatson K, Hassemer G, Funez LA, Simonsen HT, Medema MH, Staerk D, Nilsson N, Lovato P, Dorrestein PC, Rønsted N (2019) Assessing specialized metabolite diversity in the cosmopolitan plant genus euphorbia L. Front Plant Sci 10
9. Kruse LH, Bennett AA, Mahood EH, Lazarus E, Park SJ, Schroeder F, Moghe GD (2022) Illuminating the lineage-specific diversification of resin glycoside acylsugars in the morning glory (Convolvulaceae) family using computational metabolomics. Hortic Res 9: uhab079. https://doi.org/10.1093/hr/uhab079
10. Moghe GD, Leong BJ, Hurney SM, Jones AD, Last RL (2017) Evolutionary routes to biochemical innovation revealed by integrative analysis of a plant-defense related specialized metabolic pathway. elife 6:e28468. https://doi.org/10.7554/eLife.28468
11. Moghe G, Irfan M, Sarmah B (2023) Dangerous sugars: structural diversity and functional significance of acylsugar-like defense compounds in flowering plants. Curr Opin Plant Biol 73:102348. https://doi.org/10.1016/j.pbi.2023.102348
12. Eich E (2008) Solanaceae and Convolvulaceae: secondary metabolites: biosynthesis, chemotaxonomy, biological and economic significance (a handbook). Springer
13. Pereda-Miranda R, Rosas-Ramírez D, Castañeda-Gómez J (2010) Resin glycosides from the morning glory family. In: Kinghorn AD, Falk H, Kobayashi J (eds) Fortschritte der Chemie organischer Naturstoffe/Progress in the chemistry of organic natural products, vol 92. Springer, Vienna, pp 77–153

14. Anaya AL, Calera MR, Mata R, Pereda-Miranda R (1990) Allelopathic potential of compounds isolated fromIpomoea tricolor cav. (Convolvulaceae). J Chem Ecol 16:2145–2152. https://doi.org/10.1007/BF01026926
15. Pereda-Miranda R, Mata R, Anaya AL, Wickramaratne DB, Pezzuto JM, Kinghorn AD (1993) Tricolorin a, major phytogrowth inhibitor from Ipomoea tricolor. J Nat Prod 56: 571–582. https://doi.org/10.1021/np50094a018
16. Jackson DM, Peterson JK (2000) Sublethal effects of resin glycosides from the periderm of sweetpotato storage roots on Plutella xylostella (Lepidoptera: Plutellidae). J Econ Entomol 93:388–393. https://doi.org/10.1603/0022-0493-93.2.388
17. Schymanski EL, Jeon J, Gulde R, Fenner K, Ruff M, Singer HP, Hollender J (2014) Identifying small molecules via high resolution mass spectrometry: communicating confidence. Environ Sci Technol 48:2097–2098. https://doi.org/10.1021/es5002105
18. Sumner LW, Amberg A, Barrett D, Beale MH, Beger R, Daykin CA, Fan TW-M, Fiehn O, Goodacre R, Griffin JL, Hankemeier T, Hardy N, Harnly J, Higashi R, Kopka J, Lane AN, Lindon JC, Marriott P, Nicholls AW, Reily MD, Thaden JJ, Viant MR (2007) Proposed minimum reporting standards for chemical analysis. Metabolomics 3:211–221. https://doi.org/10.1007/s11306-007-0082-2

Chapter 14

Extraction, Isolation, and Structural Elucidation of Acylated Triterpene Saponins from Asteraceae Family

Soohyun Um, Jaeyoun Lee, Mira Oh, and Seung Hyun Kim

Abstract

Saponins represent specialized (secondary) metabolites primarily sourced from plants, typically characterized by an aglycone component of triterpenoids or steroids, often referred to as sapogenin, coupled with sugar moieties. Their structural intricacy and diversity, along with their manifold pharmacological properties, have garnered significant interest among researchers. Notwithstanding this interest, the study of saponins has been encumbered by challenges in their isolation, purification, and structural characterization. Fortunately, recent advancements in mass spectrometry (MS) coupled with spectral databases have facilitated the development of efficient dereplication tools. In this context, we present a detailed protocol exemplifying the isolation and structural elucidation of saponins derived from two Asteraceae plants, *Brachyscome angustifolia* and *Tripolium pannonicum*. This protocol predominantly relies on nuclear magnetic resonance (NMR) spectroscopy, supplemented by dereplication utilizing tandem mass spectrometry (MS/MS)-based molecular networking (MN), which has accelerated the identification and characterization of novel acylated triterpene saponins.

Key words Triterpenoid saponin, *Brachyscome angustifolia*, *Tripolium pannonicum*, Molecular networking (MN), Tandem MS (MS/MS), Nuclear magnetic resonance (NMR) spectroscopy

1 Introduction

Saponins are widely distributed specialized (secondary) metabolites found in plants, primarily in dicotyledons, with recent discoveries extending their presence to monocotyledons and microorganisms, and sporadically in animals such as sea cucumbers. Despite being considered challenging in natural product chemistry due to purification difficulties and structural complexity, the chemical and pharmacological diversity of saponins has positioned them as a significant research focus. In numerous instances, saponins have emerged as key bioactive compounds in traditional medicinal plants, playing roles in expectorants such as platycodins from *Platycodon grandiflorum* [1] and hederacoside C from *Hedera helix* [2], wound healing asiaticoside from *Centella asiatica* [3], and

Young Hae Choi et al. (eds.), *Natural Product Isolation and Identification: Methods and Protocols*, Methods in Molecular Biology, vol. 2895, https://doi.org/10.1007/978-1-0716-4350-1_14,

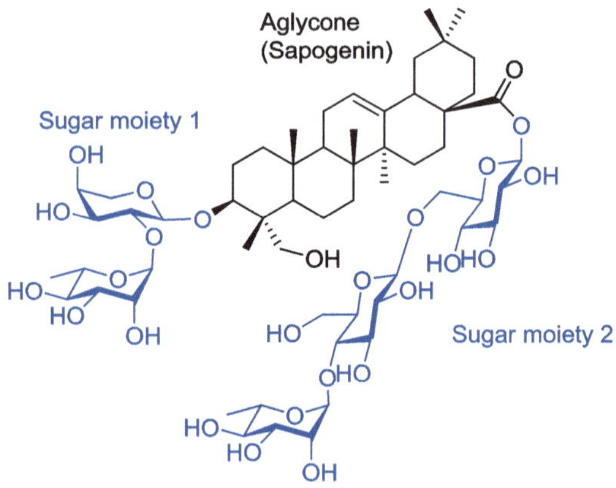

Fig. 1 Chemical structure of a typical saponin (hederacoside C from the leaves of *H. helix*)

antidiabetic saponins from *Gymnema* species [4]. Moreover, the discovery of immune-modulating ginsenosides in ginseng [5] has spurred investigations into similar immune-boosting saponins present in various natural products like *Astragalus* species [6]. Their unique property of persistent foaming upon agitation in water has further sparked interest in their industrial applications, particularly in the quest for environmentally friendly, biodegradable, and nontoxic surfactants [7].

The term "saponin" shares its etymology with soap, derived from its ability to produce lather when agitated in water. Similar to other surfactants, saponins exhibit an amphiphilic nature, possessing both lipophilic and hydrophilic groups within the same molecule (Fig. 1) [8]. The lipophilic component, termed sapogenin, comprises a polycyclic, usually tetra- or pentacyclic, triterpene or steroid skeleton derived from six isopropyl groups. Structural diversity arises from the arrangement and cyclization of these isopropyl chains, along with the positioning of functional groups such as OH, COOH, and CH_3 groups. Conversely, the hydrophilic groups of saponins consist of sugars. Based on the number of carbons to which a sugar attaches on the sapogenin, saponins can be classified as monodesmosides, bisdesmosides, tridesmosides, etc. The composition and arrangement of sugars binding to a single carbon on the sapogenin exhibit considerable variability, with sugar linkages occurring in linear or branched configurations [9]. Moreover, both sapogenin and sugars can undergo acylation, exemplified by QS-21, a prominent plant-derived vaccine adjuvant [10]. These structural characteristics render saponins highly diverse, thereby contributing to variations in their physicochemical properties and pharmacological activities.

The considerable molecular weight and structural complexity inherent in saponins pose challenges for their structural elucidation. However, recently developed LC-MS/MS analysis has emerged as a valuable tool to overcome these obstacles in combination with nuclear magnetic resonance (NMR) techniques. The substantial molecular weight of saponins facilitates their identification via MS, enabling structure-guided isolation. Furthermore, molecular networking (MN) techniques generate clusters centered on specific aglycones, providing information about the types of sugar moieties present. This recent advancement in LC-MS/MS-based MN demonstrates its efficacy as a dereplication strategy for saponin discovery.

This chapter will outline an example protocol for the separation and structural elucidation of high molecular weight saponins. Detailed experimental procedures will be provided for the isolation and identification of saponins from two Asteraceae plants, *Brachyscome angustifolia* [11] and *Tripolium pannonicum* [12], serving as illustrative examples. Each method is described independently for each plant to ensure clarity and comprehensiveness.

2 Materials

2.1 Chemicals and Solvents

MS grade high-purity solvents (>99%), such as water and AcCN, are used throughout the experimental process of the experiment to ensure reliable analysis. The use of high-purity reagents also increases analytical precision. All glassware is well cleaned, repeatedly rinsed with deionized water and HPLC grade MeOH, and oven-dried. The minimum purity required for accurate chemical identification is maintained at 99% or greater by the reference standard (*see* **Note 1**).

2.2 Instrumentation and Software

The experimenter used the following instrumentation. However, the protocol can be applied to instruments and conditions with similar specifications: an Agilent 1290 UHPLC system equipped with a 1290 Infinity binary pump (Agilent Technologies) was used for the analytical LC-MS HPLC measurements. Mass spectrometry was performed using Agilent 6530 iFunnel QTOF-MS (Agilent Technologies) equipped with an ESI source. An Agilent 1260 HPLC system with a quaternary pump, Infinity autosampler, and PDA detector was used for preparative HPLC-PDA, and an Agilent 1100 series capillary LC system coupled with Waters Micromass ZQ Mass Spectrometer was used for preparative LC-MS. NMR spectra were captured using Bruker 900-MHz, Bruker 800-MHz, and JEOL 600-MHz NMR spectrometers. Data processing was performed using MestReNova ver. 12.0.1 program and MassHunter Data Acquisition and Qualitative Analysis software, representing a comprehensive suite of advanced analytical tools for detailed compound analysis. An X-ray diffraction system (Bruker AXS) equipped with a sealed tube X-ray source (50 kV, 1 mA) and monochromatic Mo Kα radiation (λ = 0.71073 Å) was used to determine the single crystal structure.

2.3 Solvent for Extraction, Purification, and NMR Analysis of *Brachyscome angustifolia* Shoots

1. Dried plant materials: dry shoots of *B. angustifolia* (180.00 g) at 40 °C for 48 h and grind the sample to powder (17.5 g).
2. Extraction solvents: 6.0 L of HPLC grade MeOH. Extract three times with MeOH by an ultrasonicator at 40 °C for 3 h.
3. MeOH gradient for solid phase extraction (SPE): begin by sterilizing four 1-L flasks. To prepare the 40% solution, add 400 mL of HPLC grade MeOH to one flask, then fill to 1 L with deionized water. Adjust MeOH volumes to prepare 60%, 80%, and 100% solutions in separate flasks, ensuring complete mixing for a continuous gradient.
4. Chromatographic elution: to prepare the eluent solvent for the YMC RP-18 column, mix MeOH and water in a 3:1 ratio. Combine 750 mL MeOH and 250 mL deionized water in a 1-L flask and mix well.
5. Preparative HPLC and LC-MS: prepare MeOH:water mixture (3:1, v/v) for elution of SPE60 fraction on the YMC RP-18 column. To isolate the compounds of interest, elute the saponin-enriched YMC subfractions (YMC3-6) with a 42% aqueous MeOH solution using preparative LC-MS.
6. NMR solvent: dissolve the isolated compound in 600 μL of CD_3OD and then transfer to an NMR tube for analysis.

2.4 Solvent for Extraction, Purification, and NMR Analysis of *Tripolium pannonicum* Herbs

1. Dried plant materials: dry all parts of *T. pannonicum* (380.12 g) at 60 °C for 12 h, and grind the sample to powder (34.2 g).
2. Extraction solvents: 12.0 L of HPLC grade MeOH. Extract twice with MeOH by an ultrasonicator at 40 °C for 3 h.
3. MeOH gradient for SPE: prepare five 1-L graduated cylinders. Adjust MeOH volumes to make 20%, 40%, 60%, 80%, and 100% aqueous MeOH solutions in separate measuring cylinders, and transfer to solvent bottles with lids. To make the 20% MeOH solvent, add 200 mL of HPLC grade MeOH to a graduated cylinder, then fill to the 1 L mark with deionized water.
4. Fractionation with Sephadex LH-20: mix Sephadex LH-20 resin with 100% MeOH, and pack into an open column. Dilute 12.8 g of *T. pannonicum* crude extract in 10 mL of MeOH, and load to a Sephadex LH-20 column using 100% MeOH as an eluent.
5. LC-MS solvent: to make the elution solvent for the LC-MS/MS, prepare 0.1% formic acid in MS grade water as solvent A, and 0.1% of formic acid in MS grade AcCN as solvent B. Measure each solvent with a graduated cylinder and pipette, then transfer the solvent bottles, and degas with an ultrasonicator for 15 min.

6. Preparative LC-MS: to isolate the compounds, elute the fraction from Sephadex LH-20 with a 50% aqueous AcCN solution using preparative LC-MS with a J'sphere ODS-H80 (20.0 i.d. × 250 mm, 4 μm) column.
7. NMR solvent: dissolve the isolated compound in 600 μL of CD_3OD and then transfer to a 5-mm NMR tube for analysis.

3 Method

3.1 Extraction of *Brachyscome angustifolia* Shoots

In general, saponin extraction has been considered a time-consuming and laborious process. Compared to conventional methods such as reflux or Soxhlet extraction, ultrasound-assisted extraction (UAE) can be efficient, particularly when plants are extracted for isolation and purification of saponins in the laboratory [13] (*see* **Notes 2** and **3**).

1. Spread *B. angustifolia* shoots (180 g) on stainless steel trays, and dry at 40 °C for 48 h to ensure effective moisture evaporation and prevent compound degradation.
2. After drying, use a high-speed blender to grind the shoots into a fine powder to increase the surface area for optimal solvent exposure and consistent extraction.
3. Soak the powdered shoots in 2.0 L of MeOH in an ultrasonic bath for 3 h over three cycles. This process breaks down the cell walls to thoroughly extract the phytochemicals.
4. Using a rotary evaporator, remove MeOH under reduced pressure and controlled temperature at 40 °C to preserve bioactive compounds, yielding 17.5 g extract.

3.2 Extraction of *Tripolium pannonicum* Herbs

1. After washing of the soil from the roots, place the whole part of *T. pannonicum* on trays and dry at 60 °C for 12 h with a dehydrator.
2. Homogenize the dried *T. pannonicum* into a fine powder for 5 min using a high-speed blender. Homogenization is performed to increase the surface area for optimal solvent exposure and consistent extraction.
3. Transfer the fine powder from the blender to extraction bottles, and pour 12 L of MeOH into the bottles, stirring by hands to remove air bubbles as the solvent is poured in. Close the caps, place the extraction bottles in an ultrasonic bath, and sonicate for 3 h over two cycles at 40 °C.
4. Filter the extract, and concentrate in vacuo using a rotary evaporator to remove extraction solvent at 40 °C. Dissolve the dried extract in MeOH, and transfer to weighed 50-mL

vials to measure the weight of the extract. Dry the sample with SpeedVac for 6 h, and 12.8 g of crude extract was prepared for further processing.

3.3 Isolation and Purification of Saponins from the Extracts of *Brachyscome angustifolia* Shoots

1. Process the 17.5 g crude MeOH extract through an SPE C18 cartridge. Elute sequentially with 40%, 60%, 80%, and 100% MeOH gradients to obtain four fractions: SPE40, SPE60, SPE80, and SPE100.
2. Divide the SPE60 fraction on a YMC-Triart C18 semi-prep column (10.0 × 150 mm, 5 μm) using a 30–40% aqueous AcCN gradient for 50 min into eight subfractions.
3. Analyze the fractions on a preparative LC-MS with a YMC-Triart C18 column (2.0 i.d. × 150 mm, 5 μm), eluting with 42% aqueous MeOH, to isolate the compounds angustisides A–C (**1**–**3**): **1** (2.8 mg, 6.7 min), **2** (1.2 mg, 5.69 min), and **3** (2.4 mg, 4.76 min).

3.4 Isolation and Purification of Saponins of *Tripolium pannonicum* Herbs

To isolate and purify saponins, a highly polarized fraction of the extract is generally prepared to remove other undesired compounds and yield a saponin-enriched fraction. Popular methods for obtaining a highly polarized fraction include liquid–liquid extraction (LLE), SPE, and HP-20 or XAD resins. All three methods have their pros and cons. Therefore, it is possible to use a combination of the three methods. Another option is to apply SPE to the desired fraction after LLE or HP-20 open column chromatography (*see* **Notes 4** and **5**).

1. Process the 12.8 g crude MeOH extract through an Extract-Clean C18 SPE cartridge (1000 mg/75 mL). Sequentially elute twice with 40 mL of 20%, 40%, 60%, 80%, and 100% MeOH gradients. Then divide into five fractions: SPE20, SPE40, SPE60, SPE80, and SPE100.
2. Mix Sephadex LH-20 resin with MeOH, and pack into an open column. Dilute the SPE60 fraction of *T. pannonicum* in 10 mL of MeOH, and load to the Sephadex LH-20 column using 100% MeOH as the eluent. Each fraction is eluted and obtained for 15 min, resulting in eight fractions over 120 min: Fraction 1 (0–15 min); 2 (15–30 min); 3 (30–45 min); 4 (45–60 min); 5 (60–75 min); 6 (75–90 min); 7 (90–105 min); and 8 (105–120 min).
3. Purify the fraction using a preparative LC-MS with a J'sphere ODS-H80 (20.0 i.d. × 250 mm, 4 μm) eluting with 50% aqueous AcCN, to isolate the compounds pannosides A-E (**4**–**8**): **4** (3.2 mg, 21.2 min), **5** (2.6 mg, 26.9 min), **6** (2.1 mg, 35.7 min), **7** (1.8 mg, 18.4 min), and **8** (1.4 mg, 23.5 min).

3.5 Structural Elucidation of Angustiside C (3) from Brachyscome angustifolia Shoots

Saponins are often difficult to elucidate structures due to their large molecular weight and numerous peaks in the NMR spectrum, especially in the proton, which tend to overlap significantly. Therefore, it is important to utilize the features of saponins, which are composed of two distinct parts: aglycone and sugar. Important information about the type of aglycone and the type and number of sugars in a saponin can be obtained by 1D NMR. To determine how they are connected, 2D NMR is necessary. Finally, the determination of the three-dimensional structure is also critical (*see* **Note 6**).

1. Molecular formula confirmation: utilize HR-ESI-MS to determine the molecular formula of angustiside C (**3**) as $C_{95}H_{154}O_{51}$, with an m/z of 2109.9483 $[M - H]^-$ (calculated for $C_{95}H_{153}O_{51}$, 2109.9384).
2. Analysis of sapogenin moiety: compare the ^1H and ^{13}C NMR data and 2D NMR spectra to establish the chemical composition of **3** (Tables 1 and 2).
   - ^1H NMR (Fig. 2): identify seven methyl signals at chemical shifts of δ_H 1.05, 1.01, 0.98, 0.97, 0.96, 0.89, and 0.83. Observe a proton signal with an olefinic chemical shift at δ_H 5.17. Observe two signals indicating the presence of oxygenated methine groups at chemical shifts of δ_H 4.32 and 3.19. Note three methine signals at chemical shifts of δ_H 2.33, 1.36, and 0.74. Additionally, detect nine methylene signals.

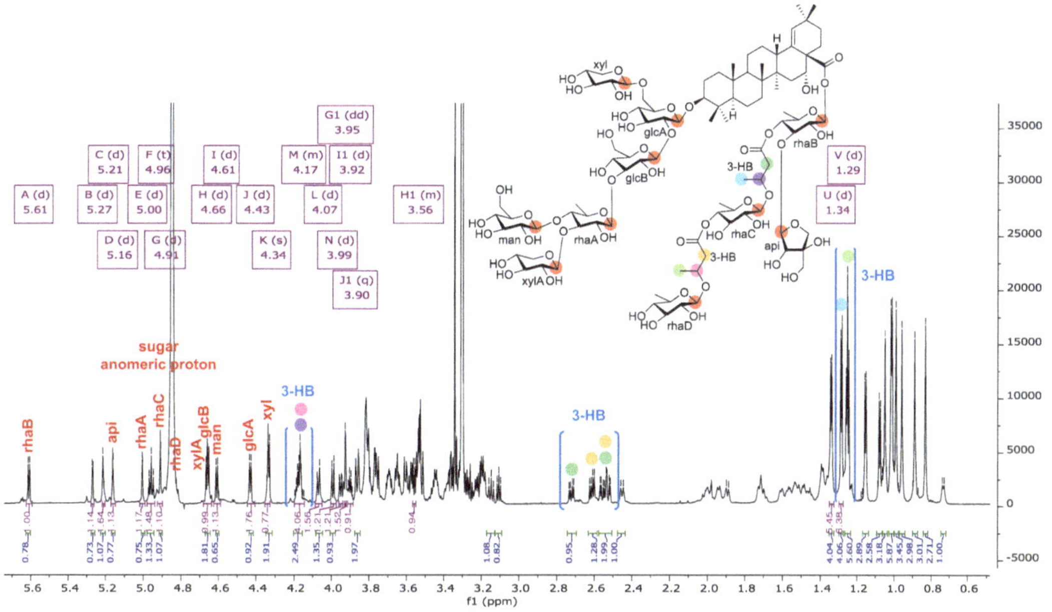

Fig. 2 ^1H NMR spectrum of angustiside C (**3**) (900 MHz, CD_3OD)

Table 1
^{1}H and ^{13}C NMR spectroscopic data of the aglycone of angustiside C (3)

No.	Type	δ_H[a]	Mult (*J* in Hz)	δ_C[b]
1	CH_2	0.99	m	40.4
		1.71	m	
2	CH_2	1.70	m	27.4
		1.94	m	
3	CH	3.18[c]	m	91.5
4	C			40.7
5	CH	0.73	m	57.5
6	CH_2	1.39	m	19.4
		1.53	m	
7	CH_2	1.39	m	35.9
		1.48	m	
8	C			42.5
9	CH	1.34	br. d (6.0)	51.7
10	C			38.2
11	CH_2	1.24	m	22.5
		1.60	m	
12	CH_2	1.37	m	26.6
		1.51	m	
13	CH	2.44	dd (12.0, 2.0)	42.2
14	C			43.7
15	CH_2	1.35	m	37.7
		1.89	dd (15.0, 3.5)	
16	CH	4.33	m	73.8
17	C			54.2
18	C			136.5
19	CH	5.21	s	135.7
20	C			32.7
21	CH_2	1.45	m	35.2
		1.56	m	
22	CH2	1.94	m	27.5
		2.00	m	

(continued)

Table 1 (continued)

No.	Type	δ_H[a]	Mult (*J* in Hz)	δ_C[b]
23	CH_3	1.05	s	28.4
24	CH_3	0.83	s	16.9
25	CH_3	0.89	s	17.5
26	CH_3	1.02	s	17.5
27	CH_3	0.96	s	17.8
28	C			176.7
29	CH_3	0.99	s	30.4
30	CH_3	1.01	s	30.7

Measured in CD_3OD
[a]900 MHz
[b]225 MHz
[c]Overlapped signal

- ^{13}C NMR: observe a total of 30 carbon signals, including eight quaternary carbons (one of which is a carbonyl), six methines (two of which are oxygenated), nine methylenes, and seven methyl groups.
- Interpret the 2D NMR spectra for compound **3** to establish the sapogenin structure (Fig. 3). Utilize the ^{1}H–^{1}H COSY spectrum to determine the spin-coupling relationships of H_2-1/H_2-2/H-3, H-5/H_2-6/H_2-7, H-9/H_2-11/H_2-12/H-13, H_2-15/H-16, and H_2-21/H_2-22. In addition, use the HMBC spectrum to determine long-range heteronuclear couplings, such as H_3-23 and H_3-24 with C-3/C-4/C-5 and others. Determine the presence of an olefinic quaternary carbon at δ_C 137.4 (C-18) and a methine carbon at δ_C 135.1 (C-19) on the E-ring of the oleanane skeleton. Identify the chemical structure that suggests an olean-18-ene skeleton with a hydroxyl group at C-16.

3. Analysis of sugar moiety: by examining the ^{1}H NMR spectrum, identify 10 anomeric proton signals within the chemical shift range of δ_H 4.33–5.61. These signals correspond to 10 sugar units attached to the aglycone moiety in compound **3**. Using COSY and HMBC data, elucidate the assignments of monosaccharide residues within compound **3** bearing two β-D-glucose units (glcA and glcB), one β-D-xylose (xyl), one acetylated β-D-xylose (xylB), one β-D-mannose (man), one α-L-rhamnose (rhaA), two β-L-rhamnose units (rhaB and rhaC), one β-L-xylose (rhaD), and one α-D-apiose (api) unit.

Table 2
^{1}H and ^{13}C NMR spectroscopic data of the sugar moieties of angustiside C (3)

No.	Type	3 δ_H[a]	Mult (*J* in Hz)	δ_C[b]
glcA				
1	CH	4.43	d (7.5)	105.4
2	CH	3.56	m	81.8
3	CH	3.53	m	78.4
4	CH	3.30	m	71.7
5	CH	3.44	m	76.8
6	CH_2	3.69	m	69.8
		4.06	dd (11.5, 2.0)	
glcB				
1	CH	4.65	d (8.0)	104.6
2	CH	3.20	m	76.4
3	CH	3.32–3.36[c]	m	78.0
4	CH	3.19	m	72.0
5	CH	3.24	m	78.5
6	CH_2	3.59	m	63.2
		3.80	m	
xyl				
1	CH	4.33	d (6.5)	105.2
2	CH	3.58	m	72.5
3	CH	3.53	m	74.3
4	CH	3.80	m	69.5
5	CH_2	3.51	m	66.6
		3.85	m	
rhaA				
1	CH	5.00	d (2.0)	102.5
2	CH	4.15	m	71.4
3	CH	3.94	dd (9.5, 3.0)	82.9
4	CH	3.64	m	79.1
5	CH	3.76	m	69.3

(continued)

Table 2
(continued)

		3		
No.	**Type**	δ_H[a]	**Mult (*J* in Hz)**	δ_C[b]
6	CH_3	1.26	d (7.0)	18.8
xylA				
1	CH	4.66	d (8.0)	105.3
2	CH	3.11	dd (9.5, 8.0)	75.7
3	CH	3.38	m	77.6
4	CH	3.32	m	71.2
5	CH_2	3.14	t (11.0)	67.1
		3.82	m	
31	CH_3			
32	C			
man				
1	CH	4.61	d (8.0)	104.9
2	CH	3.27	m	75.5
3	CH	3.32–3.36[c]	m	78.0
4	CH	3.81	m	72.7
5	CH	3.58	m	72.5
6	CH_2	3.69	m	62.4
		3.86	m	
rhaB				
1	CH	5.61	d (8.0)	95.1
2	CH	3.32–3.36[c]	m	77.9
3	CH	3.92	m	80.5
4	CH	5.27	d (3.5)	74.6
5	CH	3.89	m	71.0
6	CH_3	1.08	d (7.5)	17.0
3-HB				
33	C			172.2
34	CH_2	2.56	dd (16.0, 6.0)	42.3
		2.71	dd (16.0, 6.5)	
35	CH	4.17	m	72.8
36	CH_3	1.34	d (6.0)	22.1

(continued)

Table 2
(continued)

		3		
No.	**Type**	δ_H[a]	**Mult (*J* in Hz)**	δ_C[b]
rhaC				
1	CH	4.91	d (2.0)	101.5
2	CH	3.80	m	70.4
3	CH	3.44	m	71.5
4	CH	4.96	t (10.0)	75.7
5	CH	3.80	m	68.1
6	CH_3	1.15	d (6.0)	18.0
3-HB				
37	C			172.7
38	CH_2	2.53	dd (15.5, 5.0)	42.9
		2.61	dd (15.5, 7.5)	
39	CH	4.17	m	73.1
40	CH_3	1.29	d (6.0)	22.0
rhaD				
1	CH	4.84	d (7.5)	102.1
2	CH	3.76	m	72.5
3	CH	3.58	m	72.4
4	CH	3.36	m	74.1
5	CH	3.65	m	70.2
6	CH_3	1.25	d (6.5)	18.1
api				
1	CH	5.16	d (2.5)	112.8
2	CH	3.92	m	78.8
3	C			80.8
4	CH_2	3.75	m	75.5
		3.99	d (10.0)	
5	CH_2	3.52	m	65.4
		3.83	m	

Measured in CD_3OD
[a]900 MHz
[b]225 MHz
[c]Overlapped signal

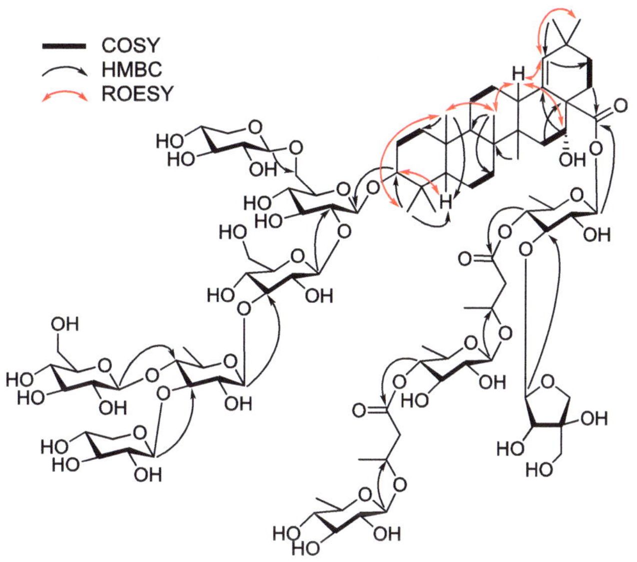

Fig. 3 Key ^1H-^1H-correlated spectroscopy (COSY), heteronuclear multiple bond coherence (HMBC), and rotating-frame nuclear Overhauser spectroscopy (ROESY) correlations of angustiside C (**3**)

4. Detection of 3-hydroxybutyrate (3-HB) and linear acyl chain sequence: compound **3** contains two 3-HB units, each characterized by the presence of a carbonyl, a methine, a methylene, and a methyl group. Investigate the linear acyl chain sequence of **3**, which is composed of fatty acid monomers and monosaccharide stacks (sugar–3-HB–sugar–3-HB–sugar).

3.6 Acid Hydrolysis of Angustiside C (3) from Brachyscome angustifolia Shoots

1. To obtain sapogenin, sugars, and 3-HB residues, hydrolyze 1 mg of compound **3** in 1 mL of 6 N HCl at 115 °C for 12 h. Then rapidly cool the reaction vial by placing it into ice water for 3 min.
2. Evaporate the reaction solvent in vacuo, and completely remove residual HCl by adding 0.5 mL of water and removing the solvent three times.
3. Dry and lyophilize the hydrolysate for 15 h.

3.7 Relative and Absolute Configurations of Angustiside C (3) from Brachyscome angustifolia Shoots

3.7.1 Sapogenin Moiety of Angustiside C (3)

1. Deduce the relative configuration of the aglycone moiety (angustic acid) of compound **3** by analyzing rotating-frame nuclear Overhauser spectroscopy (ROESY) correlations. Specifically, examine the correlations between H-24 (δ_H 0.83) and H-25 (δ_H 0.89), H-25 and H-26 (δ_H 1.01), H-13 (δ_H 2.33) and H-26, H-13 and H-16 (δ_H 4.32), H-13 and H-19 (δ_H 5.17), and H-19 and H-30 (δ_H 0.97). These correlations indicate the *R**-configuration of the hydroxyl group at C-3 and the *S**-configuration of the hydroxyl group at C-16.

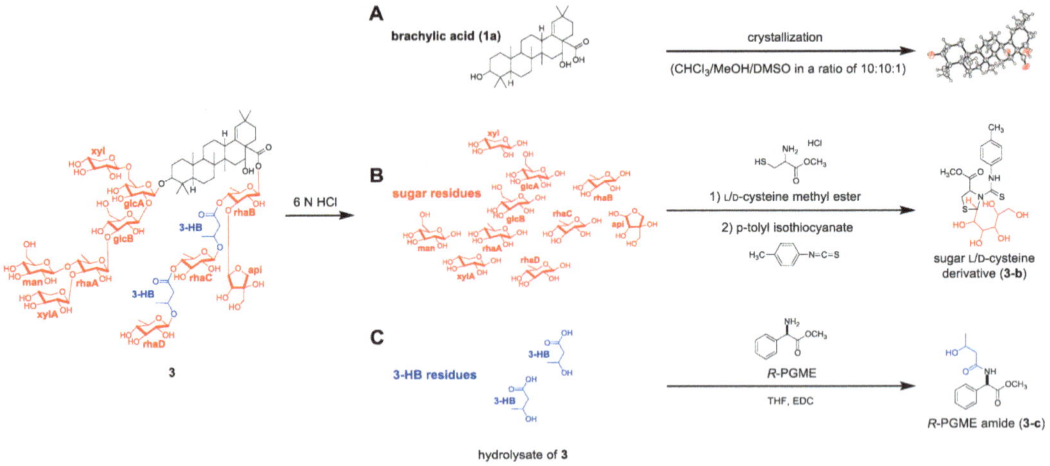

Fig. 4 Scheme of the chemical reactions with sapogenin, sugar residues, and 3-HB residues in angustiside C (**3**)

2. To determine its absolute configuration, dissolve the dried and lyophilized acid hydrolysate of **3** in MeOH and then purify by the preparative HPLC system using a YMC-Triart C18 column (10.0 i.d. × 150 mm, 5 μm), eluting with 70% aqueous AcCN. Set the flow rate at 3 mL/min. Collect the peak eluted at approximately 9.3 min.
3. To prepare a single crystal of the sapogenin, use the slow evaporation method. Dissolve 3.8 mg of purified sapogenin in a mixture of three solvents ($CHCl_3$-MeOH-DMSO = 10: 10:1), and evaporate slowly at room temperature until crystal grows.
4. Determine the single crystal structure of the sapogenin at 223 K using an X-ray diffraction system equipped with a sealed tube X-ray source (50 kV, 1 mA), and monochromatic Mo Kα radiation ($\lambda = 0.71073$ Å) corresponding to graphite (Fig. 4).

3.7.2 Sugar Moiety of Angustiside C (3)

1. Analyze the relative configurations of 10 sugar units focusing on the *J*-value coupling constants. For the two β-D-glucose units (glcA and glcB), note the large *J* coupling constants (7.5 for both), indicating β-configurations. The β-D-xylose (xyl) and acetylated β-D-xylose (xylB) configurations can be determined by COSY and HMBC analyses, which also confirm the β-configuration for the β-D-mannose (man). The α-L-rhamnose (rhaA) displays a small *J* coupling constant of 2.0 (δ_H 5.06), suggesting an α-configuration, while the two β-L-rhamnose units (rhaB and rhaC) and β-L-xylose (rhaD) show large *J* coupling constants, suggesting β-configurations. The α-D-apiose (api) unit has a small *J* coupling constant of 2.5 (δ_H 5.16), confirming its α-configuration.

2. After drying, dissolve the acid hydrolysate containing sugar residues in 0.4 mL of pyridine (99.5%) and divide equally into two 5-mL vials. Treat each vial with 5 mg of L- or D-cysteine (99.0%) methyl ester hydrochloride, and heat at 60 °C for 1 h. Then, add 2 mg of *p*-tolyl isothiocyanate (97.0%) to each vial and heat at 60 °C for 1 h (Fig. 4).
3. React the authentic sugar standards, D-glucose, D-xylose, L-rhamnose, and D-mannose, with each of L- and D-cysteine methyl ester, respectively, and then with p-tolyl isothiocyanate using the same method described above.
4. Determine the absolute configuration of sugars by comparing the retention time of the sugar derivatives of **3** with the retention times of the derivatives of authentic standard sugars. Analyze the reaction mixtures by HR-ESI-MS using a YMC-Hydrosphere C18 column (4.6 × 250 mm, 2 μm), eluting with 25% aqueous AcCN at a flow rate of 0.8 mL/min.

3.7.3 3-Hydroxybutyrate of Angustiside C (3)

1. Dissolve the dried acid hydrolysate containing 3-HB residues as described in Subheading 3.5 above in 2 mL of tetrahydrofuran (THF), and divide equally into two 15-mL vials.
2. Add 10 mg each of *R*-PGME, *S*-PGME, EDC, and 4-DMAP. Stir the reaction mixtures at room temperature for 4 h (*see* **Note 1**).
3. Determine the absolute configuration of 3-HB by comparing the retention time of 3-HB derivatives with authentic standards of *S*-3-HB and *R*-3-HB. Dissolve 30 mg each of *R*- and *S*-3-HB in 2 mL of THF.
4. Treat with 10 mg each of *S*- or *R*-PGME, EDC, and 4-DMAP (*see* **Note 1**). Stir the reaction mixtures for 4 h at room temperature, and confirm by HR-ESI-MS using a YMC-Triart C18 column (4.6 i.d. × 250 mm, 2 μm), eluting with the following gradient solvent: 10-50% aqueous AcCN (0-40 min) at the flow rate of 0.8 mL/min (Figs. 4 and 5).

3.8 Structural Elucidation of Pannoside A (4) from *Tripolium pannonicum* Herbs

1. Molecular formula confirmation: use HR-ESI-MS to determine the molecular formula of pannoside A (**4**) as $C_{60}H_{90}O_{26}$, with an *m/z* of 1249.5588 $[M + Na]^+$ (calculated for $C_{60}H_{90}O_{26}Na$, 1249.5623).
2. Analysis of sapogenin moiety: compare the ^1H and ^{13}C NMR data and 2D NMR spectra to determine the chemical composition of **1** (Tables 3 and 4).
   - ^1H NMR: identify six methyl signals at chemical shifts of δ_H 1.38, 1.29, 1.15, 0.94, 0.91, and 0.83. Observe nine proton signals indicating the presence of methylene groups at chemical shifts between δ_H 2.10 and 1.10. In addition, three

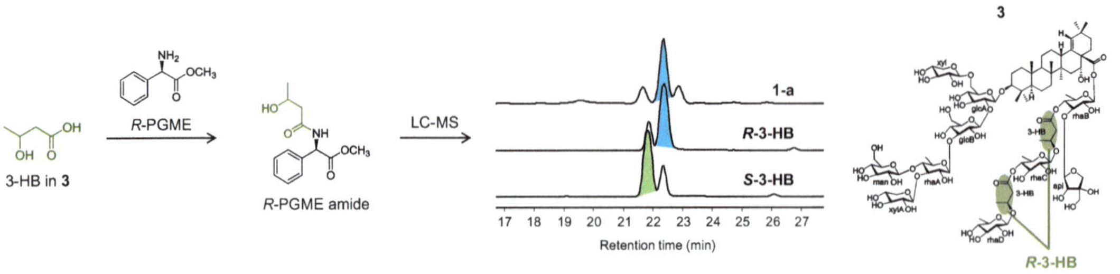

Fig. 5 HPLC-MS chromatogram of *S*- and *R*-PGME amide of 3-HB in angustiside C (**3**)

methine proton signals were detected at δ_H 2.82, 1.60, and 1.59, with two oxygenated methine proton signals at δ_H 4.33 and 4.07, and one proton signal with an olefinic chemical shift at δ_H 5.28.

- ^{13}C NMR: observe a total of 30 carbon signals, including six methyl groups, nine methylenes, six methine groups (including two oxygenated and one olefinic), and nine quaternary carbons (including one carbonyl).
- Interpret the 2D NMR spectra for pannoside A (**4**) to establish the sapogenin structure. Use the ^{1}H–^{1}H COSY spectrum to determine the spin coupling between H_2-1/H-2/H-3, H-5/H_2-6/H_2-7, H_2-11/H-12, H_2-15/H_2-16, H-18/H_2-19, and H_2-21/H_2-22. In addition, interpret the HMBC spectrum such as from H_3-24 to C-3, C-4, and C-23, and from H_3-25 to C-1 and C-10, indicating heteronuclear couplings of ring A; from H_3-26 to C-7, C-8, and C-9, and from H_2-11 to C-9, C-12, and C-13, indicating couplings of rings B and C; from H_3-27 to C-8, C-13, C-14, and C-15 indicating couplings of ring D; and from H_3-29 and H_3-30 to C-20, from H_2-19 to C-17 and C-18, and from H_2-22 to carbonyl C-28, indicating couplings of ring D. Identify the chemical structure indicating an olean-13-ene skeleton with a hydroxyl group at C-2 and C-3 (Fig. 6).

3. Analysis of sugar moiety: by examining the ^{1}H NMR spectrum, identify three anomeric proton signals of chemical shift at δ_H 5.45, 5.38, and 4.36. These signals correspond to three sugar units attached to the aglycone moiety in compound **4**. Using extensive COSY and HMBC data, elucidate the assignments of monosaccharide residues within compound **4** bearing one β-D-glucuronic acid units (gluA), one β-D-rhamnose (rha), and one α-L-rhamnose (rhaA) unit.
4. Detection of 3-HB and linear acyl chain sequence: compound **4** contains two 3-hydroxybutyrate (3-HB) units, each characterized by the presence of a carbonyl, a methine, a methylene, and a methyl group.

Table 3
^{1}H and ^{13}C NMR spectroscopic data of aglycone of pannoside A (4)

Position		1		
		δ_H	Mult (*J* in Hz)	δ_C
1	CH_2	1.27	d (6.5)	45.1
		2.10	m	
2	CH	4.33	m	70.8
3	CH	4.07	m	86.7
4	C			53.6
5	CH	1.60	m	53.2
6	CH_2	1.10	m	21.8
		1.66	m	
7	CH_2	1.33	dd (6.5, 2.0)	33.7
		1.45	m	
8	C			41.1
9	CH	1.59	m	49.7
10	C			37.6
11	CH_2	1.94	m	24.8
		2.03	m	
12	CH	5.27	t (4.0)	123.7
13	C			144.9
14	C			43.3
15	CH_2	1.21	m	29.2
		1.60	m	
16	CH_2	1.61	m	24.4
		2.07	m	
17	C			48.1
18	CH	2.82	m	42.3
19	CH_2	1.16	m	47.5
		1.73	m	
20	C			31.7
21	CH_2	1.25	m	35.1
		1.41	m	

(continued)

Table 3 (continued)

Position		1		
		δ_H	Mult (*J* in Hz)	δ_C
22	CH_2	1.61	m	33.3
		1.78	m	
23	C			182.0
24	CH_3	1.38	s	13.9
25	CH_3	1.29	s	17.4
26	CH_3	0.83	s	17.9
27	CH_3	1.15	s	26.3
28	C			178.2
29	CH_3	0.94	s	24.2
30	CH_3	0.91	s	33.6

Measured in CD_3OD
[a]200 MHz (^{13}C NMR)
[b]800 MHz (^{1}H NMR)
[c]Overlapped signal

3.9 Relative and Absolute Configurations of Pannoside A (4) from *Tripolium pannonicum* Herbs

3.9.1 Sapogenin Moiety of Pannoside A (4)

1. Deduce the relative configuration of the aglycone moiety (medicagenic acid) of compound **4** by analyzing ROESY correlations.
2. Examine the correlations of H-3 with H-5, H-5 with H-9, and H-9 with H_3-27; and the correlations of H_3-25 with H_3-26, H_3-26 with H-18, and H_3-25 with H_3-24. These correlations indicate the *R**-configuration of the hydroxyl group at C-3 and the *S**-configuration of the hydroxyl group at C-2 (Fig. 6).

3.9.2 Sugar Moiety of Pannoside A (4)

Analyze the relative configurations of 10 sugar units focusing on coupling constants. Confirm one α-configuration from a small *J*-value (2.0 for H-1_{rhaA} δ_H 5.38), and two β-configurations from large coupling constants of anomeric protons (7.5 for H-1_{gluA} δ_H 4.36, and 8.0 for H-1_{rha} δ_H 5.45).

3.9.3 3-Hydroxybutyrate of Pannoside A (4)

1. Hydrolyze pannoside A (**4**) in 6 N HCl at 115 °C for 12 h, cool rapidly, evaporate to remove HCl, and freeze dry to obtain a dried hydrolysate for further analysis.
2. Dissolve the dried hydrolysate containing 3-HB residues in 2 mL of THF. Add 10 mg each of *S*-PGME or *R*-PGME, EDC, and 4-DMAP (*see* **Note 1**). Stir the reaction mixtures for 4 h at room temperature.

Table 4
^{1}H and ^{13}C NMR spectroscopic data of the sugar and 3-HB moieties of pannoside A (4)

		1		
Position		δ_H	**Mult (*J* in Hz)**	δ_C
gluA				
1	CH	4.36	d (7.5)	105.1
2	CH	3.26	m	75.1
3	CH	3.37	m	77.6
4	CH	3.58	m	76.4
5	CH	3.41	m	73.8
6	C			176.7
rha				
1	CH	5.45	d (8.0)	95.1
2	CH	3.80	dd (9.5, 8.0)	75.4
3	CH	3.96	dd (9.5, 3.5)	74.3
4	CH	5.10	dd (3.5, 1.0)	75.5
5	CH	3.87	dd (6.5, 1.0)	71.1
6	CH_3	1.08	d (6.5)	16.7
rhaA				
1	CH	5.38	d (2.0)	99.2
2	CH	5.33	dd (3.5, 2.0)	71.4
3	CH	4.98	dd (10.5, 3.5)	73.5
4	CH	3.50	m	71.1
5	CH	3.91	m	70.5
6	CH_3	1.29	d (6.0)	18.3
rhaA-1′	C			171.8
rhaA-2′	CH_3	2.07	s	20.8
rhaA-3′	C			172.5
rhaA-4′	CH_3	1.99	s	21.0
3-HB				
1′	C			171.9
2′	CH_2	2.72	dd (16.0, 6.0)	41.5
		2.82	m	

(continued)

Table 4
(continued)

Position		1		
		δ_H	Mult (*J* in Hz)	δ_C
3′	CH	5.31	m	68.9
4′	CH_3	1.34	d (6.5)	20.1
1″	C			172.6
2″	CH_2	2.42[a]	dd (6.5, 4.0)	45.2
3″	CH	4.14	qdd (6.5, 6.0)	65.7
4″	CH_3	1.19	d (6.0)	23.4
1‴	C			
2‴	CH_2			
3‴	CH			
4‴	CH_3			
1⁗	C			
2⁗	CH_2			
3⁗	CH			
4⁗	CH_3			

Measured in CD_3OD
[a]200 MHz (^{13}C NMR)
[a]800 MHz (^{1}H NMR)
[c]Overlapped signal

3. Determine the absolute configuration of 3-HB by comparing the retention time of 3-HB derivatives with authentic standards of *S*-3-HB and *R*-3-HB. Dissolve 30 mg each of *R*- and *S*-3-HB in 2 mL of THF. Treat with 10 mg each of *S*- or *R*-PGME, EDC, and 4-DMAP (*see* **Note 1**). Stir the reaction mixtures for 4 h at room temperature, and confirm by HR-ESI-MS with a YMC-Triart C18 column (2.0 × 150 mm, 5 μm), eluting with the following gradient solvent: 10-50% aqueous AcCN (0-40 min) at the flow rate of 0.5 mL/min.

3.10 LC-MS/MS Analysis of the Extracts of *Tripolium pannonicum* Herbs

1. To prepare LC-MS samples, reconstitute the dried extract with MeOH to a concentration of 250 μg/mL. Filter the solution through a 0.45-μm syringe filter, before transferring the sample to the LC-MS vial.
2. Replace the LC column with a YMC-Triart C18 column (2.0 × 150 mm, 5 μm), and adjust the column temperature at 30 °C.

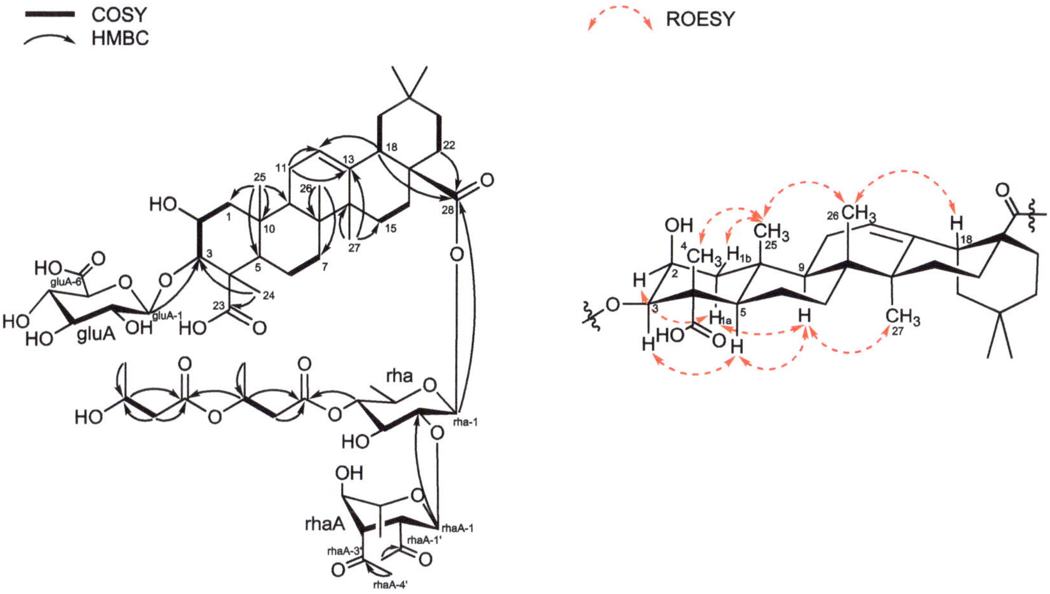

Fig. 6 Key ^{1}H-^{1}H-correlated spectroscopy (COSY), heteronuclear multiple bond coherence (HMBC), and rotating-frame nuclear Overhauser effect correlation spectroscopy (ROESY) correlations of pannoside A (**4**)

3. Purge solvents A and B at 3 mL/min for 10 min to remove gas from the solvent lines and pump.
4. Calibrate and tune the TOF mass using the reference tune mix, then set the instrument mode to high resolution (4 GHz) and mass range of 100–3200 *m/z*.
5. Set the ion source to Dual AJS ESI. Run the MS experiment with the following parameters: capillary voltage +3.5 kV (positive ion mode); nebulizer pressure 35 psig; capillary voltage 3500 V; fragmentor voltage 175 V; skimmer voltage 65 V; OCT 1 RF Vpp 750 V; drying gas temperature 300 °C; drying gas flow rate 8 L/min; sheath gas temperature 350 °C; sheath gas flow rate 11 L/min.
6. Set up an LC analysis method as follows: A: 0.1% formic acid in water; B: 0.1% formic acid in AcCN; a gradient system: 0–20 min (10–100% B), 20–25 min (100% B), and 25–30 min (10% B); flow rate: 0.3 mL/min; and injection volume: 20 μL.

3.11 Molecular Networking

Natural product isolation and structural elucidation are time-consuming and laborious procedures. Therefore, when isolating and purifying a single compound from a natural product, dereplication becomes an essential step to pre-check reported compounds in advance and set up isolation targets. By comparing experimental spectral data with databases, dereplication allows rapid identification of known metabolites [14]. LC-MS/MS combined with MN is the most prevalently used technique for dereplication. The MN method generates networks of clustered nodes by calculating similarities between compounds using tandem mass spectrometry (MS/MS) fragmentation [15, 16].

First described by Dorrestein's group in 2012, Global Natural Product Social Molecular Networking (GNPS) provides users with an interactive online platform for creating molecular networks and sharing, analyzing, and comparing tandem MS data of small molecules (https://gnps.ucsd.edu/ProteoSAFe/static/gnps-splash.jsp), allowing natural product chemists to obtain information about the chemical features of complex mixtures in advance [17] (*see* **Note 7**).

All the parameters described below are for pannoside analogs in *T. pannonicum*. These parameters may need to be fine-tuned depending on the characteristics of the target compounds.

1. Use the software MassHunter Qualitative Analysis B.06.00 for data acquisition and qualitative analysis to process LC-MS/MS data.
2. Convert the data to the mzML file format using MS-Convert. Then, upload mzML file to the WinSCP account linked to GNPS.
3. Access the GNPS platform (https://gnps.ucsd.edu/ProteoSAFe/static/gnps-splash.jsp), and select the "Create Molecular Network" tab. Process the molecular network analysis with the following conditions: precursor ion mass tolerance, 2.0 Da; fragment ion mass tolerance, 0.5 Da; cosine score for building molecular networks, 0.7; minimum number of matched peaks, 6; and minimum cluster size, 2.
4. After analyzing the molecular network-based tandem mass, visualize the data using Cytoscape software version 3.9.1. Open the GraphML file in Cytoscape, and set the node label to "parent mass," then check for clustering of nodes with similar tandem mass patterns. Specify one of these clusters as the target, which is likely to be a saponin node (Fig. 7).

4 Notes

1. The reagents used in the original method are as follows: 6 N hydrochloric acid (HCl), tetrahydrofuran (THF), *R*-, *S*-phenylglycine methyl ester (*R*-, *S*-PGME) 98.0%, ethyl-(*N′*,*N′* dimethylamino)-propylcarbodiimide hydrochloride (EDC), 4-dimethylaminopyridine (4-DMAP), authentic standards of *S*-3-hydroxybutyrate (*S*-3-HB), and *R*-3-hydroxybutyrate (*R*-3-HB).

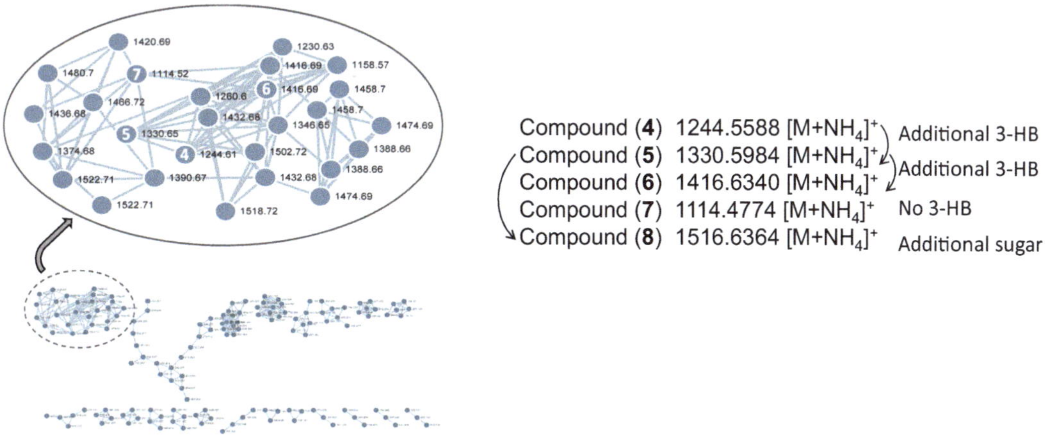

Fig. 7 Global Natural Product Social Molecular Networking (GNPS) cluster assigned to pannoside analogs in *Tripolium pannonicum*

2. Ultrasound disrupts the cell structure and facilitates the release of intracellular molecules into the solvent due to the mechanical effect of acoustic cavitation [18]. In this way, high yields of saponins can be extracted using less solvent in a relatively shorter time than conventional extraction methods [7].
3. Although saponins are quite polar, organic solvents such as MeOH or EtOH are often used since primary metabolites such as sugars and amino acids can be extracted more efficiently in water. Considering that MeOH can cause methylation of certain functional groups, it may be necessary to check that the isolated compound is not an artifact when elucidating its structure, particularly for a methoxylated compound.
4. LLE is one of the most popular methods for fractionating total extracts of natural products. It typically involves suspending the total extract in water and then separating the compounds according to their polarity using sequentially more polar solvents, starting with less polar solvents. Saponins are mainly isolated from the remaining water fraction after fractionation with organic solvents such as n-hexane, $CHCl_3$, and ethyl acetate (EtOAc) to remove the less polar compounds. It is also possible to extract n-butanol (n-BuOH) from the above residual fraction. However, when targeting highly polar saponins with many sugars, four or more linked sugars, n-BuOH may not be able to completely extract them from the residual water fraction [13]. In this case, the water layer remaining after the organic solvent fractionation can be applied to multiple chromatograms [19]. Although LLE is difficult to perform a precise fractionation, it can be used with large amounts of the sample and no sample loss due to adsorption.

5. SPE is also a popular method for obtaining saponin-rich fractions. By loading the extract onto an SPE cartridge, generally packed with C18 resin, and sequentially eluting with increasing proportions of organic solvent, usually MeOH, from water, it is possible to fractionate the extract with a finer resolution than LLE [20]. SPE can also significantly remove polysaccharides, which interfere with the separation of highly polar substances when eluting with water, to a considerable extent, and pigments by adsorption to the resin. In SPE, the steps are conditioning with MeOH, equilibration with water, loading of the sample dissolved in water, washing with water, solvent elution with increasing proportions of organic solvent, and finally washing with organic solvent, typically using MeOH. Usually, 5–10 cartridge volumes is considered sufficient for conditioning or equilibration. The volume of solvent to be eluted also depends on the volume of the cartridge. In general, most manufacturers provide a guide for selecting the SPE cartridge size and elution solvent volume.
6. Saponin is characterized by distinct methyl signals, superimposed methylene signals, and signals from the anomeric protons of sugars and oxygenated methine in ^{1}H NMR. The signals for methyl groups (δ_H 0.7 to 1.3) provide the most useful information for deducing the triterpene skeleton. In oleanane, the methyl groups are all observed as singlet, while in ursane, lanostane, and cycloartane, they appear as doublet ($J = \sim 6.50$ Hz). In hopane and lupane, isopropyl or isoprenyl groups are observed. The anomeric protons show typical chemical shifts within δ_H 4.0–5.5 in the ^{1}H NMR spectrum, and the coupling constants are well distinguished by sugar types and their glycosidic linkages [21].
7. Saponin is relatively easy to identify its cluster in MN. It has a large molecular weight compared to other clusters, generally 500 ~ 1200 amu, and forms a network according to the difference of sugars centered on a specific sapogenin, and sapogenin and its glycosides with a similar structure to this sapogenin form a close network. In *T. pannonicum*, clusters of aglycone congeners, sugar congeners, and similar structures were obtained from MS. By finding an undescribed one of these clusters and conducting targeted isolation using MS as a detector, new triterpene saponins were isolated, and the MS fragmentation pattern was also utilized as useful information on the number, type, and sequence of sugars during structural elucidation [12].

References

1. Ryu J, Lee HJ, Park SH, Kim J, Lee D, Lee SK, Kim YS, Hong JH, Seok JH, Lee CJ (2014) Effects of the root of *Platycodon grandiflorum* on airway mucin hypersecretion in vivo and platycodin D (3) and deapi-platycodin on production and secretion of airway mucin in vitro. Phytomedicine 21:529–533
2. Hong G, Kim YI, Park SJ, Lee SY, Kim JW, Yoon SH, Lee KS, Byun MK, Kim HR, Chung J (2021) Effects of a mixture of ivy leaf extract and *Coptidis* rhizome on patients with chronic bronchitis and bronchiectasis. Int J Environ Res Public Health 18:4024
3. Shukla A, Rasik AM, Jain GK, Shankar R, Kulshrestha DK, Dhawan BN (1999) In vitro and in vivo wound healing activity of asiaticoside isolated from *Centella asiatica*. J Ethnopharmacol 65:1–11
4. Kahksha, Alam O, Naaz S, Sharma V, Manaithiya A, Khan J, Alam A (2022) Recent developments made in the assessment of the antidiabetic potential of *Gymnema* species – from 2016 to 2020. J Ethnopharmacol 286: 114908
5. Luong Huynh D, Nguyen NH, Nguyen CT (2021) Pharmacological properties of ginsenosides in inflammation-derived cancers. Mol Cell Biochem 476:3329–3340
6. Yakubogullari N, Coven FO, Cebi N, Coven F, Coven N, Genc R, Bedir E, Nalbantsoy A (2021) Evaluation of adjuvant activity of astragaloside VII and its combination with different immunostimulating agents in Newcastle disease vaccine. Biologicals 70:28–37
7. Rai S, Acharya-Siwakoti E, Kafle A, Devkota HP, Bhattarai A (2021) Plant-derived Saponins: a review of their surfactant properties and applications. Science 3:44
8. Liao Y, Li Z, Zhou Q, Sheng M, Qu Q, Shi Y, Yang J, Lv L, Dai X, Shi X (2021) Saponin surfactants used in drug delivery systems: a new application for natural medicine components. Int J Pharm 603:120709
9. El Aziz MMA, Ashour AS, Melad ASG (2019) A review on saponins from medicinal plants: chemistry, isolation, and determination. J Nanomed Res 7:282–288
10. Guzmán L, Villalón K, Marchant MJ, Tarnok ME, Cárdenas P, Aquea G, Acevedo W, Padilla L, Bernal G, Molinari A, Corvalán A (2020) In vitro evaluation and molecular docking of QS-21 and quillaic acid from *Quillaja saponaria* Molina as gastric cancer agents. Sci Rep 10:10534
11. Oh M, Um S, Jeong J-H, Ko H-J, Kim SH (2023) 3-Hydroxybutyrate-containing triterpenoid saponins from *Brachyscome angustifolia* and their immunogenic activity. Phytochemistry 212:113724
12. Um S, Lee J, Lee Y, Namkung W, Kim SH (2023) Pentacyclic triterpenoids saponins pannosides A-E from *Tripolium pannonicum*. Front Mar Sci 10:1117407
13. Krzyzanowska J, Kowalczyk M, Oleszek W (2014) Analysis of plant saponins. In: Hostettmann K, Chen S, Marston A, Stuppner H (eds) Handbook of chemical and biological plant analytical methods volume II. Wiley, Chichester, pp 687–708
14. Allard PM, Péresse T, Bisson J, Gindro K, Marcourt L, Pham VC, Roussi F, Litaudon M, Wolfender JL (2016) Integration of molecular networking and in-silico MS/MS fragmentation for natural products dereplication. Anal Chem 88:3317–3323
15. Duncan KR, Crüsemann M, Lechner A, Sarkar A, Li J, Ziemert N, Wang M, Bandeira N, Moore BS, Dorrestein PC, Jensen PR (2015) Molecular networking and pattern-based genome mining improves discovery of biosynthetic gene clusters and their products from *Salinispora* species. Chem Biol 22:460–471
16. Guthals A, Watrous JD, Dorrestein PC, Bandeira N (2012) The spectral networks paradigm in high throughput mass spectrometry. Mol BioSyst 8:2535–2544
17. Watrous J, Roach P, Alexandrov T, Heath BS, Yang JY, Kersten RD, Voort M, Pogliano K, Gross H, Raaijmakers JM, Moore BS, Laskin J, Bandeira N, Dorrestein PC (2012) Mass spectral molecular networking of living microbial colonies. Proc Natl Acad Sci USA 109:E1743–E1752
18. Pingret D, Fabiano-Tixier AS, Chemat F (2013) Ultrasound-assisted extraction. In: Rostagno M, Prado J (eds) Natural product extraction: principles and applications. RSC Publishing, Cambridge, pp 89–112
19. Oleszek W, Jurzysta M, Ploszynski M, Colquhoun IJ, Price KR, Fenwick GR (1992) Zahnic acid tridesmoside and other dominant saponins from alfalfa (*Medicaco sativa* L.) aerial parts. J Agric Food Chem 40:191–196
20. Xu M, Wan Z, Yang X (2021) Recent advances and applications of plant-based bioactive saponins in colloidal multiphase food systems. Molecules 26:6075
21. Nguyen NL, Vo TH, Lin YC, Liaw CC, Lu MK, Cheng JJ, Chen MC, Kuo YH (2021) Arenarosides A–G, polyhydroxylated oleanane-type saponins from *Polycarpaea arenaria* and their cytotoxic and antiangiogenic activities. J Nat Prod 84:259–267

Chapter 15

Analysis and Identification of Polyphenols in *Fragaria*, *Rubus*, and *Ribes* Species by Nontargeted UHPLC-PDA-MS Metabolite Profiling

Simon D. A. Pont and J. William Allwood

Abstract

Profiling of secondary metabolites within *Fragaria* sp. (strawberry), *Rubus* sp. (raspberries and blackberries), and *Ribes* sp. (blackcurrants and redcurrants) is of key importance with respect to the nutritional (antioxidant) benefits associated with a range of polyphenol classes within such fruits and how they vary between growth environment and season, as well as their deployment as quantitative trait loci (QTL) within breeding populations. This chapter presents an ultra-high-performance liquid chromatography (UHPLC) approach applying aqueous functionalized C18 chromatography combined with mass spectrometry (MS) for the characterization and relative quantification of a wide range of polyphenols such as the pelargonidins, being rich in strawberry; cyanidins, procyanidins, and anthocyanins, being rich in raspberry; and delphinidins and prodelphinidins, being particularly rich in blackcurrant, as well as a range of colorless flavonoids, and amino and organic acids.

Key words *Fragaria*, *Rubus*, *Ribes*, Ultra-high-performance liquid chromatography (UHPLC), Mass spectrometry (MS), Tandem MS (MS/MS), Data-dependent analysis (DDA)

1 Introduction

The production of fleshy and soft fruit crops is of great economic importance with global production being 910 Mt in 2021 (fao.org). Fruit quality is dependent on both organoleptic and nutritional properties which are directly related to biochemical composition, especially being influenced by low molecular weight (MW) organic compounds, namely, primary metabolites such as sugars, amino acids, organic acids, water, and lipid-soluble vitamins, and secondary metabolites such as polyphenols, alkaloids, isoprenoids, and carotenoids [1]. Metabolic changes in fruit occur during fruit development and ripening, as well as during commercial storage and transport. Historically, detailed measurements of metabolites were performed via targeted methodologies, although

Young Hae Choi et al. (eds.), *Natural Product Isolation and Identification: Methods and Protocols*, Methods in Molecular Biology, vol. 2895, https://doi.org/10.1007/978-1-0716-4350-1_15,

since the advent of modern metabolomics approaches, central metabolism has been largely measured through proton nuclear magnetic resonance (^{1}H-NMR) spectroscopy and gas chromatography (GC)-MS [2–6], whereas secondary metabolism has been measured through LC-MS [4, 6–9].

During initial soft fruit growth, compounds serving as secondary metabolite and volatile precursors, such as aromatic amino acids, are being synthesized. In later growth, color develops, and softening begins, this is followed by the accumulation of ascorbic acid within the mesocarp and lipids within seeds. At the final stages of ripening, sugars and polyphenols accumulate, and the fruits further soften and darken in appearance [8]. Ripening of the fruit is a complex process under tight transcriptional phytohormone and metabolic control, under the given environmental conditions [6–8]. Flavor appreciation of soft fruits is largely driven by sweetness, acidity, and aroma [1, 6], although a high value for visual appreciation and nutrition with respect to antioxidant benefits is driven by pelargonidins (being rich in strawberry), cyanidins, procyanidins, and anthocyanins (being rich in raspberry), and delphinidins and prodelphinidins (being rich in blackcurrant) [8–10].

The measurement of these groups of polyphenols via LC-MS methods is of key importance with regard to assessments of fruit quality between growth environment and season [6], as well as for their deployment as quantitative trait loci (QTL) within breeding populations [9]. The following protocol presents a mature ultra-high-performance liquid chromatography (UHPLC) approach applying aqueous functionalized C18 chromatography combined with UV/Vis spectroscopy with photodiode array (PDA) and mass spectrometry (MS)-based detectors for the characterization and relative quantification of a wide range of key soft fruit polyphenols, which has been successfully applied across a wide range of soft fruits including strawberry, raspberry, blackberry, blackcurrant, blueberry, and honeyberry [6–12].

2 Materials

Unless otherwise stated, all solvents are minimally of HPLC purity grade, with LC-MS grade being desirable. Formic acid is of MS grade. All glassware, measuring cylinders, and LC bottles are pre-cleaned with a range of polar and nonpolar organic solvents (water, acetonitrile, propan-2-ol, MeOH, chloroform), followed by a rinse in HPLC grade MeOH and three further rinses in HPLC grade water, before oven drying. Reference standards applied to confirmation of compound identification should be of 99% purity or greater (*see* **Note 1**); stock solutions are prepared in MeOH to a concentration of 10 ppm.

2.1 Sample Extraction and Reconstitution Solvents

1. Fresh fruit extraction solvent: 99.5% (v/v) MeOH:0.5% (v/v) formic acid. Using a clean 1-L measuring cylinder, accurately measure 995 mL of HPLC grade MeOH, transfer to a clean LC bottle, and add 5 mL of MS grade formic acid. Mix by gentle swirling by hand.
2. Freeze-dried fruit extraction solvent: 80% (v/v) MeOH: 19.5% (v/v) water: 0.5% (v/v) formic acid. Using a clean 1-L measuring cylinder, accurately measure 800 mL of HPLC grade MeOH and 195 mL of HPLC grade water, transfer to a clean LC bottle, and add 5 mL of MS grade formic acid. Mix by gentle swirling by hand.
3. Reconstitution solvent (without internal standard): 20% (v/v) MeOH:80% (v/v) water. Using a clean 1-L measuring cylinder, accurately measure 200 mL of HPLC grade MeOH and 800 mL of HPLC grade water, and transfer to a clean LC bottle (*see* **Note 2**). Mix by gentle swirling by hand.
4. Reconstitution solvent (with internal standard): 20% (v/v) MeOH: 80% (v/v) water: 25 ppm (w/v) morin hydrate. Dissolve 50 mg of morin hydrate in 100 mL of MeOH (500 ppm). Dilute 20 mL morin hydrate in 500 ppm stock with 320 mL of HPLC grade water and 60 ml of HPLC grade MeOH. Use clean 100- and 500-mL measuring cylinders to measure solvent volumes, and transfer to a clean LC bottle (*see* **Note 2**). Mix by gentle swirling by hand.

2.2 LC Solvents

1. LC solvent A: HPLC grade water: 0.1% (v/v) MS grade formic acid. Using a clean 1-L measuring cylinder, accurately measure 1 l of HPLC grade water, transfer to a 1-l LC bottle, remove 1 mL of water and add 1 mL of MS grade formic acid, and mix by gentle swirling by hand.
2. LC solvent B: HPLC grade acetonitrile: 0.1% (v/v) MS grade formic acid. Using a clean 1-L measuring cylinder, accurately measure 1 l of HPLC grade acetonitrile, transfer to a 1-l LC bottle, remove 1 mL of acetonitrile and add 1 mL of MS grade formic acid, and mix by gentle swirling by hand.
3. Degas solvents A and B by placing in an ultrasonicator on degas setting for 20 min.
4. LC wash solvent: 80% (v/v) acetonitrile: 20% (v/v) water. Using a clean 1-L measuring cylinder, accurately measure 800 mL of HPLC grade acetonitrile and 200 mL of HPLC grade water, transfer to a 1-l LC bottle, and mix by swirling gently by hand.

2.3 Instrumentation

The protocol assumes the use of a Dionex U3000 biocompatible UHPLC system interfaced with a Thermo LTQ-Orbitrap XL MS (Thermo Fisher); however, the protocols can be easily replicated with alternative biocompatible HPLC or UHPLC systems with minimally 600 bar pressure capacity and with a high-resolution sub 5 ppm mass accuracy MS system with MS/MS capabilities such as modern Q-ToF/MS systems, although injection volumes will likely require re-optimization.

3 Methods

3.1 Preparation of Soft Fruits (See Note 3 and 4)

1. Freeze-drying fruits: place the representative fruit sample within a foil punnet(s), and cover loosely with a foil sheet that is pierced multiple times with a needle. Freeze-dry at −40 °C for 72 > 120 h (based upon freeze-drying with a Christ Gamma 1–16 LSC) (*see* **Note 5**). Store as whole freeze-dried fruits at −80 °C within airtight packaging.
2. After removing fruits from −80 °C, freeze-dry for a further 24 h to remove any potential moisture absorbed during storage. *Ribes* species require for whole fruits to be roughly broken up and freeze-dried for 24 h.
3. Freeze-dried fruits are homogenized with a Retsch ZM2000 mill at 10,000 Hz. The milling time per fruit is approximately 5 s, although the total cycle time is dependent on the volume of fruit to be homogenized.
4. Fresh fruits are homogenized to a fine powder under liquid nitrogen with a mortar and pestle by hand.

3.2 Extraction of Soft Fruits

1. Weigh 300 mg (298–302 mg) of freeze-dried soft fruit homogenate, or 3000 mg (2995–3005 mg) of fresh weight fruit homogenate, into a 15-mL Sarstedt brand centrifuge tube. All tubes are pre-labeled with an alcohol-resistant marker (on side of the tube and on lids). The precise sample weights are recorded. Empty tubes are also applied to blank sample extractions (*see* **Note 6**).
2. Working on ice, to the sample add 6 mL of extraction solvent, 80% (v/v) MeOH:19.5% (v/v) water:0.5% (v/v) formic acid for freeze-dried soft fruit samples, or 99.5% (v/v) MeOH:0.5% (v/v) formic acid for fresh weight fruits. The extraction buffer is pre-chilled at −20 °C minimally overnight. Each sample is then vortexed for 15 s maintaining that the pellet of material is well mixed.
3. Shake for 30 min at full power (speed 10) with a Heidolph MultiReax shaker housed within a cold room (5 °C).
4. Centrifuge for 10 min at 4000 rpm (2900×*g*) and 3 °C to pellet material.
5. Transfer 1 mL of supernatant to 2-mL microtube. Up to four aliquots can be prepared for each extract.

6. Dry down by speed vacuum concentration with a miVac concentrator system set at 30 °C, alcohol, 4-h dry time (*see* **Note 7**).
7. A common quality control (QC) or standard reference sample is included in each analytical batch to monitor instrument performance and errors of measurement for each metabolite.
   (a) If all samples are available at the commencement of the analytical experiments, create a mixed pool of equal weights of all samples. This is extracted with high levels of replication. This QC analyzed throughout the sequences and batches provides a measure of both instrument error and extraction error.
   (b) Alternatively, if materials are limited, create a pooled extract QC: with the remaining sample extract, create a QC mixture containing equal volumes of extract for all samples included within the experiment. This QC only provides a measure of instrument error.
   (c) If comparisons are to be made between growth seasons where an averaged QC mix cannot be created, then a representative reference sample or reference mixture is prepared for the first analytical batch and applied to all subsequent analytical batches across the years of the experiment (*see* **Note 8**).
8. Store all dried samples at −80 °C for up to 4 weeks in an airtight box sealed with a paraffin film.

3.3 Reconstitution of Soft Fruit Extracts

For blackcurrants, reconstitute the dried extract in 1 mL of 20% (v/v) MeOH: 80% (v/v) water, with or without 25 ppm morin hydrate (*see* **Note 2**). For redcurrant, raspberry, and strawberry, reconstitute the dried extract in 0.5 mL. If the experimental aim is to directly compare between *Fragaria, Ribes*, and *Rubus* sp., then it is the best to apply an intermediate reconstitution volume of 0.75 mL to all fruit types.

1. Add the reconstitution solvent, vortex for 20 s, and shake for 30 min on an IKA VXR Basic Vibrax shaker in a cold room (5 ° C).
2. Centrifuge for 3 min at 3 °C and 15,000 rpm (21,130×*g*) to pellet debris.
3. Transfer 0.4 mL (avoiding transferring particulate matter) to a Thomson 0.45-μm PTFE filter vial, filter, and then transfer 0.2 mL to a Thermo microvial (300 μL) and apply a cap with septa.

3.4 UHPLC-PDA-MS and DDA MS/MS Instrument Setup and Analysis

3.4.1 UHPLC and PDA Preparation

1. The Dionex U3000 UHPLC is configured to operate with a 20-μL nano viper loop, a 250-μL syringe, 15-μL needle, and 50-μL buffer tubing volume.
2. The Dionex U3000 PDA is tuned and calibrated according to the manufacturer's recommended procedures.
3. The LC column (Waters CORTECS C18 T3 1.6 μm particle 2.1 × 100 mm, 120 Ä) is pre-fitted with a guard column (Waters CORTECS C18 T3 1.6 μm particle 2.1 × 5 mm guard cartridge) and placed within the column oven at 30 °C (allow 1 h for the column core to reach stable temperature) (*see* **Note 9**).
4. Purge each of the solvents A and B for 5 min at 3 mL/min maintaining that no air is entrapped within the solvent filters or lines.
5. Purge the fresh needle wash by performing a 5-mL needle wash several times. Maintain that no air is entrapped within the solvent filter, solvent lines, syringe, and needle.
6. Prior to sample analysis, condition the LC column and new guard cartridge with solvents A and B for a minimum of 70 min at a flow rate of 300 μL/min, starting with 100% B for 40 min, 50% B for 20 min, and 2% B for 10 min.
7. Create an LC method employing an injection volume of 10 μL in partial-loop mode with an external and internal needle wash of 200 μL. The UHPLC gradient program is as follows: flow rate, 300 μL/min; hold 2% B for 0–2 min, 2–5% B for 2–5 min, 5–38% B for 5–21 min, and 38–100% B for 21–36 min; hold 100% B for 36–39 min and 100–2% B for 39–39.1 min; and hold 2% B for 39.1–44 min (equilibration) (*see* **Note 9**).
8. The UHPLC column eluent is first monitored by the Dionex U3000 PDA detector where spectra are collected in wavelength/absorbance mode from 200 to 600 nm with a filter bandwidth and wavelength step of 1 nm. The filter rise time is 1 s, and the sample rate is 5 Hz. Additionally apply three channel set points: Channel A, 280 nm; Channel B, 365 nm; and Channel C, 520 nm, with a bandwidth of 9 nm and a sample rate of 10 Hz.

3.4.2 LTQ-Orbitrap XL MS Preparation and Analytical Batch Sequence Design

The Thermo LTQ-Orbitrap XL MS system is first calibrated, after which the system is tuned in an electrospray ionization (ESI) positive mode to morin ion 303 *m/z* and operating in an ESI negative mode to morin ion 301 *m/z*, to create a set of optimized ionization and transmission values within two tune files, respective of ESI polarity, later applied to the LC-MS profiling methods. The aim is to optimize ESI transmission conditions for flavonoid compounds.

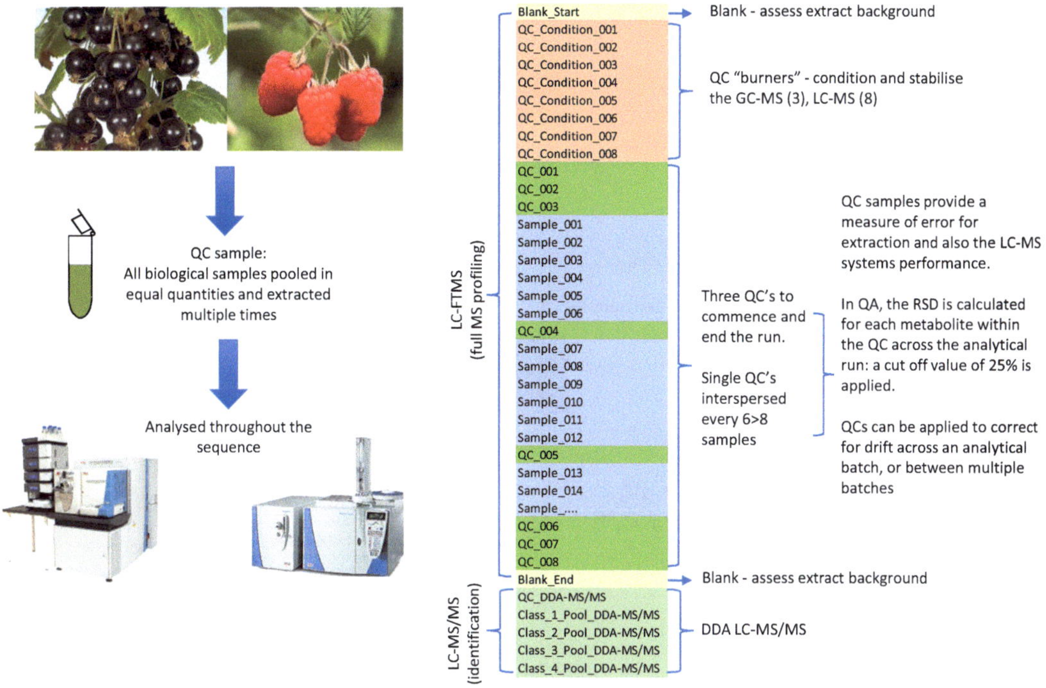

Fig. 1 Example sequence to illustrate how to set up a sample run for soft fruit analysis inclusive of quality assurance samples and blank controls. *QA* quality assurance, *QC* quality control, *DDA* data-dependent analysis, *RSD* relative standard deviation

1. Calibrate the LTQ-Orbitrap XL MS system in each ESI positive and negative mode following the manufacturer's recommended calibration procedures and ESI source conditions by applying Pierce LTQ ESI positive and ESI negative calibration solutions. A dedicated Hamilton syringe and PEAK transfer line are recommended for use with Pierce calibration solutions.
2. Fill a secondary Hamilton 500-μL syringe with a solution of 1 ppm morin in MeOH, displace entrapped air, and connect to the heated (H)ESI source with a dedicated PEAK transfer line. Infuse the morin solution at 50 μL/min under the following conditions: +4.0 kV ESI+; −3.5 kV ESI−; sheath gas: 10; aux gas: 5; sweep gas: 0; capillary temperature: 275 °C; and HESI probe temperature: 30 °C.
3. Run an automated tune optimization procedure for the capillary and tube lens voltages to obtain maximum signal intensity for morin in each positive and negative polarities. Our system optimal values are as follows: ESI+ capillary voltage: +43 V; ESI + tube lens voltage: +100 V; ESI− capillary voltage: −20 V; and ESI− tube lens voltage: −105 V.
4. Create two tune files, one for ESI positive and the other for ESI negative mode, with the following values: automatic gain

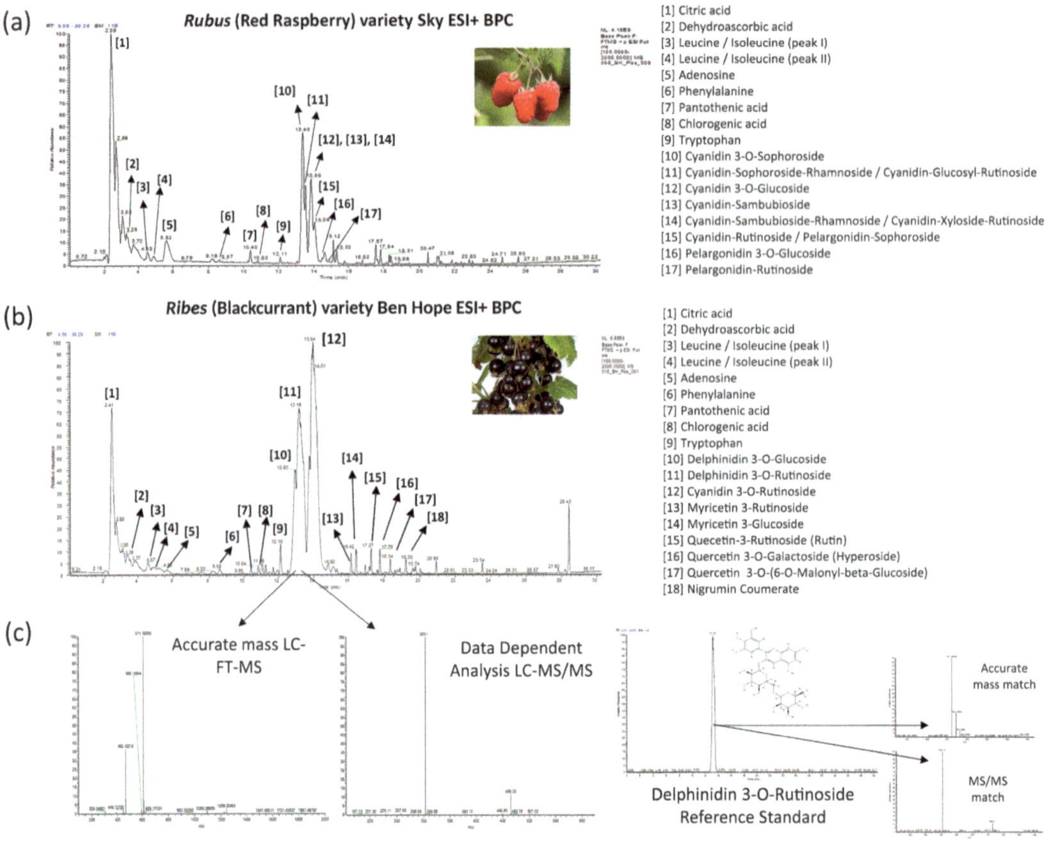

Fig. 2 Example LC-MS ESI positive mode base peak chromatograms (BPC) and LC-MS/MS identification of delphinidin-3-O-rutinoside in blackcurrant. (**a**) LC-MS ESI+ base peak chromatogram of *Rubus* (Red Raspberry) variety, Sky; (**b**) LC-MS ESI+ base peak chromatogram of *Ribes* (blackcurrant) variety, Ben Hope; (**c**) LC-ESI+ FT-MS accurate mass spectrum and LC-ESI+ MS/MS spectrum of delphinidin-3-O-rutinoside in a blackcurrant extract, alongside those matching for a reference standard and its EIC providing a retention time and MS/MS match between the reference standard and blackcurrant extract (MSI level 1) [17]

control (AGC): 1×10^5 and 5×10^5 for the LTQ and FT-MS, respectively; multipole RF: 400; ESI spray voltage: −3.5 kV (ESI−) and +4.0 kV (ESI+); sheath gas: 60; aux gas: 30; sweep gas: 0; capillary voltage: −20 V (ESI−) and + 43 V (ESI+); tube lens voltage: −105 V (ESI−) and +100 V (ESI +); capillary temperature: 275 °C; and HESI probe temperature: 100 °C.

5. Post-calibration, flush the ESI probe and needle at high flow (300 μL/min) with 80% acetonitrile to displace any remaining calibrant solution.

6. Clean the ion transfer tube (ITT), and spray cone sequentially in organic solvents, (i) 80% (v/v) MeOH: 19% (v/v) water: 1% (v/v) formic acid > (ii) water > (iii) MeOH, each with sonication for 20 min. Finally, dry the ITT and spray cone under a stream of nitrogen gas.

7. After replacing the ITT and spray cone, ion source housing, and reconnecting the LC outlet from the rotor seal to the HESI probe, open the tune file created in Subheading 3.4.2 **step 4** for the intended ESI polarity. Turn the LTQ-Orbitrap XL system to operate for minimally 2–3 h before commencement of sample analysis to maximally stabilize system temperatures and gas flows.
8. Set up two analytical methods for each ESI polarity, one applied to LC-FTMS profiling across all extracts, and a second method for data-dependent analysis (DDA) LC-MS/MS applied to QCs and experimental class-based sample pools for compound identification.
   (a) Take the UHPLC method (developed in Subheading 3.4.1 **steps** 7 and **8**, and set up the MS system to collect spectra in the FT detector in full scan profile mode (*m/z* 100–2000) at a mass resolution of 30,000 (FWHM defined at *m/z* 400), permitting a scan rate of 0.4 s (*see* **Note 10**). Set the divert valve to waste for 0–0.8 min, to source for 0.8–35 min, and to waste for 35–40 min. Create an LC-MS method for each ESI positive and ESI negative mode with the respective tune files created in Subheading 3.4.2 **step 4**.
   (b) Using the methods developed in Subheading 3.4.2 **step 8a**, create a secondary method for DDA MS/MS. A full FT-MS1 scan event is defined in profile mode and followed by IT-MS2 scan events in centroid mode for the three most intense ions defined in MS1. Set a dynamic exclusion count of 10 s to prevent repeated collection of the same MS/MS spectra, maximizing data capture for lower intensity ion MS/MS spectra. Helium is applied as the collision gas for collision-induced dissociation (CID) at a normalized collision energy (CE) of 45, a trapping window width of 2 (+/−1) *m/z* is applied, an activation time of 30 ms and activation Q of 0.25 are applied, only single charged ions are selected for DDA, and [M + 1], [M + 2], and [M + 3] isotope ions are also excluded. Further methods applying low CE (20) and high CE (55) may also be applied and benefit annotation efforts.
9. Analyze the samples in a completely randomized order as two independent analytical blocks respective of ESI polarity. For each LC-FTMS analytical block, perform eight injections of QC sample for column and MS system conditioning (*see* **Note 11**), inject a further three QCs, and perform six injections of experimental sample followed by a QC injection. Repeat until all "experimental" samples are analyzed, and finally inject three QCs at the end of the analytical block (*see* **Note 12**). Analyze

the control blank sample at the start (prior to the QC "conditioning" injections) and end of the analytical block, thus providing a measure of the sample background. The QC sample and/or sample class-based pools are finally subjected to the DDA MS/MS method(s). An example sequence template is illustrated in Fig. 1.

10. Samples are analyzed in ESI positive mode first. Once the analytical block is completed, the ITT and spray cone are cleaned according to the procedures outlined above in Subheading 3.4.2 **step 6**.
11. Replace LC solvents A and B with freshly prepared solvents, purge solvents A and B (Subheading 3.4.1 **step 4**), and condition the column (Subheading 3.4.1 **step 6**). The ESI negative mode tune file created in Subheading 3.4.2 **step 4** is opened and the MS system is placed in operate for minimally 2–3 h to stabilize the system temperatures, gas flows, and the FT-MS detector (requiring 80 min for an LTQ-Orbitrap XL), prior to commencing an identical analytical block for the ESI negative mode analyses. Example annotated UHPLC chromatograms are illustrated in Fig. 2a, b.

3.5 UHPLC-PDA-MS Data Processing, Compound Annotation, and Identification

1. The UHPLC-PDA-MS data are first centroided and converted to mzML using MSConvert (http://proteowizard.sourceforge.net/).
2. The data are uploaded to and deconvolved with XCMS Online (https://xcmsonline.scripps.edu/), with the following parameters: feature detection method: CentWave; mass error: 3 ppm, minimum and maximum peak width: 10 and 60 s, respectively; mzdiff: 0.01; S/N threshold: 6; integration method: 1; prefilter peaks: 3; prefilter intensity: 100; noise filter: 0; RT (retention time) correction method: Obiwarp; profstep: 1; alignment, minfrac: 0.5; *m/z* width: 0.015; bw: 5; min samp: 1; max samp: 100; annotation: search for isotopes +adducts; mz absolute error: 0.015; and ppm error: 5.
3. XCMS generates a Microsoft Excel-based XY matrix containing paired RT and *m/z*, along with the extracted ion chromatogram (EIC) area for each profiled sample. Peaks that elute from 0 to 0.8 min (polar front) and from 39 to 44 min (the equilibration period), or those that are dominant within blank sample extracts (minimum 3× signal level of blank within true sample to maintain peak), are excluded [13].
4. Applying PutMedID operated within Taverna Workbench 1.7.2 [14–16], peak-to-peak Pearson correlations (workflow 1) are calculated (+/−5-s moving RT window), and peaks that show high levels of correlation (0.8>) are grouped as *m/z* features associated with the same compound (i.e., an *m/z* group).

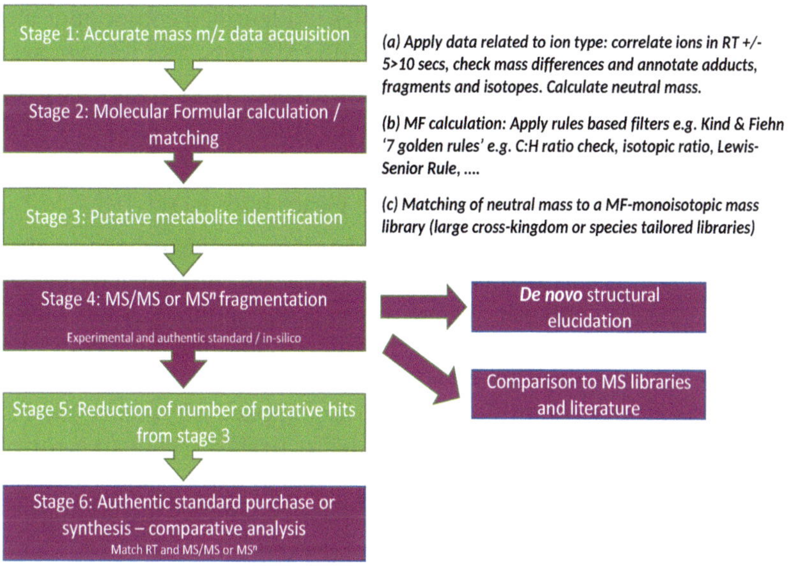

Fig. 3 Flow diagram illustrating the steps applied in the compound annotation and identification workflow. *RT* retention time, *MS/MS* tandem MS or MS^2, MS^n multistage mass spectrometry, *MF* molecular formula

5. Within PutMedID operated within Taverna Workbench 1.7.2 workflow 2, accurate mass differences between ions in each *m/z* group are calculated to allow the annotation of the parent *m/z*, as well as adduct and isotope ions, and the neutral accurate mass is next calculated (5 ppm mass error) and in turn matched to possible molecular formula(s) and in workflow 3 to metabolite name(s) (*see* **Note 13**).
6. The Manchester Metabolomics Database (MMD: http://dbkgroup.org/MMD/) is applied as the molecular formula and compound name library; it contains many of the soft fruit-relevant polyphenols (*see* **Note 14**).
7. In most cases, the process of matching accurate mass to molecular formula provides an MSI level 2 (single molecular formula) or 3 (multiple molecular formula) identification [17].
8. To increase confidence in the assigned putative identification (s), MS/MS data and UV response are also considered. Experimental MS/MS spectra are matched to reference spectra (MassBank Europe (https://massbank.eu/MassBank/), FooDB (https://foodb.ca/), and LipidMaps (https://www.lipidmaps.org/)) and to the literature (UV/Vis and MS). Where both an accurate mass match and an MS/MS match to a database reference spectrum are achieved, the identification is regarded as an MSI level 2+. If a reference standard is available, RT, accurate mass full MS, and MS/MS, match between the reference standard and the sample (Fig. 2c) confirms identification (MSI level 1) [17]. The annotation and identification workflow steps are illustrated in Fig. 3.

9. Comprehensive tables of identified compounds alongside their accurate mass *m/z*, MS/MS ions, and RTs are available.
   (a) For blackcurrant, please refer to ref. [6] (https://www.ebi.ac.uk/metabolights/editor/MTBLS773/metabolites) and references [7–8].
   (b) For Raspberry, please refer to ref. [9].
10. Finally, if applied, the data are normalized to the morin peak area to create peak ratios (or alternatively scaled to the TIC intensity where morin is not applied). A QC filter is applied to remove irreproducible compounds (*see* **Note 15**). The QC samples analyzed throughout the entire sequence (excluding conditioning QCs) are applied to quality-assure each target peak, with a high-quality peak showing an error of lower than 25% (typically 5 > 10%) relative standard deviation (RSD) across all QC and reference samples throughout the sequence. Only quality-assured peaks are taken forward to statistical analysis and modeling.

4 Notes

1. Selection of appropriate polyphenol reference standards is dependent upon the soft fruit species being studied, many, but not all, are available from ExtraSynthese (Rhône, FR), LGC (Bellshill, UK), and Merck/Sigma-Aldrich (Darmstadt, DE).
2. If the intended experiment is of small scale and all analyses can be performed within a single small analytical batch (100–120 or less injections), then application of an internal standard is not necessary. However, if the intended experiment is to involve the profiling of hundreds to thousands of samples across multiple analytical batches, then it is recommended to include an internal standard within the reconstitution solvent. Here, we add morin hydrate at a concentration of 25 ppm.
3. It must first be considered that the fruit homogenate applied to analysis should be representative of the fruits across an entire plant with respect to studies of physiology, or across a number of plants co-harvested with respect to QTL analyses and studies of seasonal or environmental impact on fruit nutritional or flavor quality.
4. If fruits are to be stored only for short periods of time prior to analysis (< 4 weeks at −20 °C or − 80 °C), then it is possible to analyze fresh weight fruits. However, if fruits are to be harvested over sustained periods of time or across multiple seasons (2–4 seasons), then it is best to freeze-dry fruits prior to storage at −80 °C. For such long-term studies, considerations should

be made upon the stability of freeze-dried fruits at −80 °C versus the stability of collecting data across several years on a given analytical platform.

5. *Fragaria* and *Rubus* species tend to have thinner epicarp and freeze-dry more rapidly (~72 h), whereas *Ribes* species have thicker epicarp and require 120 h.
6. It is recommended to use only high-quality polypropylene pipette tips, centrifuge, and microcentrifuge tubes. Regardless of the manufacturer, all plastic tubes and pipette tips should be of the same type and from a single manufacturer's production lot. It is recommended to test different manufacturer's tubes, and to perform regular production batch/lot testing, to maintain that the plastics provide a low-bleed background. In our experience, Sarstedt brand centrifuge tubes, Eppendorf brand safelock microcentrifuge tubes, and Fisher brand 1- and 0.2-mL pipette tips provide a low-bleed background.
7. Dried soft fruit extracts tend to retain water/form syrups which are later difficult to reconstitute; it is recommended that the dried extracts are further placed in a freeze-drier for 24 h at −40 °C, removing residual water held within the syrup and leaving a pellet with a "fluffy" appearance, which will later reconstitute with ease.
8. For QC approaches (a) and (b), it is recommended that extraction error, as well as instrument error, be assessed, by applying a reference sample mixture (approach c) extracted minimally in triplicate for each analytical batch.
9. The UHPLC method has been developed over many years of application; the presented UHPLC method provides optimal resolution; however, previous HPLC methods suited to lower pressure 600 bar HPLC systems are detailed below. The MS method should also be adapted so that the LC flow is directed to waste and not the HESI source during the different equilibration period timings applied in the different method variants.
   (a) The original method and gradient applied the Synergi C18 Hydro-RP 80 Ä, 150 × 2.0 mm, 4 μm particle size column pre-fitted with a guard cartridge and holder (Security Guard AQ C18 4 × 2.0 mm and Security Guard cartridge holder) (Phenomenex), housed within the column oven at 30 °C. The following method was applied: injection volume 10 μL in partial-loop mode with an external and internal needle wash of 200 μL. The UHPLC gradient program was as follows: flow rate: 300 μL/min; hold 2% B for 0–2 min, 2–5% B for 2–5 min, 5–45% B for 5–25 min, and 45–100% B for 25–26 min; hold 100% B for 26–29 min and 100–2% B for 29–30 min; and hold 2% B for 30–35 min (equilibration).

(b) A high-throughput, although lower resolution, method has also been applied where it has been necessary to analyze high numbers of samples within breeding populations. This method and gradient applied the Thermo Hypersil Gold C18 50 × 2.1 mm 1.9 μm particle size analytical and 10 × 2.1 mm guard column fitted with a Uniguard 2-mm column holder (Thermo Fisher), housed within the column oven at 30 °C. The following method was applied: injection volume 10 μL in partial-loop mode with an external and internal needle wash of 200 μL. The UHPLC gradient program was as follows: flow rate: 400 μL/min; hold 2% B for 0–1 min, 2–5% B for 1–2.5-min, 5–45% B for 2.5–13 min, and 45–100% B for 13–14 min; hold 100% B for 14–17 min and 100–2% B for 17–18 min; and hold 2% B for 18–23 min (equilibration).

10. All samples are first profiled in full scan mode so that as many FT full scan data points as possible at resolution 30,000 are obtained, thus providing a smooth peak shape for quantification and maximum data points for reliable mass accuracy.

11. In our experience, a minimum of 6 > 8 injections of a mixed QC sample (analytical system-dependent) are required before both chromatographic and MS systems are fully stabilized with respect to minimal chromatographic shift and variance in MS signal intensity. Analysis upon a Thermo Orbitrap XL system indicated that 8 injections were required, whereas analysis upon an Agilent TOF/MS system only required 6 injections to reach full stability.

12. The maximum number of injections within an analytical block is both MS system- and sample concentration-dependent. It is recommended to test your specific LC-MS system to define the maximum number of injections before the observation of reduced signal. In our experience, with a sample such as blackcurrant which typically has the highest levels of polyphenols, analytical blocks should be limited to approximately 120 injections (including QC injections) prior to operator intervention to clean the ITT and spray cone following the procedure outlined in Subheading 3.4.2 **step 6**. Periodically, typically every 1000 injections, the MS system will require venting and the S and Exit lenses cleaned following the manufacturer's recommended procedures.

13. The PutMedID workflow makes assumptions on how compounds ionize; it does not account for the fact that certain compounds carry endogenous positive or negative charges, therefore not forming adducts in ESI, and being detected as

m/z that directly match to the natural compounds MW. For compounds that carry a positive charge, for example, delphinidin-3-*O*-rutinoside ($C_{27}H_{31}O_{16}^{+}$), we must remove a proton from the molecular formula within the library (i.e., $C_{27}H_{30}O_{16}$), therefore tricking the workflow that an $[M + H]^{+}$ ion is formed. Likewise, for compounds that carry a negative charge, for example, glucobrassicin ($C_{16}H_{19}N_{2}O_{9}S_{2}^{-}$), we must add a proton to the molecular formula within the library (i.e., $C_{16}H_{20}N_{2}O_{9}S_{2}$), therefore tricking the workflow that an $[M\text{-}H]^{-}$ ion is formed.

14. Although the Manchester Metabolomics Database was created from across a large number of web resources for metabolite and lipid databases and already contains large numbers of plant- and fruit-relevant compounds, it must be considered that there is a lot of further information available in the literature on the LC-MS analysis of soft fruits and their annotated metabolites. It is recommended that researchers conduct thorough searches of literature and compile information on metabolite names and their molecular formula; this information can then be easily formatted and added to the molecular formula and compound name libraries applied within PutMedID.
15. Since the QC samples are analyzed multiple times with high frequency throughout the analytical sequence, they provide a measure of analytical performance and errors, without the need to analyze each individual sample in duplicate or triplicate.

Acknowledgement

JWA and SP would like to acknowledge support from the Scottish Government Rural and Environmental Science and Analytical Services (RESAS) program.

Author Contribution SP developed the methods applied to soft fruit preparation, co-developed the extraction protocols, and provided critical evaluation of the protocols for fruit preparation and extraction. JWA co-developed the extraction protocols, developed the LC-MS analysis and annotation protocols, and wrote this chapter.

References

1. Allwood JW, Gibon Y, Ossorio S, Araújo WL, Vallarino JG, Pétriacq P, Moing A (2021) Chapter One – Developmental metabolomics to decipher and improve fleshy fruit quality. In: Pétriacq P, Bouchereau A (eds) Advances in botanical research, vol 98, pp 3–34. https://doi.org/10.1016/bs.abr.2020.09.016
2. Biais B, Allwood JW, Deborde C, Xu Y, Maucourt M, Beauvoit B, Dunn WB, Jacob D, Goodacre R, Rolin D, Moing A

(2009) ^{1}H-NMR, GC-EI-TOFMS, and data-set correlation for fruit metabolomics: application to spatial metabolite analysis in melon. Anal Chem 81:2884–2894. https://doi.org/10.1021/ac9001996

3. Spraul M, Schütz B, Rinke P, Koswig S, Humpfer E, Schäfer H, Mörtter M, Fang F, Marx UC, Minoja A (2009) NMR-based multi parametric quality control of fruit juices: SGF profiling. Nutrients 1(2):148–155. https://doi.org/10.3390/nu1020148
4. Moing A, Aharoni A, Biais B, Rogachev I, Meir S, Brodsky L, Allwood JW, Erban A, Dunn WB, Lorraine Kay L, de Koning S, de Vos CHR, Jonker H, Deborde C, Maucourt M, Bernillon S, Gibson Y, Hansen TH, Husted S, Goodacre R, Kopka J, Schjoerring JK, Rolin D, Hall R (2011) Spatial and temporal metabolic profiling using multiple analytical platforms highlights the crosstalk between primary and secondary metabolites and mineral elements in melon fruit. New Phytol 190:683–696. https://doi.org/10.1111/j.1469-8137.2010.03626.x
5. Mori K, Beauvoit B, Biais B, Chabane M, Allwood JW, Deborde C, Maucourt M, Goodacre R, Cabasson C, Moing A, Rolin D, Gibon Y (2019) Central metabolism is tuned to the availability of oxygen in developing melon fruit. Front Plant Sci 10:594. https://doi.org/10.3389/fpls.2019.00594
6. Pott DM, Durán-Soria S, Allwood JW, Pont S, Gordon SL, Jennings N, Austin C, Stewart D, Brennan RM, Masny A, Sønsteby A, Krüger E, Jarret D, Vallarino JG, Usadel B, Osorio S (2023) Dissecting the impact of environment, season and genotype on blackcurrant fruit quality traits. Food Chem 402:134360. https://doi.org/10.1016/j.foodchem.2022.134360
7. Jarret DA, Morris J, Cullen DW, Gordon SL, Verrall SR, Milne L, Hedley PE, Allwood JW, Brennan RM, Hancock RD (2018) A transcript and metabolite atlas of blackcurrant fruit development highlights hormonal regulation and reveals the role of transcription factors. Front Plant Sci 9:1235. https://doi.org/10.3389/fpls.2018.01235
8. Allwood JW, Woznicki T, Xu Y, Foito A, Aaby K, Sungurtas J, Freitag S, Goodacre R, Stewart D, Remberg SF, Heide OM, Sønsteby A (2019) Application of HPLC-PDA-MS metabolite profiling to investigate the effect of growth temperature and day length on blackcurrant fruit. Metabolomics 15:12. https://doi.org/10.1007/s11306-018-1462-5
9. McDougall GJ, Allwood JW, Dobson G, Austin C, Verrall S, Alexander CJ, Hancock RD, Graham J, Hackett CA (2023) Quantitative trait loci mapping of polyphenol metabolites from a 'Latham' x 'Glen Moy' red raspberry (***Rubus idaeus*** L) cross. Metabolomics 19:71. https://doi.org/10.1007/s11306-023-02033-7
10. Haugeneder A, Trinkl J, Härtl K, Hoffman T, Allwood JW, Schwab W (2018) Answering biological questions by analysis of the strawberry metabolome. Metabolomics 14:145. https://doi.org/10.1007/s11306-018-1441-x
11. Grace MH, Xiong J, Esposito D, Ehlenfeldt M, Lila MA (2019) Simultaneous LC-MS quantification of anthocyanins and non-anthocyanin phenolics from blueberries with widely divergent profiles and biological activities. Food Chem 277:336–346. https://doi.org/10.1016/j.foodchem.2018.10.101
12. Gamble L, Pont SDA, Allwood JW, Jarret DA, Hancock RD (2023) Comparative analysis of quality and nutritional traits from *Lonicera caerulea* (honeyberry) cultivars and other berries grown in Scotland. Ann Appl Biol 182(2):171–182. https://doi.org/10.1111/aab.12805
13. Di Guida R, Engel J, Allwood JW, Weber RJM, Jones MR, Sommer U, Viant MR, Dunn WB (2016) Non-targeted UHPLC-MS metabolomic data processing methods: a comparative investigation of normalisation, missing value imputation, transformation and scaling. Metabolomics 12(5):1–14. https://doi.org/10.1007/s11306-016-1030-9
14. Brown M, Dunn WB, Dobson P, Patel Y, Winder CL, Francis-McIntyre S, Begley P, Carroll K, Broadhurst D, Tseng A, Swainston N, Spasic I, Goodacre R, Kell DB (2009) Mass spectrometry tools and metabolite-specific databases for molecular identification in metabolomics. Analyst 134:1322–1332. https://doi.org/10.1039/b901179j
15. Brown M, Wedge DC, Goodacre R, Kell DB, Baker PN, Kenny LC, Mamas MA, Neyses L, Dunn WB (2011) Automated workflows for accurate mass-based putative metabolite identification in LC/MS-derived metabolomics datasets. Bioinformatics 27:1108–1112. https://doi.org/10.1093/bioinformatics/btr079

16. Allwood JW, Weber RJM, Zhou J, He S, Viant MR, Dunn WB (2013) CASMI—the small molecule identification process from a Birmingham perspective. Metabolites 3:397–411. https://doi.org/10.3390/metabo3020397
17. Sumner LW, Amberg A, Barrett D, Beale MH, Beger R, Daykin CA, Fan TW-M, Fiehn O, Goodacre R, Griffin JL, Hankemeier T, Hardy N, Harnly J, Higashi R, Kopka J, Lane AN, Lindon JC, Marriott P, Nicholls AW, Reily MD, Thaden JJ, Viant MR (2007) Proposed minimum reporting standards for chemical analysis. Metabolomics 3:211–221. https://doi.org/10.1007/s11306-007-0082-2

Chapter 16

Identification of Primary Metabolites in ^{1}H NMR-Based Metabolomics of Plants

Hye Kyong Kim and Young Hae Choi

Abstract

Identifying metabolites in NMR-based plant metabolomics is challenging due to the complexity of plant metabolites. This complexity stems from the abundance and diverse chemical properties of compounds, which vary in concentration across plant specimens. Additionally, the lack of automated identification software complicates the analysis process. Primary metabolites such as amino acids and sugars are widespread in plants, yet their identification is not straightforward due to various stereoisomeric forms and dynamic equilibria. Our protocol offers a manual approach to identify these metabolites, particularly focusing on amino acids and sugars. Through step-by-step guidance, we aim to empower researchers to navigate plant metabolomics complexities effectively.

Key words Identification, NMR spectroscopy, Primary metabolites, Amino acid, Plant metabolomics

1 Introduction

Metabolomics has become an essential tool in plant science for more than two decades, with recent advances in analytical technology and data analysis broadening its applications significantly. These advancements have enabled the simultaneous determination of a wide range of metabolites, encompassing primary and secondary metabolites, leading to unprecedented insights into plant metabolism, physiology, and responses to environmental stimuli.

Furthermore, the development of metabolomics databases and bioinformatics tools has facilitated data interpretation and knowledge sharing within the scientific community, accelerating the pace of discovery and innovation in plant science and natural product research.

However, one of the most significant challenges in metabolomics remains the identification of peaks or signals, particularly in NMR-based approaches applied to plants. This challenge persists due to the lack of sufficiently robust automated software for

Young Hae Choi et al. (eds.), *Natural Product Isolation and Identification: Methods and Protocols*, Methods in Molecular Biology, vol. 2895, https://doi.org/10.1007/978-1-0716-4350-1_16,

identification, coupled with the intricate chemical and physical diversity of plant metabolites. Additionally, the wide concentration range of metabolites and the species-specific nature of many secondary metabolites further complicate the task [1–3].

In light of these challenges, our focus will be on providing an overview aimed at rapidly identifying commonly encountered primary metabolites in plants. Secondary metabolites will not be addressed, as we aim to concentrate on discussing primary metabolites most commonly encountered. Before delving into identification methods, we will outline standardized steps for sample preparation and analysis, with the goal of streamlining and optimizing the process.

2 Materials

2.1 NMR Solvents

1. Methanol-d_4 (CD_3OD) 99.8% (tetradeuteromethanol).
2. Sodium deuteroxide (NaOD) 99.5% (40% in D_2O).
3. TMSP-d_4: 3-(trimethylsilyl)propionic-2,2,3,3-d_4 acid sodium salt, 99 atom % D.
4. Deuterium oxide (D_2O) >99.9 atom % D.
5. 1.0 M NaOD: add 1 mL of NaOD to 9 mL of D_2O in the glass bottle and mix them well.
6. 90 mM phosphate buffer: add 1.232 g KH_2PO_4 and 10 mg TMSP-d_4 in 100 mL D_2O. After complete dissolution, adjust pH to 6.0 with 1 M NaOD (*see* **Note 1**).

2.2 Extraction Setup

1. Ultrasonicator water bath operating at a high frequency (≥2 MHz).
2. Microcentrifuge fit for microtubes.
3. 2-mL microtubes.
4. pH meter with an electrode.
5. 3-mm NMR tubes and rack.

3 Methods

All experiments can be performed in the room temperature unless specifically mentioned.

3.1 Sample Preparation for NMR

1. Weigh 30 mg of the dried sample into a 2-mL microtube (*see* **Note 2**).
2. Add 0.5 mL of MeOD followed by the addition of 0.5 mL of phosphate buffer (90 mM, pH 6.0) containing 0.1% (w/w) TMSP.

3. Vortex the mixture for 1 min to ensure thorough mixing.
4. Ultrasonicate for 10–20 min.
5. Centrifuge for 10 min using a microtube centrifuge at 14,000×*g* (*see* **Note 3**).
6. Transfer the supernatant to a 2-mL microtube.
7. Transfer 300 μL of the supernatant to a 3-mm NMR tube. Place the tubes in the 96-well rack for the Bruker SampleJet.

3.2 NMR Analysis

1. Ensure all parameters are set for your analysis and set temperature to 298 K (25 °C) (*see* **Note 4**).
2. Tune and match the NMR tube.
3. Lock the spectrometer frequency to the deuterium resonance arising from the NMR solvents (*see* **Note 5**).
4. Shim the sample (*see* **Note 6**).
5. Select suitable experiments, for ^{1}H NMR standard setups are as follows:
   (a) Pulse sequence comprising relaxation delay: 60°—acquire.
   (b) Pulse power: set to achieve a 60° flip angle, 10 kHz spectral width, water pre-sat applied during 1.5-s relaxation delay.
   (c) Processing parameters: zero-fill to 64 k data points, apply exponential line broadening of 0.3 Hz.

3.3 NMR Spectrum Processing

1. After Fourier transformation, manually phase the spectrum (zero and first phase) correction.
2. Correct the baseline and calibrate the spectrum by setting the TMSP peak at δ 0.00.

3.4 Identification of Metabolites

Some primary metabolites commonly detected in most plant samples can be categorized based on their typical chemical shift in the ^{1}H NMR spectra.

Aliphatic amino acids: typically appear in the range of δ 3.0–δ 0.5.

Organic acids: typically found in the range of δ 3.5–δ 2.0.

Sugars: typically observed in the range of δ 5.5–δ 3.0.

Aromatic amino acids: typically seen in the range of δ 8.0–δ 6.5.

Below are some representative metabolites commonly identified in plant samples using ^{1}H NMR spectra, along with descriptions of methods for their identification. Following this procedure facilitates the rapid identification of primary metabolites.

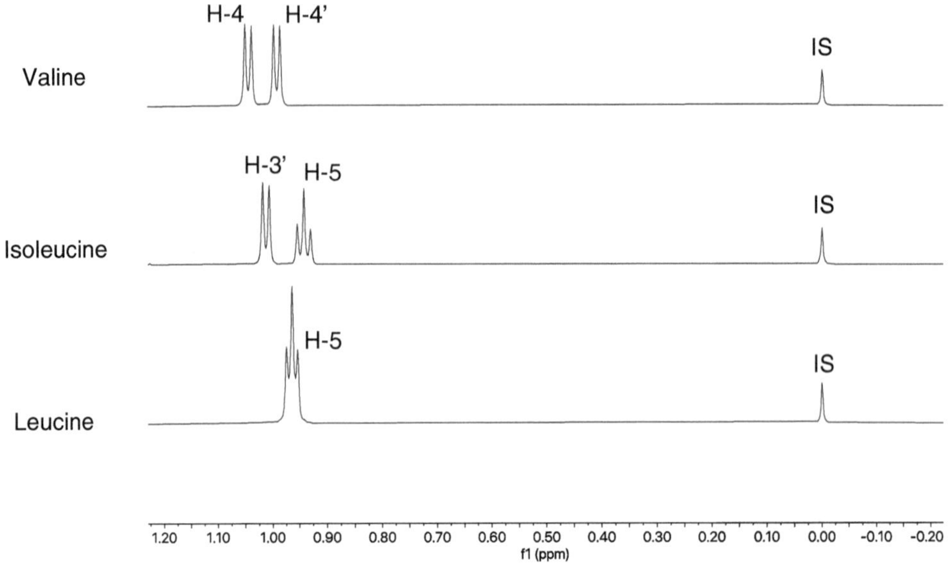

Fig. 1 Expanded ^{1}H NMR spectra (δ 1.20–δ 0.0) of valine, isoleucine, and leucine. Each number indicates the proton signals of the molecule. IS: internal standard (TMSP, 3-(trimethylsilyl)propionic-2,2,3,3-d_4 acid sodium salt). All spectra were acquired using Bruker 600-MHz NMR equipped with a cryoprobe in MeOD-KH_2PO_4 buffer in D_2O (1:1) as described in the Materials and Methods section

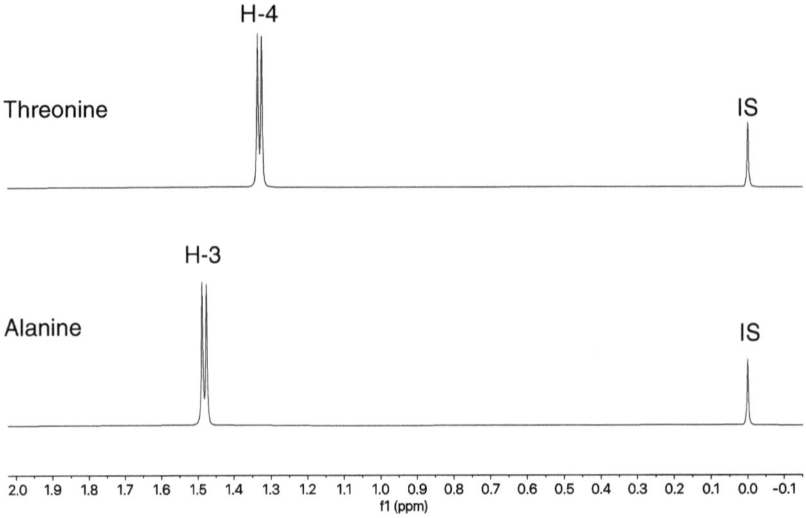

Fig. 2 Expanded ^{1}H NMR spectra (δ 2.00–δ 0.0) of threonine and alanine. Each number indicates the proton signals of the molecule. *IS* internal standard (TMSP)

1. First, examine the region of aliphatic amino acids (δ 3.0–δ 0.5).
2. Focus on the range of δ 1.2–δ 0.8 to identify characteristic signals of valine, isoleucine, and leucine (*see* **Note** 7). Valine exhibits two prominent CH_3 signals, both doublets, at δ 1.06 and δ 1.00. Leucine is characterized by a triplet at δ 0.96, while

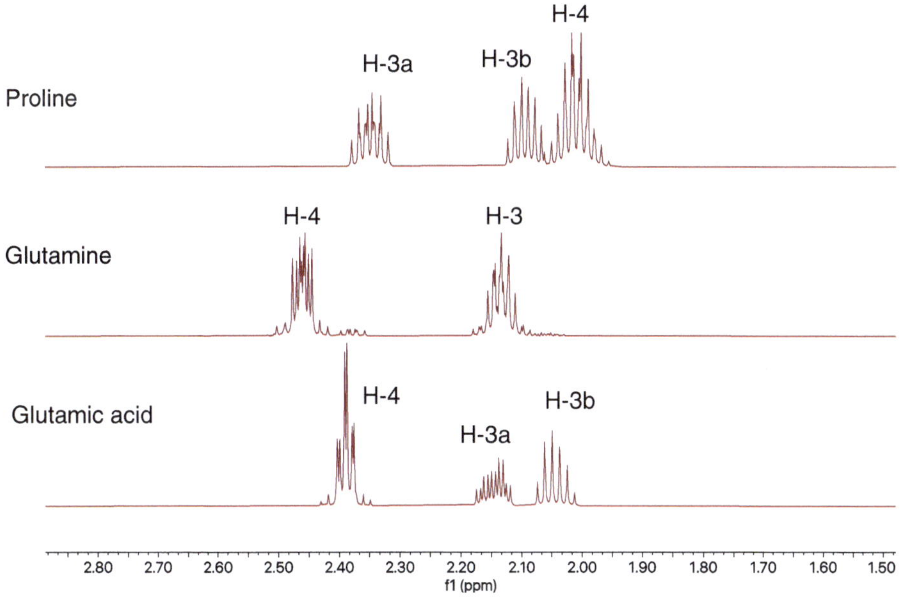

Fig. 3 Expanded ^{1}H NMR spectra (δ 2.85–δ 1.50) of proline, glutamine, and glutamic acid. Each number indicates the proton signals of the molecule

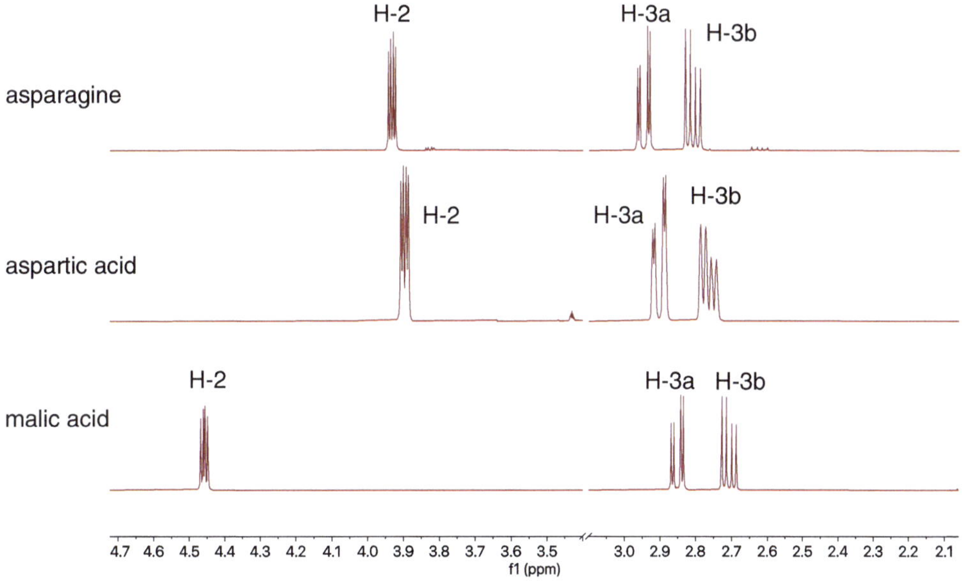

Fig. 4 Expanded ^{1}H NMR spectra (δ 4.7–δ 2.1) of asparagine, aspartic acid, and malic acid. Each number indicates the proton signals of the molecule. The solvent peak from MeOD at δ 3.3 was excluded due to its excessive size

isoleucine displays a triplet at δ 0.96 and a doublet at δ 1.03. (Fig. 1) (*see* **Note 8**).

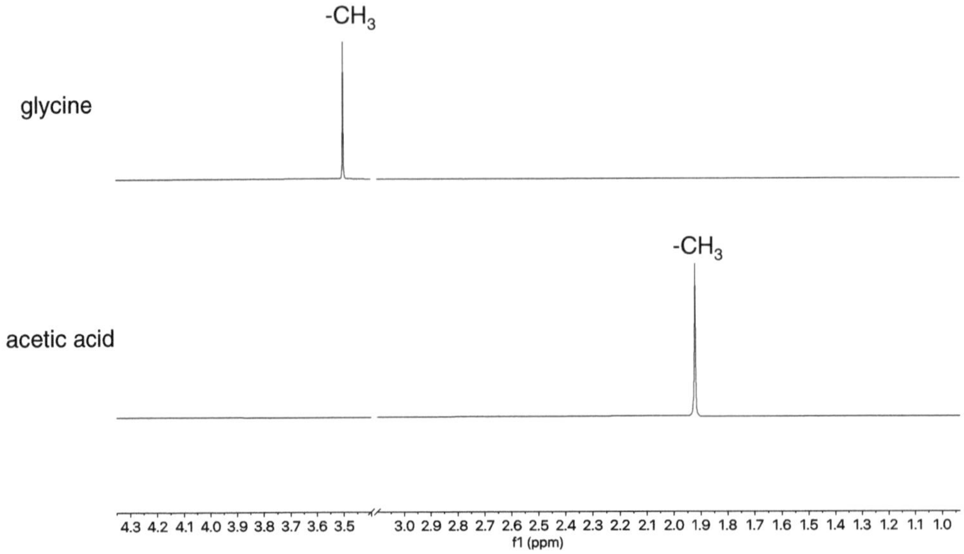

Fig. 5 Expanded ^{1}H NMR spectra (δ 4.3–δ 1.0) of glycine and acetic acid. In this area, only the methyl signals ($–CH_3$) of the molecules are observable. The solvent peak from MeOD at δ 3.3 was excluded due to its excessive size

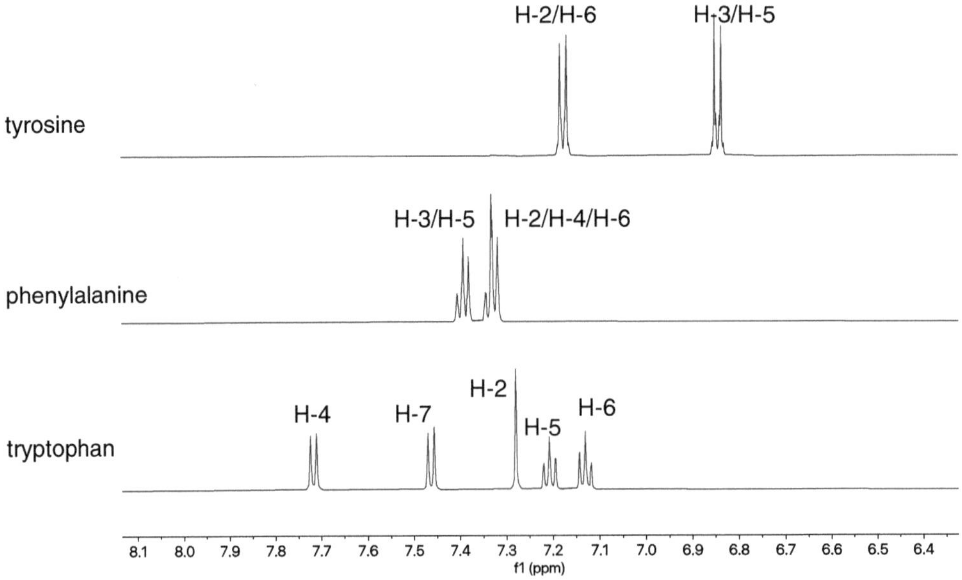

Fig. 6 Expanded ^{1}H NMR spectra (δ 8.1–δ 6.4) of tyrosine, phenylalanine, and tryptophan. Each number indicates the proton signals of the molecule

3. Threonine is detected at δ 1.32 with a doublet in the ^{1}H NMR spectrum of the plant extract. However, it may be overlapped with signals from lactic acid signals (*see* **Note 9**).
4. Alanine is detected at δ 1.48 with a doublet in ^{1}H NMR of the plant extract (Fig. 2).

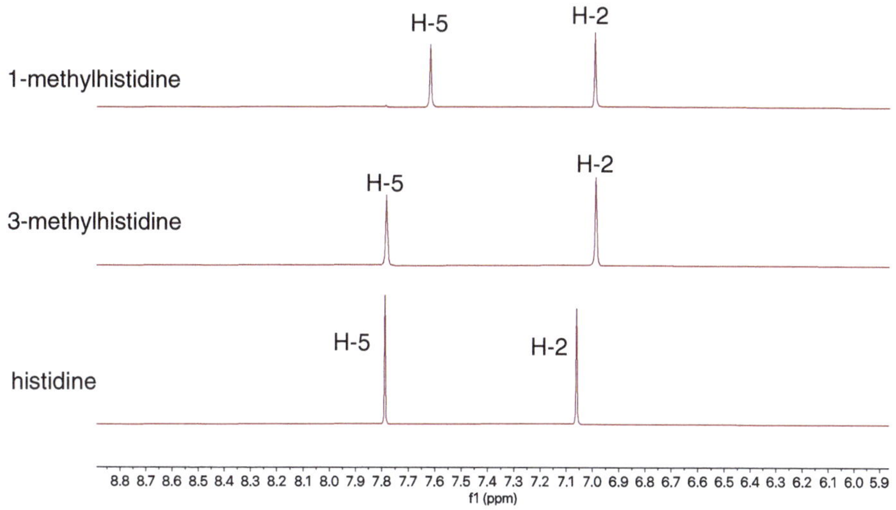

Fig. 7 Expanded ^{1}H NMR spectra (δ 8.8–δ 5.9) of 1-methylhistidine, 3-methylhistidine, and histidine. Each number indicates the proton signals of the molecule

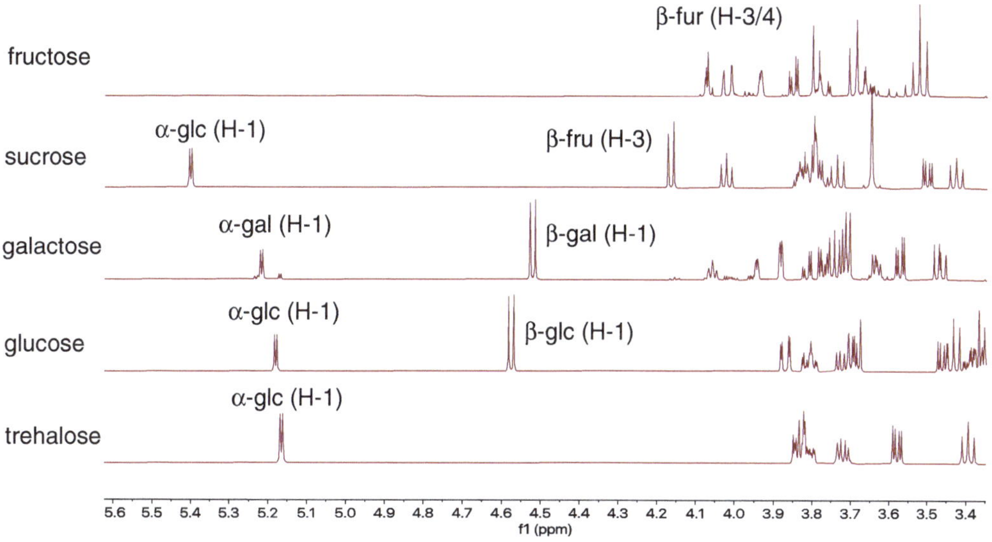

Fig. 8 Expanded ^{1}H NMR spectra (δ 5.6–δ 3.4) of common sugars (fructose, sucrose, galactose, glucose, and trehalose). The H-3 and H-4 signals of β-furanose form of fructose overlapped around δ 4.1. For full identification of fructose, *see* **Note 18**. In the sucrose spectra, anomeric proton (H-1) of α-glucose and β-glucose is very distinctive. Both galactose and glucose exhibit α- and β-anomers with a similar chemical shift. *Fur* furanose, *glc* glucose, *gal* galactose, *fru* fructose

5. The characteristic signals of glutamine, glutamic acid, and proline are typically found in the range of δ 2.6–δ 2.0. The distinguishing feature between glutamic acid and glutamine lies in the signals around δ 2.05 (Fig. 3) (*see* **Note 10**).

6. Asparagine and aspartic acid are typically detected in the range of δ 4.0–δ 2.7 with a double-doublet pattern. Distinguishing between asparagine and aspartic acid solely by ^{1}H NMR spectra can be challenging due to subtle differences in chemical shift. Consequently, ambiguous annotation is often used for these amino acids. Their spectral pattern closely resembles that of malic acid, yet they exhibit a higher chemical shift for H-2. Most H-2 signals of amino acids are detected in the range of δ 4.0–δ 3.6. (Fig. 4) (*see* **Note 11**).
7. Examine the region of δ 5.0–δ 3.0. Besides the appearance of specific amino acid signals (specially H-2 of amino acids), most organic acids are also evident in this region (*see* **Note 12**). While most of amino acids can be identified with their methyl signals appearing in the aliphatic area, shorter amino acids such as glycine can be distinguished by the singlet for their CH_3 protons around δ 3.6 (Fig. 5).
8. Acetic acid exhibits a prominent singlet signal in many plant extracts, typically appearing around δ 2.1–δ 1.8 (Fig. 5). This signal is highly sensitive to pH variations, resulting in slight shifts in its chemical shift. However, its recognition is facilitated by its prevalence as a major peak in plant extracts (*see* **Note 13**).
9. Succinic acid appears around δ 2.4, as prominent singlet from its two CH_2 groups.
10. Amino acids containing benzene rings or heterocyclic rings typically exhibit signals within the range of δ 8.0–δ 6.5. For example, tyrosine is identified at δ 7.2 and δ 6.8, both appearing as doublets, attributable to its benzene ring protons (Fig. 6). Phenylalanine is typically detected around δ 7.4, often presenting as a multiplet in NMR spectra. However, it's important to note that the observed pattern may vary depending on factors such as concentration and the magnetic power of the NMR spectrometer employed (*see* **Note 14**) (Fig. 6).
11. In the aromatic region, five distinct types of tryptophan signals are typically detected: a doublet, doublet, singlet, triplet, and triplet. The pattern of the indole ring, which is characteristic of tryptophan, is unique. However, it's worth noting that there may be numerous similar compounds present in plant extracts, such as indole alkaloids. The presence of a singlet at δ 7.3 can serve as evidence of tryptophan due to its characteristic signal (Fig. 6).
12. Histidine is typically detected at δ 7.8 and δ 7.1, with two protons in the imidazole ring, both appearing as doublets with a very small coupling constant of 1.2 Hz. If the signals at δ 7.8 were to shift upfield (displaying a lower chemical shift), it could indicate the presence of 1-methyl histidine (*see* **Note 15**) (Fig. 7).

13. Examine where most of sugar signals appeared (δ 5.5–δ 3.0). In the spectra of monosaccharides, two isoforms, namely, α- and β-forms, are commonly detected. The ratio between the α- and β-forms can vary depending on the specific sugar. For example, the ratio of α- to β-forms of glucose differs from that of arabinose. However, the ratio between the α- and β-forms of the same sugar typically remains constant. For instance, the ratio of α- to β-forms of glucose is commonly reported as 34:66 [4]. In ^{1}H NMR spectra, the α- and β-forms are distinguished by the coupling constants of the anomeric protons (H-1). Usually, only one anomeric proton is detected in the ^{1}H NMR spectra, and its coupling constant helps to differentiate between the α- and β-forms (*see* **Note 16**).
14. Chemical shifts are not reliable for sugar analysis, as they can be significantly influenced by different forms and attachments to hydroxyl groups. Coupling constants, on the other hand, provide a more reliable method for sugar analysis. Typically, the same sugar in the same form exhibits consistent coupling constants, although there may be a few exceptions (*see* **Note 17**).
15. Most common sugars in plant extract are glucose and sucrose. Glucose exhibits two anomeric signals corresponding to α- and β-glucose in solution. The anomeric proton of α-glucose appears at δ 5.20 (*d*, J = 3.8 Hz), while that of β-glucose appears at δ 4.60 (*d*, J = 7.8 Hz) (Fig. 8). Sucrose is disaccharide consisting of glucose and fructose (β-D-fructofuranosyl-(2 → 1)-α-D-glucopyranoside). The anomeric proton of α-glucose moiety in sucrose appears around δ 5.40, while the fructose moiety of sucrose appears at δ 4.16 (H-3, *d*, J = 8.5 Hz) (Fig. 8).
16. The moiety of fructose and fructose in sucrose has different ^{1}H NMR chemical shifts (*see* **Note 18**).
17. Galactose and glucose exhibit similar ^{1}H NMR chemical shift, but the β-form of galactose demonstrates a lower chemical shift compared to that of glucose (*see* **Note 19**). β-Glucose appears around δ 4.6 and β-galactose at δ 4.5 (Fig. 8).
18. Trehalose possesses only one anomeric proton at δ 5.18, which may lead to misinterpretation as glucose. However, unlike glucose, trehalose lacks the anomeric proton of the β-form at δ 4.58. Therefore, the presence of only one anomeric proton at δ 5.18 without a corresponding β-form signal at δ 4.58 indicates trehalose, not glucose. Trehalose is a common sugar found in microbes [5]. Additionally, by analyzing the ratio between the intensities of signals at δ 5.18 and δ 4.58, one can determine the ratio between trehalose and glucose (Fig. 8).

4 Notes

1. It may require a few milliliters of 1.0 M NaOD to reach pH 6.0. Add it slowly, being careful to avoid contamination with any other solvent.
2. The amount of sample required can vary depending on the nature of the sample and the quality of the analytical equipment. Typically, we use 10–50 mg of dried samples for analysis by a 600-MHz NMR spectroscopy equipped with a cryoprobe. If the volume ratio (sample: solvent = 1:3) cannot be achieved due to larger sample volumes, you can increase the volume of the extraction solvent. After centrifugation, only 250–300 μL of clear solution is necessary to transfer into the 3-mm NMR tube.
3. Variable speeds (14,000–17,000×*g*) can be used to obtain a clear supernatant. For lower-speed centrifuge, more time is required to obtain a clear supernatant. Occasionally, light samples such as leaves may float on the surface and prove challenging to remove through centrifugation. In such instances, carefully pipette out the solution, avoiding any solid material, and then repeat the centrifugation process.
4. Different modules may be present in one apparatus, requiring different parameters. For example, using a 3- or 5-mm tube may have different settings on the machine.
5. Either MeOD or D_2O can be used for this purpose; however, using MeOD is preferable as it offers superior and faster shimming.
6. **Steps 2–4** can be set up in the automation system, although it is advisable to process the initial sample manually to guarantee high-quality spectra.
7. Signals are highly prominent as they are all coming from methyl ($-CH_3$) signals.
8. The chemical shift mentioned in this chapter is a specific value measured under the conditions described in the Materials and Methods section. Please be aware that depending on the temperature and solvents used, their chemical shift may change.
9. Distinguishing between the signals of lactic acid and threonine solely based on ^1H NMR spectra is quite challenging. However, threonine is commonly found, whereas lactic acid is less frequently encountered in plant extracts. Therefore, when analyzing plant extracts, we typically anticipate the presence of threonine.
10. Glutamine and glutamic acid both have two methylene groups (CH_2), and these four methylene protons appear as closely

grouped multiplets. However, the signals of glutamic acid appear at δ 2.14 (H-3a, 1H) and δ 2.05 (H-3b, 1H), while glutamine signals appear as a multiplet at δ 2.13 (H-3, 2H).

11. Please note that the chemical shift of malic acid is sensitive to both pH and concentration. For detailed information on how these factors influence its chemical shift, refer to the book chapter that describes their variations [6].
12. Protons from sugar will also be observed in this region (δ 5.0–δ 3.0); however, their description will be provided in the last section.
13. The singlet peak of acetic acid shifts downfield in the lower pH level [7].
14. Phenylalanine exhibits three distinct sets of magnetically different protons within its benzene ring, typically appearing as a multiplet spread across the range of δ 7.5–δ 7.3 in NMR spectra. With higher resolution, it becomes feasible to detect two or even three sets of protons distinctly separated. However, it's common for these signals to appear predominantly as a single multiplet in many cases [8].
15. Histidine signals are not typically prominent in plant extracts; however, they may occasionally be detected in plant samples. Because of its small coupling constant, it often appears as a pair of singlets rather than distinct doublets. It's important to note that while histidine can be metabolized to 3-methylhistidine in animal muscle, the occurrence of 1-methylhistidine is less common. Furthermore, both methylhistidines are not frequently found in plants [9].
16. Moreover, in nature, most sugars exist predominantly in a specific stereo form, typically the D-form, such as D-glucose (L-glucose being non-natural). However, sugars like arabinose, galactose, fucose, and rhamnose exist in both L- and D-forms, necessitating separation via a chiral column (enantiomers) in an HPLC system [10]. It's important to note that this separation is not achievable in NMR. When monosaccharides are dissolved in water, they undergo cyclization to form furanose (5-membered ring) and pyranose (6-membered ring) structures. Each of these forms can exist in both α- and β-configurations. Therefore, in essence, a single sugar can generate four distinct forms: α-furanose, β-furanose, α-pyranose, and β-pyranose. The preference between furanose and pyranose structures varies depending on the individual sugar [4]. While most hexoses predominantly exist in the pyranose form, pentoses can be found in both furanose and pyranose configurations. The coupling constant of anomeric protons differs between the α- and β-forms, typically ranging from

3 to 4 Hz for the α-configuration and 7 to 8 Hz for the β-configuration.

17. *C*-glycoside (in flavonoid glycoside) and S-glycoside (in glucosinolates) have different coupling constants [11]. β-Glucose in both C-glycoside and S-glycoside is 9–10 Hz, but in normal, the coupling constant of β-glucose is 7–8 Hz.

18. It has been documented that fructose exists in numerous tautomeric forms in aqueous solution, including β-pyranose, β-furanose, α-furanose, α-pyranose, and keto tautomers. The relative abundance of these forms at 20 °C in D_2O is reported as follows: β-pyranose (68.23%), β-furanose (22.35%), α-furanose (6.24%), α-pyranose (2.67%), and keto tautomers (0.5%). Please refer to the provided reference [12] for a comprehensive assignment of these tautomers in the NMR spectra. However, within the sucrose moiety, fructose exists solely in the β-furanose form. Therefore, the characteristic signal of fructose typically appears around δ 4.17 (H-3, d, $J = 8.0$ Hz). In free fructose, the signals for H-3 and H-4 of β-furanose, along with H-3 of α-furanose, typically appear together around δ 4.10 (Fig. 8).

19. The difference between glucose and galactose lies in the configuration at C4. Due to the configuration of galactose, it exhibits smaller coupling constants between H-3 and H-4, and H-4 and H-5 compared to glucose [13].

References

1. Verpoorte R, Choi YH, Mustafa NR et al (2008) Metabolomics: back to basics. Phytochem Rev 7:525–537. https://doi.org/10.1007/s11101-008-9091-7
2. Wolfender JL, Rudaz S, Choi YH et al (2013) Plant metabolomics: from holistic data to relevant biomarkers. Curr Med Chem 20:1056–1090. PMID: 23210790
3. Wishart DS, Cheng LL, Copié V et al (2022) NMR and metabolomics—a roadmap for the future. Metabolites 12:678. https://doi.org/10.3390/metabo12080678
4. El Khadem H (ed) (2012) Carbohydrate chemistry: monosaccharides and their oligomers. Elsevier, Amsterdam
5. Zhang Y, DeBosch BJ (2020) Microbial and metabolic impacts of trehalose and trehalose analogues. Gut Microbes 11:1475–1482. https://doi.org/10.1080/19490976.2020.1750273
6. Kim HK, Choi YH, Verpoorte R (2006) Metabolomic analysis of *Catharathus roseus* using NMR and principal component analysis. In: Saito K, Dixon RA, Willmitzer L (eds) Plant metabolomics: biotechnology in agricultural and forestry, vol 57. Springer, Heidelberg, pp 261–276
7. Martin M, Labouesse J, Canioni P et al (1993) N-acetyl-L- aspartate and acetate ^{1}H NMR signal overlapping under mild acidic pH conditions. Magn Reson Med 29:692–694. https://doi.org/10.1002/mrm.1910290518
8. Govindaraju V, Young K, Maudsley AA (2000) Proton NMR chemical shifts and coupling constants for brain metabolites. NMR Biomed 13: 129–153. https://doi.org/10.1002/1099-1492(200005)13:3<129::aid-nbm619>3.0.co;2-v. Erratum in: NMR Biomed 2015 July;28(7):923–924. PMID: 10861994
9. Almeida A, Schwalbach LM, Dewaal H et al (2008) 3-methylhistidine as an indicator for protein breakdown: an experimental model in male *Capra hircu*. Scand J Lab Anim Sci 35: 259–263. https://doi.org/10.23675/sjlas.v35i4.156

10. Lopes JF, Gaspar EMSM (2008) Simultaneous chromatographic separation of enantiomers, anomers and structural isomers of some biologically relevant monosaccharides. J Chromatogr A 1188:34–42. https://doi.org/10.1016/j.chroma.2007.12.016
11. Zhang Y-Q, Zhang M, Wang Z-L et al (2022) Advances in plant-derived C-glycosides: Phytochemistry, bioactivities, and biotechnological production. Biotech Adv 60:108030. https://doi.org/10.1016/j.biotechadv.2022.108030
12. Barclay T, Ginic-Markovic M, Johnston MR et al (2012) Observation of the keto tautomer of d-fructose in D_2O using ^{1}H NMR spectroscopy. Carbohydr Res 347:136–141. https://doi.org/10.1016/j.carres.2011.11.003
13. Clemens A (2017) Re: How to differentiate between glucose and galactose in NMR? Retrieved from: https://www.researchgate.net/post/How-to-differentiate-between-glucose-and-galactose-In-NMR/59ee9583cbd5c253612eae3a/citation/download

Chapter 17

Isolation of Natural Products Using Preparative TLC

Kyu Won Kim, Kyung Ho Lee, Won Sok Ha, Yena Kim, and Young Pyo Jang

Abstract

Known for their diverse and potent physiological activities, natural products continue to be essential for the discovery and development of new drugs. This chapter explores the pivotal role of preparative thin-layer chromatography (Prep-TLC) in the isolation of natural products. This chapter begins with an understanding of the historical significance and structural complexity of natural products, and discusses the problems caused by complex mixtures present in extracts, as well as the multifunctionality, cost-effectiveness, and compatibility with different sample types of Prep-TLC to address these challenges. We then explain the problems of separating these compounds and introduce Prep-TLC as a solution. The focus shifts to the fundamental principles and methodologies of Prep-TLC, highlighting its role in achieving the efficient separation and isolation of natural products. To illustrate the successful application of Prep-TLC to isolate bioactive compounds from various natural sources, real-world case and case studies are presented. In this chapter, we aim to provide practical assistance to researchers and practitioners by explaining the principles and methodologies of Prep-TLC that facilitate the discovery of novel physiologically active compounds.

Key words Preparative TLC, Natural products, Isolation, Purification, Medicinal plants

1 Introduction

Natural products, which originated from a wide range of biological sources, including plants, fungi, and microorganisms, have been a diverse source of physiologically active compounds with a potential as a lead compound for the drug development [1]. Complex structures found in nature often become essential targets for the discovery and development of new drugs because they have unique pharmacological properties. However, the separation of biologically active compounds from natural crude extract is a formidable task that requires a strategic approach. In this chapter, we explore the important role of preparative thin-layer chromatography (Prep-TLC) as a powerful technique for the isolation of natural products, a key step in the search and utilization of these natural compounds.

Young Hae Choi et al. (eds.), *Natural Product Isolation and Identification: Methods and Protocols*, Methods in Molecular Biology, vol. 2895, https://doi.org/10.1007/978-1-0716-4350-1_17,

The isolation of natural products involves navigating through a complex matrix of compounds present in the crude extract obtained from the natural source. This mixture often includes a myriad of structurally related and unrelated compounds, necessitating the use of efficient separation techniques. While various methods such as liquid–liquid extraction, column chromatography, and solid-phase extraction are commonly employed, preparative TLC has been employed as a versatile and efficient tool for the initial separation of complex mixtures.

Thin-layer chromatography (TLC) has long been utilized as a qualitative analytical technique, allowing researchers to visualize and identify components within a mixture. However, the preparative variant of TLC extends its utility to the isolation of compounds in larger quantities [2]. This technique involves the separation of compounds on a thin layer of adsorbent material, typically silica gel or alumina, followed by the elution and collection of individual fractions for further analysis. Some examples of natural product isolation using the Prep-TLC technique are listed in Table 1, drawing upon a collection of scientific articles that showcase the versatile and impactful applications of Prep-TLC in the realm of natural product isolation.

The simplicity, cost-effectiveness, and speed of Prep-TLC make it an attractive choice for the isolation of natural products. Moreover, its compatibility with a wide range of solvents and the ability to handle a diverse set of sample types contribute to its versatility in the isolation process. This chapter aims to provide an overview of the principles, methodologies, and applications of Prep-TLC in the isolation of natural products.

2 Materials and Basic Preparations

The successful implementation of Prep-TLC for the isolation of natural products relies on the use of appropriate materials and equipment. In the past, the plates were often self-prepared in the laboratory for preparative purpose; however, commercial plates are mostly used in modern times in order to acquire robust and reproducible result. All the solvents for sample preparation and development should be at least of analytical grade.

Choosing the right chromatographic plate is fundamental to achieve successful separations.

Various considerations come into play, such as the nature of the compounds, the desired resolution, and the compatibility with the selected adsorbent. Glass and aluminum are usual materials for backing (support) type of plate. The glass plate is rigid and transparent, and offers high chemical and heat resistance, so it is mostly used for the analytical TLC plate. However, in the case of preparative purpose, the recovery of target compounds is mostly important

Table 1
Examples of Prep-TLC isolation of natural products

Compound	Plate type	Development solvent	References
Diketopiperazine	Silica gel 60 F254 0.5 mm	$CHCl_3$-MeOH (95:5)	[3]
Peptide	Silica gel 60 F254 2 mm	$CHCl_3$-MeOH (75:25)	[4]
Ephedrine	Aluminum oxide	Toluene-$CHCl_3$-EtOH-MeOH (2:5:3:1)	[5]
6-Hydroxykynurenic acid	Silica gel F254	EtOAc-formic acid-AcOH-H_2O (80:5:6:10)	[5]
Flavone glycoside	Silica gel F254	$CHCl_3$-EtOH-glacial AcOH (94:5:1)	[6]
Flavanonol	Silica gel F254	n-BuOH-AcOH-H_2O (2:2:6)	[7]
Porphyrins	Silica gel G	PE-$CHCl_3$ (1:5)	[8]
Sesamin	Silica gel 60	MeOH-acetone (1:4)	[9]
Coumarins	Silica gel 60	MeOH-acetone (1:4)	[9]
Vindoline	Silica gel 60	PE-EtOAc-acetone-EtOH (70:20:10:1)	[10]
L-dopa	Silica gel F254	BuOH-AcOH-H_2O (6:1:2)	[11]
Imide	Silica gel 60 F254	n-Hex-acetone (12:1)	[12]
Polymethoxy flavone	Silica gel	Hex-EtOAc (60:40)	[13]
Phenoxazine pigment	Silica gel	Hex-EtOAc (40:60)	[14]
Octabenzone	Silica gel	$CHCl_3$	[15]
Fatty acid alkyl ester	Silica gel 60 F254 0.5 mm	Toluene-EtOAc-diethylamine (7:2:1)	[16]
Amaryllidaceae alkaloid	Silica gel 60 F254	Toluene-Diethylamine (9:1)	[17]
Physciosporin	Silica gel 60 F254	CH_2Cl_2-MeOH (95:5)	[18]
Olivomycin A derivative	Silica gel	BuOH-AcOH-H_2O (75:3:25)	[19]
Phyllacanthone	Silica gel	Hexane-EtOAc (9:1)	[20]
Labdane diterpene	Silica gel	Hexane-EtOAc (85:15)	[21]
Oleanolic acid	Silica gel 60	Toluene-EtOAc-glacial AcOH (7:2:1)	[22]
Caffeine	Silica gel 60	$CHCl_3$-EtOAc (1:3)	[23]
Lycopene	Silica gel	Acetone-MeOH (1:1)	[24]
Acetylated oligosaccharides	Silica gel 60 F254 1 mm	CH_2Cl_2-EtOAc (7:3)	[25]
Securinine	Silica gel 60 F254 1 mm	$CHCl_3$-MeOH-25% NH_4OH (10:1:0.01)	[26]

(continued)

Table 1 (continued)

Compound	Plate type	Development solvent	References
Coumarin ester	Silica gel	Acetone-$CHCl_3$ (12:88)	[27]
Cyclic peroxides	Silica gel 60 0.5 mm	Hexane-EtOAc (9:1)	[28]
Carvone	HPTLC silica gel 60 F254 0.2 mm	Toluene-EtOAc (95:5)	[2]
Resveratrol	HPTLC silica gel 60 F254 0.2 mm	$CHCl_3$-MeOH (9:1)	[2]
Rosmarinic acid	HPTLC silica gel 60 F254 0.2 mm	Toluene-EtOAc-formic acid (5: 4:1)	[2]
Rhodomycin analogues	Silica gel 60 F254	$CHCl_3$-MeOH-25% Aq. NH_3 (85: 14:1)	[29]
Kaempferol	Silica gel 60 0.5 mm	Benzene-AcOH-H_2O (125:72:3)	[30]
Guanosine	Silica gel 60 1 mm	EtOAc-AcOH-H_2O (10:2:3)	[31]
Coumarins	Silica gel 60	EtOAc-PE (various)	[32]
Urospermal-15-O-acetate	Silica gel 60 1.5 mm	CH_2Cl_2-MeOH (95:5)	[33]
b-Carboline alkaloid	Silica gel 60 F254	$CHCl_3$-MeOH (8:2)	[34]
Roseanone	Silica gel 60 F254	$CHCl_3$-MeOH (9:1)	[35]
2-Amino-3-dodecanol	Silica gel 60 F254	$CHCl_3$-MeOH (7:3)	[36]
Norophthalmic acid derivative	Silica gel 60 F254	Ethylmethylketone-EtOH-ammonia (1:1:1)	[36]
Peptide mixture	Silica gel 60 F254 0.5 mm	EtOAc-MeOH-H_2O (7:2:1)	[37]

and rigorous chemical derivatization is not necessary. Therefore, the aluminum plate can be a good alternative to the glass plate. Common types of chromatographic plates are discussed as follows.

2.1 Selection of Absorbent Materials of the Stationary Phase

The choice of adsorbent material for the stationary phase is critical for achieving optimal separation. Silica gel and alumina are the most common adsorbents, each with its own set of characteristics.

2.1.1 Silica Gel Normal Phase

It is suitable for a broad range of compounds due to its polar nature and offers versatility and compatibility with various solvents. It is a common choice for the isolation of natural products due to its compatibility with a diverse range of compounds found in complex mixtures.

Silica gel plates, such as Silica Gel 60 F254, are widely used for their versatility in separating a range of compounds based on polarity. They are available in different dimensions and thicknesses, allowing flexibility based on sample load and separation goals. It is recommended to use F254-free plates for the purpose of minimizing the incorporation of impurities, such as when the purity of the material to be separated is important or when an activity evaluation is performed with a separated material. Silica gel is hygroscopic, absorbing moisture from the environment, which may impact reproducibility. It is most important to control the humidity during Prep-TLC experiment.

2.1.2 Silica Gel Reverse Phase

It is suitable for the separation of nonpolar compounds, providing an alternative when normal-phase silica gel is not suitable. Since the separation capacity is selective for each material, it is important to select it through a preliminary experiment, and it is disadvantageous in separating large amounts of samples because it is more expensive than a normal-phase silica gel.

2.1.3 Alumina (Aluminum Oxide)

It provides selectivity based on acidity, making it ideal for separating polar and acidic/basic compounds. Different alumina plates offer varying degrees of acidity for fine-tuning separations. Alumina is often chosen when dealing with complex mixtures containing a variety of compounds with diverse polarities.

Alumina plates, like Alumina N and Alumina C, offer selectivity based on acidity and are suitable for polar and acidic/basic compounds. Choosing the appropriate alumina plate depends on the specific separation requirements.

2.2 Sample Application

The sample is quantitatively applied as bands using a capillary of the requested volume, or an automated device suitable for application by spraying on of indicated volumes as band. Since the sample dissolved in a solvent having a low polarity is less diffused when dropping, it is preferable to resolve the sample with a solvent having a low polarity if possible.

2.3 Developing Chamber

A glass chamber having a size sufficient to hold the plate is mainly used. Since it is very important to saturate the plate sufficiently before development for separation, chambers having two troughs are generally used. The saturation method will be described again in the protocol section.

2.4 Developing Solvent

The developing solvent is prepared in advance, considering the pre-saturation process, ideally just before the experiment, in sufficient quantity. For the experiment mentioned as an example, 100 mL of a mixture containing toluene and formic acid in a ratio of 4:1 was prepared. The remaining developing solvent after use is stored in glass bottles with lids that can be tightly sealed.

2.5 Visualization

To specify the zone of target compounds, a UV lamp (254 nm and 366 nm) is usually utilized. In the case of compounds that have no chromophore, some chemical derivatizations are needed for a small part of the plate or co-developed separate plate. Derivatizing agents and a suitable hot plate for heating are needed in this case.

2.6 Recovery

Suitable stainless spatula is required to scrape off the target bands from the plate. A sonicator is recommended to extract the target substance from the scraped silica gel band. Separate the extraction solvent and silica gel using a tabletop centrifuge with a performance of 1000×*g* or higher.

2.7 Other Consumables

Assemble glassware, solvent reservoirs, syringes, filters, storage vials, and any other consumables are needed for the experiment.

3 Methods

A detailed step-by-step protocol is provided for the execution of Prep-TLC for the isolation of natural products. This protocol encompasses the preparation of materials, sample loading, development of the chromatogram, visualization, and extraction of target compounds. Following protocol is for a 20 × 20 cm silica gel plate. As a specific example, the separation of curcuminoids from turmeric (*Curcuma longa* rhizome) extract was exemplified.

3.1 Sample Preparation

1. Place 1 g of powdered turmeric into a test tube, and add 2 mL of MeOH. Extract the mixture using an ultrasonicator for 30 min. Filter the extract using a filter paper, and remove the solvent using a vacuum rotary evaporator. Dissolve the extract in ethyl acetate, and adjust the concentration to 100 mg/mL. Dissolving in low-polar solvent such as $CHCl_3$ or acetone, if possible, helps minimize the spread of the band when loading sample on the plate (*see* **Note 1**).
2. Prepare the developing solvent by mixing 80 mL of toluene and 20 mL of acetic acid which was chosen by the preliminary TLC study to get optimal separation of target compounds and store in the glass bottle with lid (*see* **Note 2**).
3. Prepare 20 × 20 cm preparative TLC plate. With a ruler and a pencil, mark the application position (15 mm from one edge) and an expected position of the developing front (e.g., 10–15 mm) on the other edge of the plate. The length of the band is set to a maximum of 18 cm.

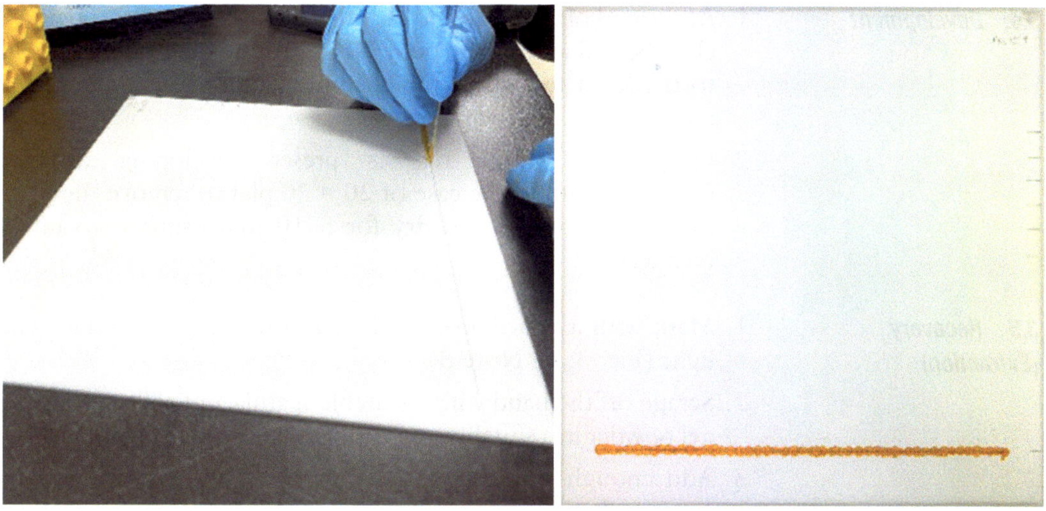

Fig. 1 Application of sample into the Prep-TLC plate. A narrower sample band is recommended for the good separation result

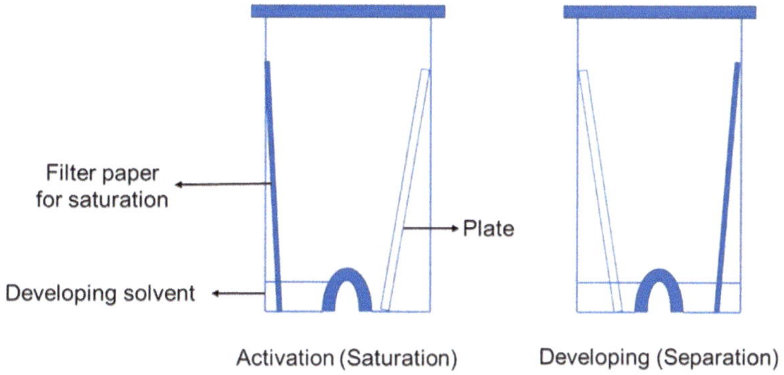

Fig. 2 Saturation and developing a Prep-TLC plate in a twin trough TLC chamber

3.2 Loading Sample

Load 200 μL of sample onto the starting line of plate with capillary or micropipette (Fig. 1). Repeatedly apply small amounts of sample to start line to minimize diffusion. If available, automatic sample loader helps to get a better separation and an accurate amount.

3.3 Saturation

1. Use a twin trough chamber. Twenty min prior to the development, place a saturation filter paper in the rear trough.
2. Pour 30 mL of the developing solvent onto the filter paper at the back of the twin trough chamber, and place the sample-loaded plate at the front empty trough with the layer facing the saturation filter paper (Fig. 2).
3. Cover the lid and allow the chamber saturation for 20 min.

3.4 Development

1. At the end of the plate and chamber saturation, gently open the chamber lid and slowly pour 20 mL of the developing solvent to the front trough of the chamber along the corners of the inner wall of the chamber.
2. Once the solvent reaches preset developing distance (120–175 mm in the case of 20 × 20 plate), remove the plate from the chamber and dry for 5–10 min using a stream of cold air.

3.5 Recovery (Extraction)

1. Mark with a pencil around the band to scrape off under UV light (Fig. 3, *see* **Note 3**).
2. Scrape off the band with a suitable spatula and collect the silica gel powder in a suitable microtube or glassware (Figs. 4 and 5).
3. Add enough solvent such as MeOH to be the powder fully immersed.
4. Sonicate it more than 10 min so that the compound can be sufficiently extracted from the silica gel.
5. Centrifuge at 1000×*g* for 5 min, and filter the supernatant through a 0.2-mm syringe filter.
6. Add enough solvent to the precipitated silica gel, and repeat three times **steps 3–5**.
7. Combine the filtrate and remove the solvents using a vacuum rotary evaporator or speed-vac. To calculate the exact amount of isolated compound, the weight of an empty Eppendorf tube or vial was weighed precisely before.
8. Calculate the yield using the following equation:

$$\text{Yield }(\%) = \frac{\text{Final amounts after evaporation (mg)}}{\text{Loaded sample volume (mL)} \times \text{Sample concentration (mg/mL)}} \times 100.$$

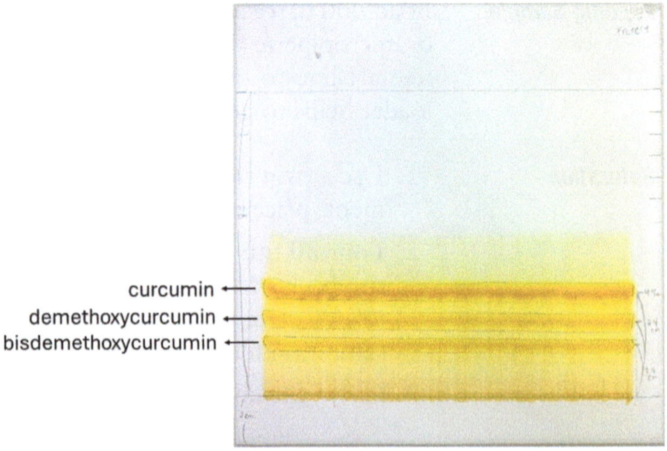

Fig. 3 Development of turmeric extract on a Prep-TLC plate. Three major bands are corresponding to curcumin, demethoxycurcumin, and bisdemethoxycurcumin, respectively, from the top

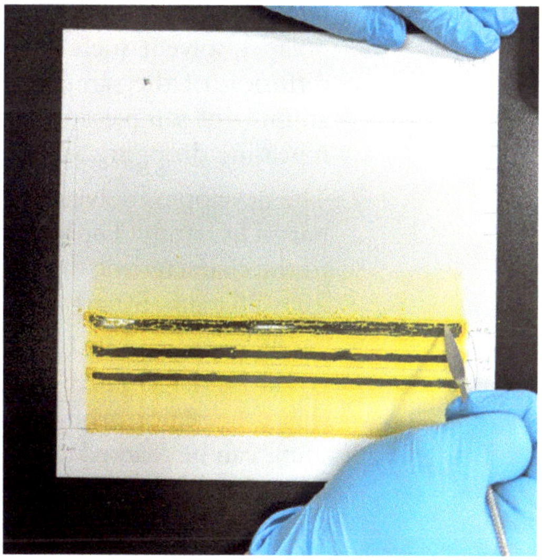

Fig. 4 Scraping out the target band from a glass TLC plate. The marked area is carefully scraped out using a stainless spatula without loss

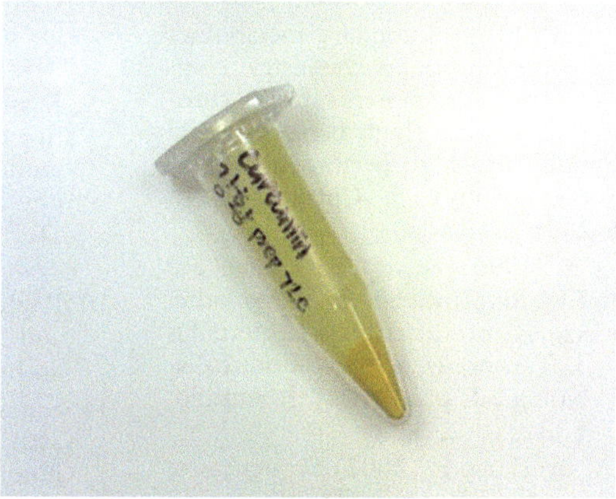

Fig. 5 Microtube containing scraped-out silica gel powder from a TLC plate. Some extracting solvents such as MeOH and $CHCl_3$ are used to extract target natural products from the silica gel thoroughly

4 Notes

1. In the case of the plant extract, the general range for sample concentration for Prep-TLC lies between 100 mg/mL and 500 mg/mL. For more purified mixture, such as reaction mixture which contains a few compounds, the concentration can be elevated up to a gram scale. Regarding the dissolving solvent for the sample, selecting a solvent with lower polarity among solvents capable of sufficiently dissolving the sample is a

method to minimize the diffusion of the sample. If a high polarity solvent such as alcohol or water should be used, the diffusion of the sample can be reduced by dropping a small amount of sample and blowing a solvent enough time, then repeating dropping again.

2. The developing solvent for Prep-TLC is chosen by the preliminary TLC study. The optimized solvent system varies according to the characteristic of the sample and TLC plate. In the case of 20 × 20 plate, 50 mL of developing solvent is enough.

3. In the case of non-UV-absorbing compounds, the Rf value could be determined during the preliminary TLC study, and then, the region around the Rf value in the Prep-TLC experiment can be marked as a slightly wider region to minimize the loss of the target compound.

References

1. Cragg GM, Newman DJ (2013) Natural products: a continuing source of novel drug leads. Biochim Biophys Acta 1830(6):3670–3695
2. Do TKT et al (2014) Secondary metabolites isolation in natural products chemistry: comparison of two semipreparative chromatographic techniques (high pressure liquid chromatography and high performance thin-layer chromatography). J Chromatogr A 1325:256–260
3. Cui CB et al (1996) Novel mammalian cell cycle inhibitors, tryprostatins A, B and other diketopiperazines produced by Aspergillus fumigatus: I. Taxonomy, fermentation, isolation and biological properties. J Antibiot 49(6):527–533
4. Jaworski A, Brückner H (2001) Sequences of polypeptide antibiotics stilboflavins, natural peptaibol libraries of the mold Stilbella flavipes. J Pept Sci 7(8):433–447
5. Khaleefa AS, Hamad MN (2020) Isolation, structural characterization and identification of major constituents in Ephedra foliata naturally growing in Iraq by TLC, GC-MS and UPLC-ESI-MS/MS. Iraqi J Pharm Sci 29(2): 48–61. (P-ISSN: 1683-3597, E-ISSN: 2521-3512)
6. Syarifah AL, Retnowati R, Soebiantoro S (2019) Characterization of secondary metabolites profile of flavonoid from Salam leaves (Eugenia polyantha) using TLC and UVSpectrophotometry. Pharm Sci Res 6(3):9
7. Sharma V, Janmeda P (2017) Extraction, isolation and identification of flavonoid from Euphorbia neriifolia leaves. Arab J Chem 10(4):509–514
8. Cardinal RA et al (1969) Isolation of porphyrins from porphyria urine by preparative thin-layer chromatography. J Chromatogr 38:6
9. Kassim NK et al (2019) Isolation of antioxidative compounds from Micromelum minutum guided by preparative thin layer chromatography-2,2-diphenyl-1-picrylhydrazyl (PTLC-DPPH) bioautography method. Food Chem 272:185–191
10. Raghu N et al (2023) Isolation, purification and characterization of vindoline from Catharanthus roseus. Int J Sci Res Arch 8(1):310–319
11. Rima et al (2023) Nutritional content and the activities of L-Dopa (L-3,4-dihydoxyphenyalanine) from Mucuna pruriens L. DC seeds of Central Java accession. Arab J Chem 16(1):13
12. Cheng MJ et al (2023) Natural product from the culture broth of Actinobacteria Actinomadura miaoliensis. Chem Nat Compd 59(2): 351–353
13. Aksit H, Ozan E, Erenler R (2023) Isolation, characterization, and antiproliferative activity of polymethoxy flavones from *Mentha dumetorum*. Nat Prod J 13(7):7
14. Lagashetti AC et al (2022) Antioxidant, antibacterial and dyeing potential of crude pigment extract of Gonatophragmium triuniae and its chemical characterization. Molecules **27**(2): 393
15. Mansour A et al (2021) Octabenzone isolated from Saharan Myrtle leaves (*Myrtus nivellei*). Nat Prod J 11(3):400–403

16. Bhadekar NS, Zodape GV (2021) Isolation and partial purification of bioactive compounds from sponge sigmadocia fibulata (schmidt) collected from west coast of Mumbai, India. Biomed Pharmacol J 14(3):1675–1684
17. Al Shammari L et al (2020) Alkaloid profiling of *Hippeastrum* cultivars by GC-MS, isolation of Amaryllidaceae alkaloids and evaluation of their cytotoxicity. Rec Nat Prod 14(2): 154–159
18. Taş İ et al (2019) Physciosporin suppresses the proliferation, motility and tumourigenesis of colorectal cancer cells. Phytomedicine 56:10–20
19. Gozari M et al (2019) An "olivomycin A" derivative from a sponge-associated Streptomyces sp. strain SP 85. 3 Biotech 9(12):439
20. de Oliveira RG et al (2018) Antibacterial activity of terpenoids isolated from *Cnidoscolus quercifolius* Pohl (Euphorbiaceae), a Brazilian medicinal plant from Caatinga biome. Eur J Integr Med 24:30–34
21. Dziwornu GA et al (2017) Isolation, characterization and antiproliferative activity of new metabolites from the South African endemic red algal species laurencia alfredensis. Molecules 22(4):513
22. Biradi M, Hullatti K (2017) Bioactivity guided isolation of cytotoxic terpenoids and steroids from *Premna serratifolia*. Pharm Biol 55(1): 1375–1379
23. Wahyuono S, Widyarini S (2016) Optimization method of caffeine isolation of merapi green coffee beans. Int J PharmTech Res 9(5): 255–259
24. Saeid A et al (2016) Effects of extraction and purification methods on degradation kinetics and stability of lycopene from watermelon under storage conditions. J Food Sci 81(11): C2630–C2638
25. Ben Youssef S et al (2016) Efficient purification and complete NMR characterization of galactinol, sucrose, raffinose, and stachyose isolated from *Pinus halepensis* (Aleppo pine) seeds using acetylation procedure. J Carbohydr Chem 35(4):224–237
26. Sparzak B, Dybowski F, Krauze-Baranowska M (2015) Analysis of *Securinega*-type alkaloids from *Phyllanthus glaucus* biomass. Phytochem Lett 11:353–357
27. Razavi SM, Janani M (2015) A new ester coumarin from Ferula Persica wild, indigenous to Iran. Nat Prod Res 29(8):717–721
28. Hoye TR et al (2015) New cytotoxic cyclic peroxide acids from Plakortis sp. marine sponge. ARKIVOC 2015(5):164–175
29. Holkar S et al (2013) Rhodomycin analogues from Streptomyces purpurascens: isolation, characterization and biological activities. Springerplus 2(1):93
30. Gautam K, Kumar P (2013) Isolation, quantification and anti-microbial screening of flavonoids from callus of Vitex negundo Linn. Int J Green Pharm 7(4):309–314
31. Fuentes E et al (2013) Protective mechanisms of guanosine from Solanum lycopersicum on agonist-induced platelet activation: role of sCD40L. Molecules 18(7):8120–8135
32. Smyth TJP et al (2012) Investigation of antibacterial phytochemicals in the bark and leaves of Ficus coronata by high-performance liquid chromatography-electrospray ionization-ion trap mass spectrometry (HPLC-ESI-MSn) and ESI-MSn. Electrophoresis 33(4):713–718
33. Jansen O et al (2012) Anti-plasmodial activity of Dicoma tomentosa (Asteraceae) and identification of urospermal A-15-O-acetate as the main active compound. Malar J 11:289
34. Bratchkova A et al (2012) β-Carboline alkaloid constituents from a Thermoactinomyces Sp. strain isolated from Livingston Island, Antarctica. Biotechnol Biotechnol Equip 26(3): 3005–3009
35. Jothy SL et al (2011) Bioassay-directed isolation of active compounds with antiyeast activity from a *Cassia fistula* seed extract. Molecules 16(9):7583–7592
36. Ivanova V et al (2010) Structural elucidation of a bioactive metabolites produced by Streptomyces avidinii SB9 strain, isolated from permafrost soil in Spitsbergen, Arctic. Biotechnol Biotechnol Equip 24(4):2092–2095
37. Maddau L et al (2009) Occurrence and characterization of peptaibols from Trichoderma citrinoviride, an endophytic fungus of cork oak, using electrospray ionization quadrupole time-of-flight mass spectrometry. Microbiology 155(10):3371–3381

Chapter 18

Sample Preparation of *Caenorhabditis elegans* for GC-MS-Based Metabolomics in Toxicity Assessment

Jesus G. Arellano Spadaro, Marjolein Wildwater, Henrie Korthout, and Hye Kyong Kim

Abstract

The nematode *Caenorhabditis elegans*, widely recognized as a model organism due to its ease of breeding and well-characterized genomes, boasts complete digestive, reproductive, and endocrine systems, as well as conserved signaling pathways shared with mammals. It has become an invaluable resource for metabolomics research, particularly in examining responses to chemical or environmental factors and toxicity assessments. In this article, we provide detailed, step-by-step protocols for cultivating *C. elegans* and conducting metabolomics analyses, specifically focusing on sample preparation for GC-MS analysis in response to toxic compound treatments. We highlight the critical choice of extraction solvent, introducing two representative systems for extracting metabolites from *C. elegans*.

Key words *Caenorhabditis elegans*, Extraction, Metabolomics, GC-MS analysis, Toxicity assessment, Extraction solvent

1 Introduction

Caenorhabditis elegans, a soil nematode, is widely utilized as a model organism due to its short life cycle, ease of breeding, and well-characterized genomes [1–4]. This nematode is particularly valuable for conducting intricate studies on development, metabolism, reproduction, and various behaviors, which can be challenging to investigate in higher model organisms [5–7]. Notably, it has emerged as a reliable model for investigating toxicity and its associated mechanisms [8]. This is attributed to its complete and metabolically active digestive, reproductive, endocrine, and sensory-neuromuscular systems, along with conserved signaling pathways shared with mammals [9]. One of our studies aimed to assess the toxicity of widely recognized harmful compounds, such as toluene and benzene. To investigate the impact of these toxic compounds on *C. elegans*, we employed a

Young Hae Choi et al. (eds.), *Natural Product Isolation and Identification: Methods and Protocols*, Methods in Molecular Biology, vol. 2895, https://doi.org/10.1007/978-1-0716-4350-1_18,

metabolomics approach as a comprehensive method to understand their overall influence on metabolism. To achieve this, a robust protocol was essential, encompassing the cultivation of nematodes and the optimization of extraction methods for subsequent analysis. Given the challenging nature of obtaining small-sized samples in large quantities, we opted for the most sensitive analytical method—namely, gas chromatography-mass spectrometry (GC-MS). Consequently, we will now outline a robust protocol for the sample preparation of *C. elegans* to facilitate metabolomics analysis using GC-MS.

2 Materials

Prepare all solutions using ultrapure water (Milli-Q water) and all reagents with analytical grade.

Prepare and store all reagents at room temperature otherwise specifically mentioned.

2.1 Potassium Phosphate Buffer

1. 1 M Potassium phosphate dibasic trihydrate (1 M $K_2HPO_4 \cdot 3H_2O$): Dissolve 228.23 g $K_2HPO_4 \cdot 3H_2O$ in 1 L Milli-Q water.
2. 1 M Potassium phosphate monobasic (1 M KH_2PO_4): Dissolve 136.09 g KH_2PO_4 in 1 L Milli-Q water.
3. Preparation of potassium phosphate buffer: Add 132 mL of 1 M $K_2HPO_4 \cdot 3H_2O$ and 868 mL of 1 M KH_2PO_4 to a 1 L glass bottle (pH 5.9 ± 0.1). Autoclave buffer for 20 min at 120 °C. Let buffer cool to room temperature and measure pH (*see* **Note 1**). Store at room temperature for up to several months.

2.2 Alkaline-Bleach Solution

1. Milli-Q water.
2. 5% Sodium hypochlorite solution.
3. 4 M NaOH: Add 200 mL of Milli-Q water to a 250 mL glass bottle. Weigh 40 g of NaOH and transfer it to the 250 mL glass bottle. Stir with a magnetic rod until dissolved. Top off with Milli-Q water until a final volume of 250 mL. Store at room temperature for up to 6 months.
4. Preparation of alkaline-bleach solution: Add 7 mL Milli-Q water, 8 mL 5% sodium hypochlorite solution, and 5 mL 4 M NaOH to a 50 mL glass bottle. Prepare fresh before use.

2.3 M9-T (Tween-20) Buffer

1. 1 M Magnesium sulfate ($MgSO_4$): Dissolve 1.2 g in 10 mL Milli-Q water (*see* **Note 2**). Use a 0.22 μm membrane filter for sterilization purposes. 10% Tween 20: Add 1 mL Tween

20 to 9 mL Milli-Q water. Filter with a 0.22 μm membrane filter.

2. M9 buffer: Weigh 3.0 g of KH_2PO_4, 7.5 g of $Na_2HPO_4 \cdot 2H_2O$ (sodium phosphate dibasic dihydrate), and 5.0 g of NaCl (sodium chloride). Add to a 1 L glass bottle and fill up to 1 L Milli-Q water. Autoclave M9 buffer for 20 min at 120 °C. Let buffer cool to room temperature.
3. Preparation of M9-T buffer: Add 1 mL of filtered 1 M $MgSO_4$ and 5 mL of 10% Tween 20 to 1 L M9 buffer and mix well. Store at room temperature for up to several months.

2.4 Nematode Growth Medium (NGM) Plates

1. 1 M Calcium chloride ($CaCl_2$): Dissolve 1.47 g in 10 mL Milli-Q water (*see* **Note 2**). Filter with a 0.22 μm membrane filter.
2. 1 M $MgSO_4$: Prepare as described in Subheading 2.3, **item 1**.
3. 0.13 M Cholesterol: Dissolve 50 mg in 10 mL of 100% Ethanol (EtOH) (*see* **Note 2**). Filter with a 0.22 μm membrane filter.
4. Potassium phosphate buffer: As described in Subheading 2.1: "Potassium phosphate buffer."
5. NGM agar: Weigh 2.5 g of peptone, 3.0 g of NaCl, and 17 g of agar. Add to a 1 L glass bottle and add 975 mL Milli-Q water. Autoclave for 20 min at 120 °C and let it cool to approximately 65 °C (*see* **Note 3**).
6. Preparation of NGM plates: Add 25 mL of potassium phosphate buffer (complete), 1 mL of 1 M MgSO4, 1 mL of 1 M $CaCl_2$, and 1 mL of 0.13 M cholesterol to the 975 mL autoclaved NGM agar which was left to cool down to 65 °C. Mix well and pipette 25 mL per 9 cm culture petri dish. Store NGM plates at room temperature for up to 2 weeks (*see* **Note 4**).

2.5 S-medium

1. 1 M $MgSO_4$: Prepare as described in Subheading 2.3, **item 1**.
2. 1 M $CaCl_2$: Prepare as described in Subheading 2.4, **item 1**.
3. 0.13 M Cholesterol: Prepare as described in Subheading 2.4, **item 3**.
4. 25 mg/mL Nystatin: Dissolve 50 mg in 2 mL Milli-Q water (*see* **Note 5**).
5. 10,000 U/mL Penicillin-Streptomycin (Pen-Strep) (*see* **Note 6**).
6. 1 M Potassium citrate tribasic monohydrate: Dissolve 3.24 g in 10 mL in Milli-Q water (*see* **Note 2**). Filter with a 0.22 μm membrane filter.
7. Trace metals: Weigh 1.86 g Disodium ethylenediaminetetraacetic acid (EDTA), 0.69 g $FeSO_4 \cdot 7H_2O$, 0.2 g $MnCl_2 \cdot 4H_2O$, 0.29 g $ZnSO_4 \cdot 7H_2O$, 0.025 g $CuSO_4 \cdot 5H_2O$. Transfer to a 1 L glass bottle and fill up to 1 L with Milli-Q water. Filter with a

0.22 μm membrane filter when needed. Store in the dark at room temperature.

8. S-Basal: Weigh 6 g KH_2PO_4, 1.3 g $K_2HPO_4 \cdot 3H_2O$, 5.85 g NaCl. Add to a 1 L glass bottle and fill till 1 L with Milli-Q water. Autoclave for 20 min at 120 °C. Let buffer cool to room temperature. Store at room temperature for up to several months.
9. Preparation of S-medium: Add 3 mL of 1 M $MgSO_4$, 3 mL of 1 M $CaCl_2$, 1 mL of 0.13 M cholesterol, 0.5 mL of 25 mg/mL nystatin, 0.5 mL of 10,000 U/mL penicillin-streptomycin, 10 mL of 1 M potassium citrate tribasic monohydrate, 10 mL trace metals to 972 mL of S-basal to prepare 1 L of S-medium (complete) (*see* **Note 7**). Prepare fresh and store in the dark until use (*see* **Note 8**).

2.6 *Escherichia Coli* OP50 Bacterial Strain Culture (OP50 Culture)

1. Luria Bertani (LB) broth: Weigh 8 g of Lennox L Broth Base. Add to a 1 L glass bottle and add 400 mL Milli-Q water. Autoclave for 20 min at 120 °C and let it cool to room temperature.
2. OP50 culture (complete): Pick 1 colony of overnight OP50, *Escherichia coli* strain, and inoculate at 37 °C at 150 rpm for 8–10 h, until OD600 = 0.700–1.000 (*see* **Note 9**). Store OP50 culture at 4 °C for up to 2 weeks.

2.7 Preparation of Test Solution

1. Stock solution: Weigh the amount of test samples (either pure compound or extracts) and dissolve them in either dimethyl sulfoxide (DMSO) or EtOH to obtain the desired concentration of the stock solution, usually 1–10 mg/mL (*see* **Note 10**).

2.8 Sample Extraction Solvent

1. Chloroform, MeOH, and water.
2. 80% MeOH solution (v/v, solvent A): Mix 80 mL MeOH and 20 mL water in the bottle.
3. 50% MeOH solution (v/v): Mix 10 mL MeOH and 10 mL water in the glass bottle (*see* **Note 11**).

2.9 Internal Standard Solution for GC-MS Analysis

1. 0.20 mg/mL methyl palmitate solution

   For solvent A: Weigh 2 mg of methyl palmitate into a 10 mL volumetric flask and fill it with pyridine up to the line.

   For solvent B: Weigh 2 mg of methyl palmitate into a 10 mL volumetric flask and fill it with pyridine up to the line (polar fraction).

   Weigh 2 mg of methyl palmitate into a 10 mL volumetric flask and fill it with chloroform up to the line (non-polar fraction).

2.10 Derivatization

Use *N,O*-Bis(trimethylsilyl)trifluoroacetamide (BSTFA) with 1% trimethylchlorosilane (TMCS). Commercially available.

3 Methods

Carry out all experiments at room temperature unless otherwise specified.

3.1 Synchronize L1 C. elegans Culture

1. One week before starting experiments, seed NGM plates by transferring 400 μL OP50 culture into NGM plates (*see* **Note 12**).
2. Pick 10 L4s, N2 Bristol *C. elegans* strain, and transfer to OP50 seeded NGM plates (*see* **Note 13**). Incubate at 16 °C for 7 days.
3. Following the expansion of the *C. elegans* culture after 7 days, harvest the nematodes by washing nematodes off the plates with 3–4 mL M9-T buffer and transfer them to a 15 mL tube. You can transfer a maximum of 10 plates per falcon tube (*see* **Note 14**). Centrifuge at 1400×*g* for 1 min and remove the supernatant to have a pellet of nematodes. Continue harvesting the other plates in the same 15 mL tube. Caution: The nematodes pellet should not have a volume of more than 1.5 mL before the bleach step, **step 4**.
4. Bleach nematode pellet by adding 10 mL alkaline-bleach solution and vortex for 45 s. Centrifuge the bleached nematodes at 1400×*g* for 1 min remove the supernatant. Bleach one more time by adding 10 mL Alkaline-bleach solution, vortex for 45 s. and centrifuge at 1400×*g* for 1 min.
5. After the second bleaching step, you should obtain a small pellet consisting of *C. elegans* eggs. Wash the pellet by adding 10 mL of M9-T solution, resuspending the egg pellets by gently inverting the tube, centrifuging at 1400×*g* for 1 min, and then removing the supernatant. Repeat this washing procedure two more times.
6. After the last wash with M9-T, remove the supernatant and resuspend the eggs with 2 mL M9-T.
7. Transfer the 2 mL egg suspension to a 6 cm falcon petri dish and incubate overnight at 20 °C.

3.2 Compound Treatment in the Liquid Media

1. Prepare 8 × concentrated OP50 culture by centrifuging the OP50 culture which has an $OD_{600} = 0.700–1.000$ at 2500×*g* for 15 min. Remove all the LB supernatant and resuspend with S-medium.
2. Add 30 mL of the 8× concentrated OP50 and 70 mL of S-medium to a 250 mL Erlenmeyer flask.

3. Take the overnight *C. elegans* suspension from the 20 °C incubator (*see* **Note 15**). Transfer to a 15 mL tube and count the concentration of nematodes per μL (*see* **Note 16**). In a maximum volume of 100 μL, add approximately 10,000 L1 *C. elegans* to the 250 mL flask.
4. Add the compound solution at different time points (Day-0, Day-1, and Day-2) at different concentrations (*see* **Note 17**).
5. After incubation with compounds, collect the ±10,000 nematodes per sample into a pre-weighed 2 mL microtube (*see* **Note 18**).
6. Store all tubes at −80 °C before freeze-drying.

3.3 Preparations for the Extraction

3.3.1 Freeze-Drying

1. Seal the microtube tube containing nematodes with parafilm after opening the lid and make a small hole with a needle.
2. Dry nematodes using a freeze dryer (*see* **Note 19**).
3. Take out the microtube tube from the freezer and weigh the biomass of nematodes.

3.3.2 Extraction

1. Add a glass bead (5 mm) to each tube (*see* **Note 20**).
2. Place the tubes in the Tissue Lyser for 5 min with 30 cycle/s to disrupt the tissues of the nematodes.
3. Add extraction solvent, (A) 1 mL of 80% MeOH or (B) for two-phase solvent: 0.5 mL of chloroform and 0.5 mL of 50% MeOH (*see* **Note 21**).
4. Sonicate for 15 min in an ultrasonication bath.
5. Centrifuge them for 15 min at 1400×*g*.
6. Collect supernatant (solvent A). Or collect each phase separately by pipetting (solvent B).
7. Repeat the extraction and combine each approximately 1 mL of extracts in the glass vial.
8. Dry them using Speed Vac at 35 °C (*see* **Note 22**).

3.4 GC-MS Analysis

3.4.1 Non-polar Phase (Solvent B)

1. Add 100 μL of internal standard solution (methyl palmitate, 0.20 mg/mL in chloroform) to the vial.
2. Dissolve them completely, if necessary ultrasonicate for 1 min to facilitate dissolving.
3. Centrifuge them for 10 min at 1400×*g*.
4. Take supernatant to the glass vial insert (*see* **Note 23**).
5. Analyze by GC-MS.

3.4.2 Polar Phase (Solvent B) or Extract from Solvent A

1. Add 100 μL of internal standard solution (methyl palmitate, 0.20 mg/mL in pyridine) to the vial.
2. Dissolve them completely, if necessary ultrasonicate for 1 min to facilitate dissolving.
3. Add 100 μL of BSTFA containing 1% TMCS.
4. Vortex them well and incubate them at 80 °C for 60 min.
5. Centrifuge them for 10 min at 1400×*g*.
6. Take supernatant to the glass vial insert.
7. Analyze by GC-MS.

3.5 GC-MS Condition

1. Set up GC-MS condition as follows.
2. GC-MS equipment: 7890A chromatograph, 7693 automatic samplers, 5975C mass single-quadrupole detector.
3. GC column: DB-5 (30 m × 0.25 mm, 0.25 μm film, J&W Science, Folsom, CA, USA).
4. Carrier gas: Helium (99.9% purity), flow rate: 1.0 mL/min.
5. Oven temp: 60 °C for 1 min, increase to 290 °C at 7 °C/min, hold for 5 min, increase to 310 °C at 5 °C/min, hold for 3 min.
6. Sample injection volume: 1 μL.
7. Injection temperature: 280 °C, splitless mode.
8. Interface temperature: 280 °C.
9. MS source temperature: 230 °C, MS Quadrupole temperature:150 °C.
10. Ionization (EI) energy: 70 eV.
11. Data process: Mass Hunter (B.07, Agilent), AMDIS (V. 2.63, Agilent, Folsom, CA, USA), and MS search (V. 2.0, Agilent).
12. Identification of compounds: The National Institute of Standards and Technology (NIST) library.

3.6 Data Analysis

1. Use raw data (.D) for further metabolomics data analysis.
2. Convert raw data to mzXL or mz file using Proteowizard or equivalent software.
3. Preprocess including deconvolution, alignment, and normalization using GNPS or MZmine 3 (*see* **Note24**).
4. Curate the data for multivariate data analysis (such as adding group info or sample name).
5. Perform multivariate data analysis using MetaboAnalyst 5.0 [10] such as principal component analysis (PCA), partial least squares-discriminant analysis (PLS-DA), and orthogonal projections to latent structures-discriminant analysis (OPLS-DA) (*see* **Note 25**) (Expected result of PCA was shown in Fig. 1).

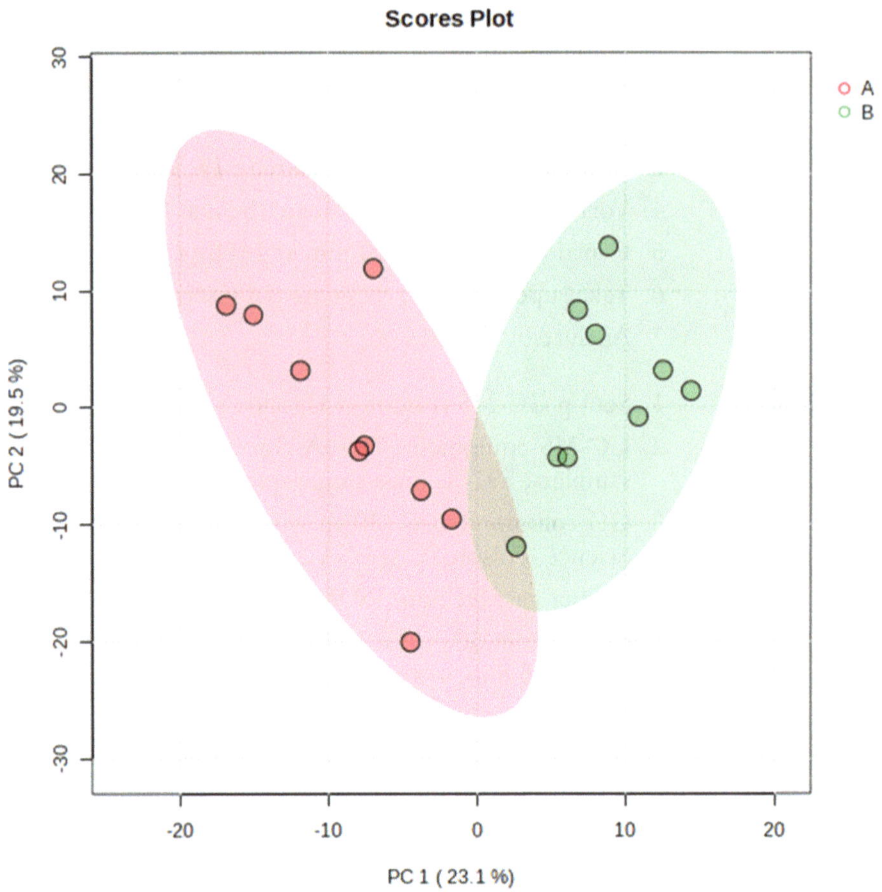

Fig. 1 Score plot from PCA. Two groups, control group vs. treated group, are well separated. Group A (red): control groups (treated with EtOH), group B (green): Naphthalene treated groups

6. Analyze principal components (PCs) in case of PCA or variable importance in projection (VIP) score in case of OPLS-DA model to obtain important metabolites between group separation (In this case, changes metabolites in treated groups compared to control groups) (*see* **Note 26**).
7. To visualize the different metabolites in different groups, use heatmap analysis in the MetaboAnalyst 5.0 (*see* **Note 27**) (Expected result of heatmap is shown in Fig. 2).

3.7 Examples of Data

3.7.1 GC-MS Chromatogram

1. Typical total ion current (TIC) chromatogram of solvent A extract (Fig. 3).
2. TIC of non-polar phase from solvent B extract (Fig. 4).
3. TIC of polar phase from solvent B extract (Fig. 5).

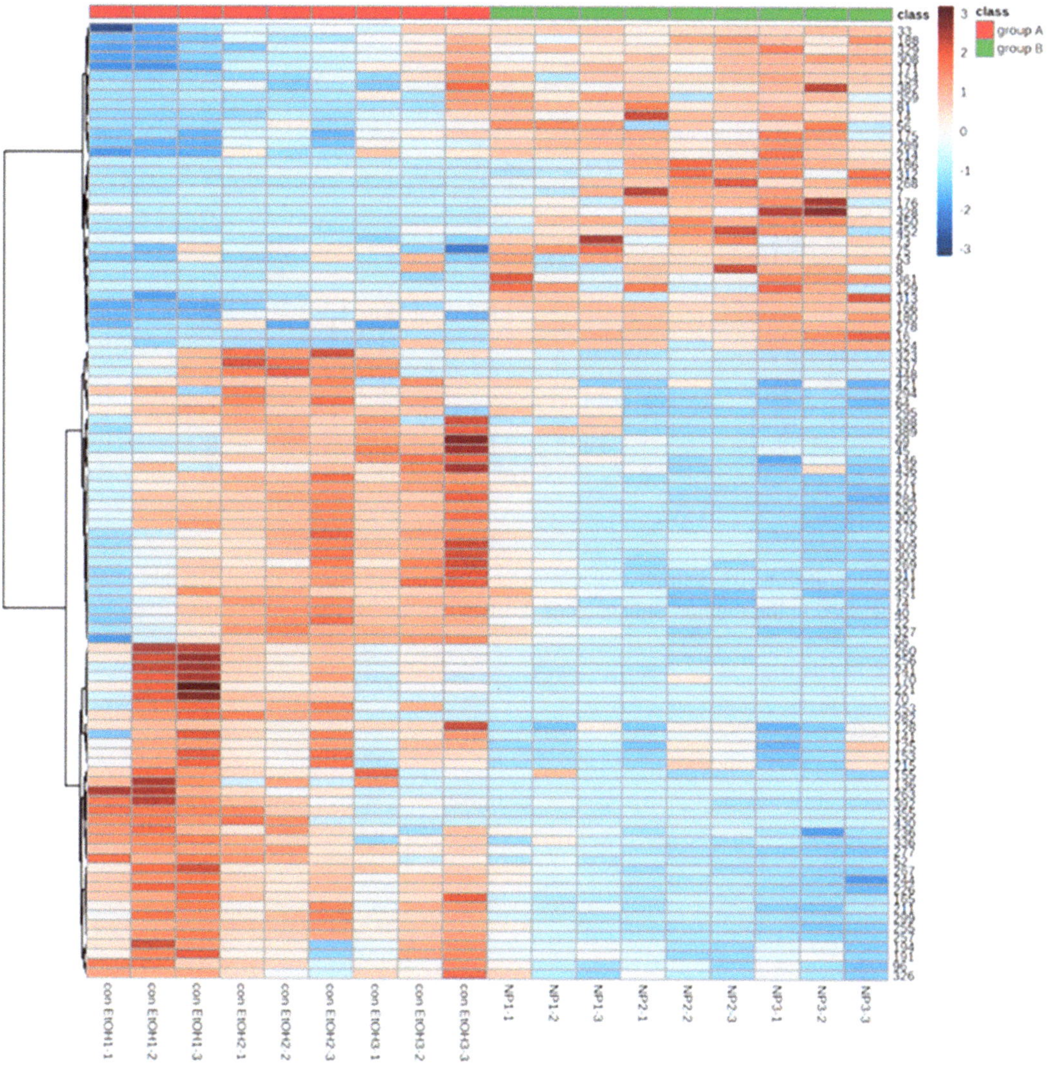

Fig. 2 Heatmap analysis of the non-polar phase from the control group and treated group. Group A (red): control group (treated with EtOH as a carrier), group B (green): Naphthalene treated group

3.7.2 The List of Compounds Identified in the GC-MS Analysis

1. Identified peaks from *C. elegans* extract using solvent A (Table 1).
2. Identified peaks from *C. elegans* extract using solvent B (non-polar phase) (Table 2).
3. Identified peaks from *C. elegans* extract using solvent B (polar phase) (Table 3).

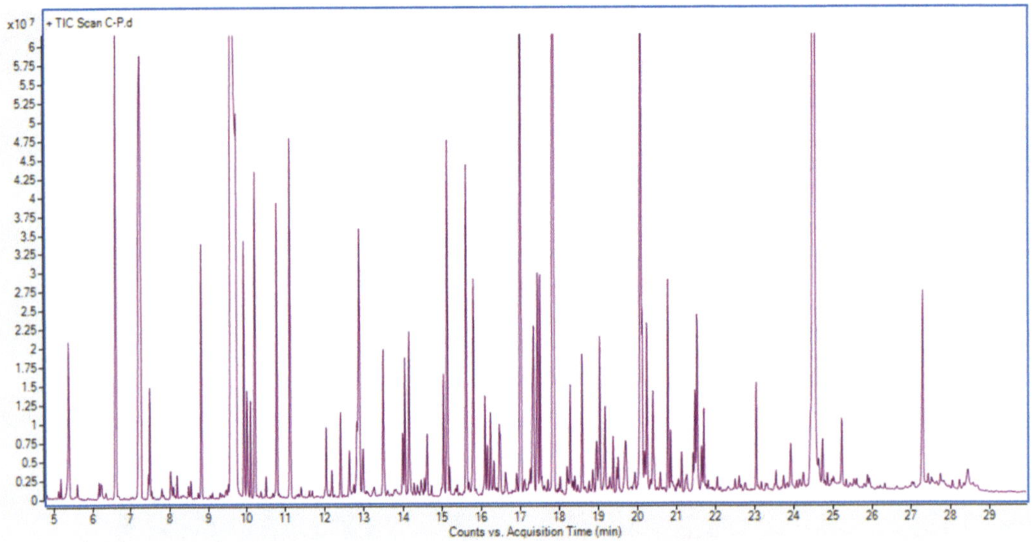

Fig. 3 TIC chromatogram of 80% extract (Solvent A) of *C. elegans* cultures. Extraction and analysis were performed as described in the method Subheadings 3.3 and 3.4

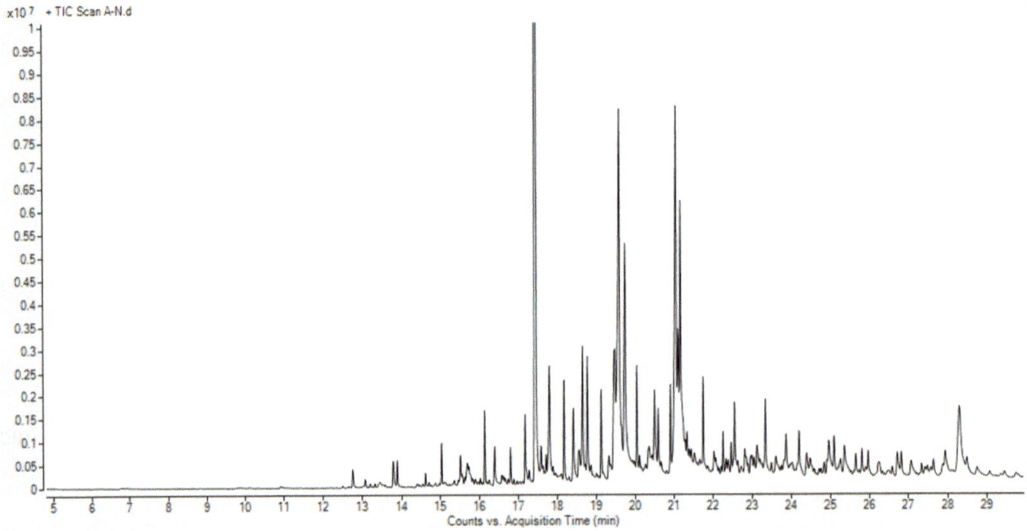

Fig. 4 TIC chromatogram of the non-polar phase of *C. elegans* extract. Extraction was performed using solvent B (Chloroform-MeOH-Water = 2:1:1) as described in the method Subheadings 3.3 and 3.4

4 Notes

1. It is possible that the pH changes slightly by ±0.1, but this is very rare and if it happens it is within an acceptable range.
2. To ensure effective filtration through a 0.22 μm membrane filter, it is advisable to prepare a minimum volume of 10 mL for solutions for $MgSO_4$, $CaCl_2$, potassium citrate tribasic

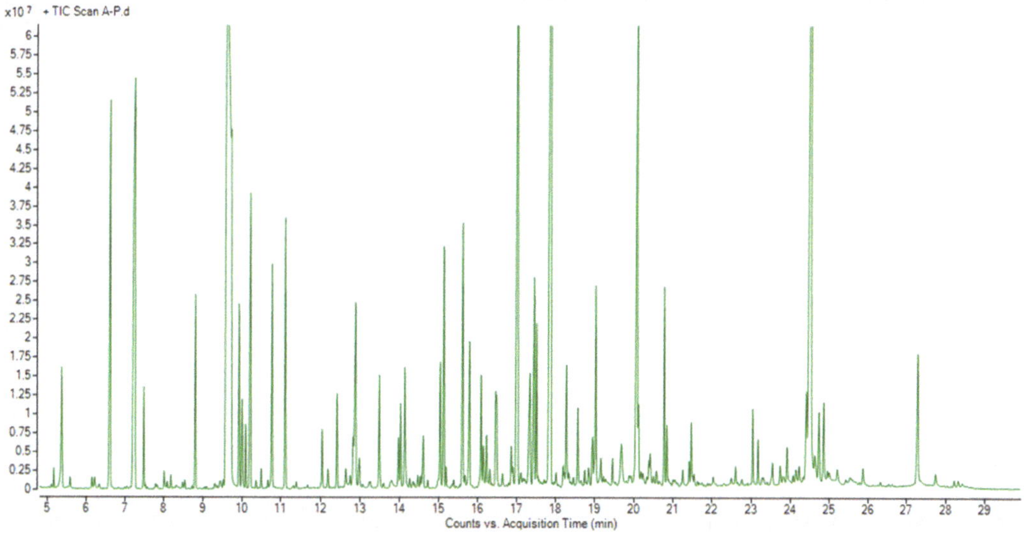

Fig. 5 TIC chromatogram of the polar phase of *C. elegans* extract. Extraction was performed using solvent B (Chloroform-MeOH-Water = 2:1:1) as described in the method Subheadings 3.3 and 3.4

monohydrate, and cholesterol solution. The cholesterol solution should be stored at −20 °C, while the other solutions can be stored at room temperature.

3. After autoclaving the NGM agar, it is best to cool down the agar in an incubator pre-set to 65 °C to prevent solidification. The chemicals $CaCl_2$, $MgSO_4$, and cholesterol, which are added after autoclaving, degrade at high temperatures.
4. The 25 mL NGM agar plates can be stored in an airtight container upside up to 4 weeks. The shelf life of the plates depends on the dryness of the plates. If the plates start to dry and crack, they are unusable.
5. For nystatin, it is recommended to prepare 2 mL in Milli-Q water, no filtering is needed. Work sterile in a flow cabinet. Store at −20 °C for up to 1 year.
6. Pen-Strep can be purchased at a concentration of 10,000 U/ mL. Once the stock is thawed, it can be stored at 4 °C for up to 6 months. It is recommended to make 1.5 mL aliquots and store them at −20 °C until further use.
7. This is the recipe for preparing 1 liter of S-medium. Adjust the quantity as needed based on the number of conditions being tested.
8. It's crucial to prepare S-medium as fresh as possible for your experiments. Store it away from direct light and in a dark environment to maintain its quality.
9. To expedite the growth of the OP50 culture, inoculating 5–10 colonies is an option. The rate of growth in LB broth depends

Table 1
Compound identification in the 80% MeOH extract (Solvent A)

Cpd	Name	CAS no.	Mass (DB)	Formula	RT	Area
1	Silanamine, *N,N'*-methanetetraylbis[1,1,1-trimethyl-	1000-70-0	186.1	C7H18N2Si2	5.37	41,753,184
2	Tris(trimethylsilyl)carbamate	1000366-58-2	277.1	C10H27NO2Si3	6.151	3,548,536
3	Ethanolamine, 2TMS derivative	17165-52-5	205.1	C8H23NOSi2	6.198	6,363,923
4	D-(−)-Lactic acid, 2TMS derivative	1000333-99-1	234.1	C9H22O3Si2	6.626	135,379,721
5	L-Alanine, 2TMS derivative	27844-07-1	233.1	C9H23NO2Si2	7.264	241,207,435
6	Glycine, 2TMS derivative	7364-42-3	219.1	C8H21NO2Si2	7.488	20,515,442
7	Propanoic acid, 2-methyl-3-[(trimethylsilyl)oxy]-, trimethylsilyl ester	55530-42-2	248.1	C10H24O3Si2	8.011	7,501,671
8	2-Aminobutanoic acid, 2TMS derivative	55133-91-0	247.1	C10H25NO2Si2	8.174	4,212,174
9	Ethanimidic acid, *N*-(trimethylsilyl)-, trimethylsilyl ester	10416-59-8	203.1	C8H21NOSi2	8.527	3,447,463
10	L-Valine, 2TMS derivative	7364-44-5	261.2	C11H27NO2Si2	8.805	56,207,385
11	L-Serine, 2TMS derivative	70125-39-2	249.1	C9H23NO3Si2	9.457	3,366,488
12	Silanol, trimethyl-, phosphate (3:1)	10497-05-9	314.1	C9H27O4PSi3	9.606	644,349,318
13	L-Isoleucine, 2TMS derivative	7483-92-3	275.2	C12H29NO2Si2	9.925	54,920,802
14	L-Proline, 2TMS derivative	7364-47-8	259.1	C11H25NO2Si2	10	26,711,679
15	Glycine, 3TMS derivative	5630-82-0	291.2	C11H29NO2Si3	10.095	17,219,676
16	Butanedioic acid, 2TMS derivative	40309-57-7	262.1	C10H22O4Si2	10.217	84,078,100
17	Uracil, 2TMS derivative	10457-14-4	256.1	C10H20N2O2Si2	10.495	4,053,379
18	Serine, 3TMS derivative	64625-17-8	321.2	C12H31NO3Si3	10.781	61,884,485

(continued)

Table 1 (continued)

Cpd	Name	CAS no.	Mass (DB)	Formula	RT	Area
19	L-Threonine, 3TMS derivative	7537-02-2	335.2	C13H33NO3Si3	11.12	83,409,210
20	3-Amino-2-piperidone, 2TMS derivative	32565-12-1	258.2	C11H26N2OSi2	12.05	17,021,102
21	Aminomalonic acid, tris (trimethylsilyl)-	1000079-52-5	335.1	C12H29NO4Si3	12.199	6,516,700
22	Malic acid, 3TMS derivative	38166-11-9	350.1	C13H30O5Si3	12.43	18,075,625
23	Parabanic acid, bis-*O*-(trimethylsilyl)-	1000149-42-3	258.1	C9H18N2O3Si2	12.654	12,257,980
24	*N*-(4-[Bis(trimethylsilyl)amino] butyl)acetamide	1000378-73-1	274.2	C12H30N2OSi2	12.763	3,812,327
25	L-Aspartic acid, 3TMS derivative	55268-53-6	349.2	C13H31NO4Si3	12.844	16,416,174
26	L-5-Oxoproline, 2TMS derivative	30274-77-2	273.1	C11H23NO3Si2	12.912	91,922,110
27	*cis*-4-Aminocyclohexanecarboxylic acid, *N*-(tert-butyldimethylsilyl), *tert*-butyldimethylsilyl ester	1000384-19-1	371.3	C19H41NO2Si2	12.994	15,033,957
28	2,3,4-Trihydroxybutyric acid tetrakis(trimethylsilyl) deriv.	38191-88-7	424.2	C16H40O5Si4	13.272	3,531,724
29	Pentanedioic acid, 2-[(trimethylsilyl)oxy]-, bis (trimethylsilyl) ester	55530-62-6	364.2	C14H32O5Si3	13.51	36,873,748
30	L-Ornithine, 3TMS derivative	24595-70-8	348.2	C14H36N2O2Si3	13.985	13,851,242
31	L-Glutamic acid, 3TMS derivative	15985-07-6	363.2	C14H33NO4Si3	14.046	28,068,929
32	L-Phenylalanine, 2TMS derivative	7364-51-4	309.2	C15H27NO2Si2	14.154	42,304,632
33	Silanol, trimethyl-, pyrophosphate (4:1)	18395-45-4	466.1	C12H36O7P2Si4	14.46	4,322,771
34	Asparagine, 3TMS derivative	55649-62-2	348.2	C13H32N2O3Si3	14.623	15,813,180
35	L-Lysine, 3TMS derivative	24595-69-5	362.2	C15H38N2O2Si3	15.051	27,506,817

(continued)

Table 1
(continued)

Cpd	Name	CAS no.	Mass (DB)	Formula	RT	Area
36	2-Aminoadipic acid, 3TMS derivative	66434-50-2	377.2	C15H35NO4Si3	15.152	95,143,272
37	Phosphoric acid, 2-(trimethylsiloxy)-1-[(trimethylsiloxy)methyl] ethyl bis(trimethylsilyl) ester	31038-12-7	460.2	C15H41O6PSi4	15.2	5,189,094
38	Phosphoric acid, bis (trimethylsilyl) 2,3-bis [(trimethylsilyl)oxy]propyl ester	31038-11-6	460.2	C15H41O6PSi4	15.628	87,830,993
39	L-Glutamine, 3TMS derivative	70591-28-5	362.2	C14H34N2O3Si3	15.811	57,497,584
40	3,5-Dioxa-4-phospha-2-silaoctan-8-oic acid, 2,2-dimethyl-4,7-bis [(trimethylsilyl)oxy]-, trimethylsilyl ester, 4-oxide	31038-13-8	474.2	C15H39O7PSi4	16.096	22,092,697
41	9H-Purin-6-ol, 2TMS derivative	17962-89-9	280.1	C11H20N4OSi2	16.157	12,472,275
42	DL-Ornithine, 4TMS derivative	1000333-13-9	420.2	C17H44N2O2Si4	16.239	17,893,661
43	Citrulline, 4TMS derivative	1000017-84-6	463.3	C18H45N3O3Si4	16.32	8,100,812
44	D-(+)-Talofuranose, pentakis (trimethylsilyl) ether (isomer 1)	1000380-17-3	540.3	C21H52O6Si5	16.463	23,983,186
45	Myristic acid, TMS derivative	18603-17-3	300.2	C17H36O2Si	16.612	6,207,598
46	D-Ribose, 4TMS derivative	33648-69-0	438.2	C17H42O5Si4	16.904	4,747,176
47	Talose, 5TMS derivative	56192-85-9	540.3	C21H52O6Si5	17.019	187,285,056
48	β-D-Allopyranose, 5TMS derivative	1000380-16-2	540.3	C21H52O6Si5	17.114	4,413,429
49	Pentadecanoic acid, TMS derivative	74367-22-9	314.3	C18H38O2Si	17.257	7,944,895
50	L-Lysine, 4TMS derivative	55429-07-7	434.3	C18H46N2O2Si4	17.345	62,471,031
51	Hexadecanoic acid, methyl ester	112-39-0	270.3	C17H34O2	17.454	61,725,441

(continued)

Table 1
(continued)

Cpd	Name	CAS no.	Mass (DB)	Formula	RT	Area
52	L-Tyrosine, 3TMS derivative	51220-73-6	397.2	C18H35NO3Si3	17.522	49,207,969
53	Talose, 5TMS derivative	56192-85-9	540.3	C21H52O6Si5	17.851	258,124,982
54	1,5-Anhydrohexitol, 4TMS derivative	1000380-13-8	452.2	C18H44O5Si4	18.194	5,314,337
55	7-(Trimethylsilyl)-2,6-bis [(trimethylsilyl)oxy]-7H-purine	1000332-89-5	368.2	C14H28N4O2Si3	18.282	22,845,233
56	D-(−)-Fructofuranose, pentakis (trimethylsilyl) ether (isomer 2)	1000380-16-7	540.3	C21H52O6Si5	18.954	14,833,801
57	Uric acid, *N,O,O′,O″*-tetrakis (trimethylsilyl)-	1000071-71-2	456.2	C17H36N4O3Si4	19.036	34,138,179
58	5,8,11,14-Eicosatetraynoic acid, TBDMS derivative	1000333-54-6	410.3	C26H38O2Si	19.287	3,361,494
59	10-Heptadecenoic acid, (*Z*)-, TMS derivative	1000333-62-6	340.3	C20H40O2Si	19.375	10,452,883
60	D-(−)-Fructofuranose, pentakis (trimethylsilyl) ether (isomer 1)	1000380-16-6	540.3	C21H52O6Si5	19.463	4,477,482
61	Heptadecanoic acid, TMS derivative	55517-58-3	342.3	C20H42O2Si	19.497	6,357,404
62	Homoserine, 3TMS derivative	1177129-58-6	335.2	C13H33NO3Si3	19.687	23,412,970
63	α-D-Glucopyranose, 2,3,4,6-tetrakis-*O*-(trimethylsilyl)-, bis(trimethylsilyl) phosphate	55520-79-1	692.3	C24H61O9PSi6	19.925	5,934,506
64	Cystathionine, 4TMS derivative	1000141-33-0	510.2	C19H46N2O4SSi4	20.088	224,359,695
65	11-Octadecenoic acid, (*E*)-, TMS derivative	1000333-56-0	354.3	C21H42O2Si	20.237	36,923,040
66	Stearic acid, TMS derivative	18748-91-9	356.3	C21H44O2Si	20.393	28,735,734
67	Glyceryl-glycoside TMS ether	1000365-90-1	686.3	C27H66O8Si6	20.78	44,456,529

(continued)

Table 1
(continued)

Cpd	Name	CAS no.	Mass (DB)	Formula	RT	Area
68	α-D-Glucopyranosiduronic acid, 3-(5-ethylhexahydro-2,4,6-trioxo-5-pyrimidinyl)-1,1-dimethylpropyl 2,3,4-tris-O-(trimethylsilyl)-, methyl ester	55556-81-5	648.3	C27H52N2O10Si3	21.045	3,462,255
69	10-Nonadecenoic acid, (*Z*)-, TMS derivative	1000333-64-9	368.3	C22H44O2Si	21.133	8,551,028
70	Glucopyranose, 1,2,3,4-tetrakis-O-(trimethylsilyl)-, bis (trimethylsilyl) phosphate, D-	28176-73-0	692.3	C24H61O9PSi6	21.473	23,675,836
71	Eicosapentaenoic acid, TMS derivative	1000333-53-2	374.3	C23H38O2Si	21.527	44,708,624
72	5,8,11-Eicosatrienoic acid, (*Z*)-, TMS derivative	1000333-53-6	378.3	C23H42O2Si	21.636	10,889,691
73	Arachidonic acid, TMS derivative	113516-18-0	376.3	C23H40O2Si	21.703	17,841,098
74	D-Galactofuranose, 1,2,3,5-tetrakis-*O*-(trimethylsilyl)-, bis(trimethylsilyl) phosphate	69688-48-8	692.3	C24H61O9PSi6	22.043	3,440,186
75	Inosine, 4TMS derivative	1000332-74-7	556.2	C22H44N4O5Si4	23.034	22,880,422
76	D-(+)-Cellobiose, octakis (trimethylsilyl) ether, methyloxime (isomer 2)	1000380-09-9	947.5	C37H89NO11Si8	23.726	3,981,037
77	α-D-Galactopyranose, 2,3,4,6-tetrakis-O-(trimethylsilyl)-, bis(trimethylsilyl) phosphate	55530-81-9	692.3	C24H61O9PSi6	23.91	14,345,985
78	Maltose, 8TMS derivative, isomer 1	1000380-08-1	918.4	C36H86O11Si8	24.229	3,689,528
79	D-(+)-Trehalose, octakis (trimethylsilyl) ether	1000380-43-8	918.4	C36H86O11Si8	24.502	423,664,493
80	Guanosine, 5TMS derivative	1000367-43-9	643.3	C25H53N5O5Si5	24.724	12,562,505
81	D-Ribofuranose, 1,2,3-tris-*O*-(trimethylsilyl)-, bis (trimethylsilyl) phosphate	69744-64-5	590.2	C20H51O8PSi5	25.206	18,315,525

(continued)

Table 1
(continued)

Cpd	Name	CAS no.	Mass (DB)	Formula	RT	Area
82	Adenosine, *N*-(trimethylsilyl)-2′,3′-bis-*O*-(trimethylsilyl)-, 5′-[bis(trimethylsilyl) phosphate]	32653-14-8	707.3	C25H54N5O7PSi5	27.297	60,398,879
83	Cholecalciferol, TMS derivative	1000332-99-6	456.4	C30H52OSi	28.451	12,481,925
84	D-Erythro-Pentofuranose, 2-deoxy-1,3-bis-*O*-(trimethylsilyl)-, bis (trimethylsilyl) phosphate	69688-43-3	502.2	C17H43O7PSi4	28.58	3,785,383

Identification was performed using the NIST library as described in the method Subheading 3.5

on the freshness of the OP50 colony. It's advisable to use a single colony to achieve a homogenous OP50 culture.

10. Previous toxicity assays conducted in *C. elegans* have shown that concentrations of EtOH [11] or DMSO [12] up to 0.5% can be safely utilized to administer compounds, with minimal impact on *C. elegans* lifespan or physiological rates. It's advisable to use a solvent concentration below 0.5%.
11. When chloroform:MeOH:water (solvent B) is in the ratio of 2: 1:1 (v/v/v), it separates into two phases, with the non-polar phase settling at the bottom and the polar phase at the top. Both phases require separate analysis. The non-polar phase can be directly analyzed using GC-MS without further processing. However, the polar phase needs to undergo derivatization to enhance detection.
12. It is recommended to seed the plates with 400 μL of OP50 bacteria 1 week before starting to grow your *C. elegans* population. This gives time for the OP50 culture to grow on the NGM plates.
13. *C. elegans* has a maximum brood size of about ±350 eggs. At the time of the bleaching, each *C. elegans* would have produced 200 offspring that would have developed into gravid adults, each carrying ±10 fertilized and viable eggs. Each NGM plate should yield approximately 20,000 eggs which will hatch to be L1s for the experiment. Do not add more than 10 L4s per plate as the nematodes will run out of OP50 culture food. Instead, prepare more NGM plates, seed with OP50, and add a maximum of 10 L4s per plate.

Table 2
The list of identified compounds in the non-polar phase of *C. elegans* extract

Cpd	Name	CAS no.	Mass (DB)	Formula	RT	Area
1	2,4-Di-tert-butylphenol	96-76-4	206.2	C14H22O	12.776	784,102
2	Diethyl phthalate	84-66-2	222.1	C12H14O4	13.788	1,409,450
3	Hexadecane	544-76-3	226.3	C16H34	13.89	1,053,874
4	Heptadecane	629-78-7	240.3	C17H36	15.044	1,526,620
5	Eicosane	112-95-8	282.3	C20H42	15.519	1,889,731
6	Tetradecanoic acid	544-63-8	228.2	C14H28O2	15.695	1,464,953
7	Octadecane	593-45-3	254.3	C18H38	16.137	2,479,209
8	Pentadecanoic acid	1002-84-2	242.2	C15H30O2	16.388	1,920,478
9	Heptadecane, 3-methyl-	6418-44-6	254.3	C18H38	16.802	1,297,250
10	Nonadecane	629-92-5	268.3	C19H40	17.182	2,794,716
11	Hexadecanoic acid, methyl ester	112-39-0	270.3	C17H34O2	17.46	75,811,033
12	Methyl 16-hydroxy-hexadecanoate	36575-67-4	286.3	C17H34O3	17.589	2,106,788
13	2-Methyl-Z-4-tetradecene	1000130-78-3	210.2	C15H30	17.637	1,077,881
14	Dotriacontane, 1-iodo-	1000406-32-4	576.4	C32H65I	17.718	1,103,206
15	*n*-Hexadecanoic acid	57-10-3	256.2	C16H32O2	17.807	7,267,638
16	l-(+)-Ascorbic acid 2,6-dihexadecanoate	28474-90-0	652.5	C38H68O8	18.105	1,027,705
17	Eicosane	112-95-8	282.3	C20H42	18.173	4,015,122
18	Heptadecanoic acid	506-12-7	270.3	C17H34O2	18.418	3,919,190
19	7-Methyl-*Z*-tetradecen-1-ol acetate	1000130-99-6	268.2	C17H32O2	18.56	2,133,226
20	*cis*-10-Heptadecenoic acid	29743-97-3	268.2	C17H32O2	18.655	6,792,093
21	Heneicosane	629-94-7	296.3	C21H44	19.124	3,400,452
22	7-Methyl-*Z*-tetradecen-1-ol acetate	1000130-99-6	268.2	C17H32O2	19.32	1,804,203
23	9,12-Octadecadienoic acid (*Z,Z*)-	60-33-3	280.2	C18H32O2	19.47	10,678,772
24	Oleic acid	112-80-1	282.3	C18H34O2	19.592	31,569,356
25	Octadecanoic acid	57-11-4	284.3	C18H36O2	19.741	17,275,614
26	Docosane	629-97-0	310.4	C22H46	20.04	4,453,871

(continued)

Table 2
(continued)

Cpd	Name	CAS no.	Mass (DB)	Formula	RT	Area
27	*cis*-10-Nonadecenoic acid	73033-09-7	296.3	C19H36O2	20.495	5,266,678
28	Hentriacontane	630-04-6	436.5	C31H64	20.909	3,372,499
29	*cis*-5,8,11,14,17-Eicosapentaenoic acid	10417-94-4	302.2	C20H30O2	21.052	24,686,674
30	8,11,14-Eicosatrienoic acid, (*Z,Z,Z*)-	1783-84-2	306.3	C20H34O2	21.106	8,029,943
31	*cis*-5,8,11,14,17-Eicosapentaenoic acid	10417-94-4	302.2	C20H30O2	21.167	17,588,514
32	2H-Pyran, 2-(7-dodecynyloxy) tetrahydro-	16695-32-2	266.2	C17H30O2	21.33	2,334,364
33	*cis*-5,8,11,14,17-Eicosapentaenoic acid	10417-94-4	302.2	C20H30O2	21.52	2,337,793
34	1-Heptatriacotanol	105794-58-9	536.6	C37H76O	21.656	893,686
35	Tetracosane	646-31-1	338.4	C24H50	21.744	4,631,010
36	Nonahexacontanoic acid	40710-32-5	999.1	C69H138O2	22.022	1,248,968
37	2-Methyltetracosane	1560-78-7	352.4	C25H52	22.253	1,353,048
38	2H-Pyran, 2-(7-dodecynyloxy) tetrahydro-	16695-32-2	266.2	C17H30O2	22.464	1,093,562
39	Pentacosane	629-99-2	352.4	C25H52	22.552	3,611,954
40	Nonahexacontanoic acid	40710-32-5	999.1	C69H138O2	22.81	1,591,022
41	Pentanoic acid, undecyl ester	81186-23-4	256.2	C16H32O2	23.115	1,654,486
42	*tert*-Hexadecanethiol	25360-09-2	258.2	C16H34S	23.19	1,140,977
43	Hentriacontane	630-04-6	436.5	C31H64	23.333	3,245,707
44	*tert*-Hexadecanethiol	25360-09-2	258.2	C16H34S	23.862	2,373,464
45	1-Hexacosanol	506-52-5	382.4	C26H54O	24.948	3,440,675
46	Hentriacontane	630-04-6	436.5	C31H64	25.084	2,076,901
47	7-Methyl-Z-tetradecen-1-ol acetate	1000130-99-6	268.2	C17H32O2	25.24	1,487,738
48	Nonacosane	630-03-5	408.5	C29H60	25.342	2,723,083

(continued)

Table 2
(continued)

Cpd	Name	CAS no.	Mass (DB)	Formula	RT	Area
49	Hexacosyl nonyl ether	1000406-38-1	508.6	C35H72O	25.627	1,171,409
50	1-Hexacosanol	506-52-5	382.4	C26H54O	25.79	1,258,030
51	Hentriacontane	630-04-6	436.5	C31H64	25.953	1,230,906
52	Nonacosane	630-03-5	408.5	C29H60	26.238	1,251,204
53	Octatriacontyl pentafluoropropionate	1000351-89-1	696.6	C41H77F5O2	26.706	1,555,642
54	Cholesterol	57-88-5	386.4	C27H46O	27.942	2,741,563
55	Cholecalciferol	67-97-0	384.3	C27H44O	28.302	10,342,392
56	Ethyl iso-allocholate	1000043-05-3	436.3	C26H44O5	28.499	2,161,768
57	Octadecane, 3-ethyl-5-(2-ethylbutyl)-	55282-12-7	366.4	C26H54	28.757	1,311,748

Extraction was performed using solvent B (Chloroform-MeOH-water = 2:1:1). Identification was performed using the NIST library as described in the method Subheading 3.5

14. The 15 mL Falcon tube can hold a maximum of 15 mL *C. elegans* suspension. It is possible to wash 5 plates at a time, centrifuge for 1 min at 1400×*g*, remove supernatant, and wash 5 more plates. The final *C. elegans* pellet, before bleaching, should not be more than 1.5 mL in volume.
15. After bleaching, eggs are resuspended in M9-T and left overnight to hatch at 20 °C. Most of the eggs will hatch if the bleaching is done correctly. Too harsh of a bleaching will kill the eggs and will not hatch. The quality of bleaching and L1s can be checked by the number of eggs that are left and the hatched L1s. A hatching of more than 90% indicates good bleaching and viable L1s.
16. Count the concentration of L1s by taking 10 μL of L1 suspension, spread on an empty petri dish, and count the total number in 10 μL. Perform this in triplicate to have an accurate concentration of L1 *C. elegans* suspension.
17. It is recommended to use a compound concentration below LD_{50}. If the compound is lethal or developmental toxic to the nematodes, there will not be enough working material for GC-MS.

Table 3
The list of identified compounds in the polar phase of *C. elegans* extract

Cpd	Name	CAS no.	Mass (DB)	Formula	RT	Area
1	L-Alanine, TBDMS derivative	1000333-28-0	203.1	C9H21NO2Si	5.166	2,996,190
2	Silanamine, *N,N'*-methanetetraylbis[1,1,1-trimethyl-	1000-70-0	186.1	C7H18N2Si2	5.37	32,728,267
3	Tris(trimethylsilyl)carbamate	1000366-58-2	277.1	C10H27NO2Si3	6.144	2,776,483
4	Ethanolamine, 2TMS derivative	17165-52-5	205.1	C8H23NOSi2	6.219	3,901,819
5	D-(−)-Lactic acid, 2TMS derivative	1000333-99-1	234.1	C9H22O3Si2	6.619	97,178,271
6	L-Alanine, 2TMS derivative	27844-07-1	233.1	C9H23NO2Si2	7.25	185,764,026
7	Glycine, 2TMS derivative	7364-42-3	219.1	C8H21NO2Si2	7.481	18,256,805
8	4-Hydroxybutanoic acid, 2TMS derivative	55133-95-4	248.1	C10H24O3Si2	8.004	4,066,685
9	L-Valine, 2TMS derivative	7364-44-5	261.2	C11H27NO2Si2	8.798	39,588,609
10	L-Serine, 2TMS derivative	70125-39-2	249.1	C9H23NO3Si2	9.43	3,153,046
11	Silanol, trimethyl-, phosphate (3:1)	10497-05-9	314.1	C9H27O4PSi3	9.606	543,962,554
12	L-Isoleucine, 2TMS derivative	7483-92-3	275.2	C12H29NO2Si2	9.912	39,139,610
13	L-Proline, 2TMS derivative	7364-47-8	259.1	C11H25NO2Si2	9.993	21,099,898
14	Glycine, 3TMS derivative	5630-82-0	291.2	C11H29NO2Si3	10.081	11,573,854
15	Butanedioic acid, 2TMS derivative	40309-57-7	262.1	C10H22O4Si2	10.21	77,846,839
16	Uracil, 2TMS derivative	10457-14-4	256.1	C10H20N2O2Si2	10.489	3,519,180
17	Serine, 3TMS derivative	64625-17-8	321.2	C12H31NO3Si3	10.774	48,161,793
18	L-Threonine, 3TMS derivative	7537-02-2	335.2	C13H33NO3Si3	11.113	56,199,413

(continued)

Table 3
(continued)

Cpd	Name	CAS no.	Mass (DB)	Formula	RT	Area
19	3-Amino-2-piperidone, 2TMS derivative	32565-12-1	258.2	C11H26N2OSi2	12.05	12,809,575
20	Aminomalonic acid, tris (trimethylsilyl)-	1000079-52-5	335.1	C12H29NO4Si3	12.199	4,212,661
21	Malic acid, 3TMS derivative	38166-11-9	350.1	C13H30O5Si3	12.43	19,341,446
22	Dihydrouracil, 2TMS derivative	74810-47-2	258.1	C10H22N2O2Si2	12.647	5,712,991
23	*N*-(4-[Bis(trimethylsilyl)amino] butyl)acetamide	1000378-73-1	274.2	C12H30N2OSi2	12.763	3,921,302
24	L-Aspartic acid, 3TMS derivative	55268-53-6	349.2	C13H31NO4Si3	12.838	12,355,620
25	L-5-Oxoproline, 2TMS derivative	30274-77-2	273.1	C11H23NO3Si2	12.899	58,122,696
26	4-Aminobutanoic acid, 3TMS derivative	39508-23-1	319.2	C13H33NO2Si3	12.994	9,235,410
27	Pentanedioic acid, 2-[(trimethylsilyl)oxy]-, bis (trimethylsilyl) ester	55530-62-6	364.2	C14H32O5Si3	13.503	25,481,098
28	α-D-Glucopyranoside, methyl 2-(acetylamino)-2-deoxy-3-*O*-(trimethylsilyl)-, cyclic methylboronate	54477-01-9	331.2	C13H26BNO6Si	13.808	4,628,807
29	L-Ornithine, 3TMS derivative	24595-70-8	348.2	C14H36N2O2Si3	13.985	9,423,521
30	L-Phenylalanine, 2TMS derivative	7364-51-4	309.2	C15H27NO2Si2	14.148	30,285,373
31	Asparagine, 3TMS derivative	55649-62-2	348.2	C13H32N2O3Si3	14.616	12,807,968
32	L-Lysine, 3TMS derivative	24595-69-5	362.2	C15H38N2O2Si3	15.051	30,785,201
33	2-Aminoadipic acid, 3TMS derivative	66434-50-2	377.2	C15H35NO4Si3	15.139	56,802,528
34	Phosphoric acid, 2-(trimethylsiloxy)-1-[(trimethylsiloxy)methyl] ethyl bis(trimethylsilyl) ester	31038-12-7	460.2	C15H41O6PSi4	15.2	4,847,324

(continued)

Table 3
(continued)

Cpd	Name	CAS no.	Mass (DB)	Formula	RT	Area
35	Phosphoric acid, bis (trimethylsilyl) 2,3-bis [(trimethylsilyl)oxy]propyl ester	31038-11-6	460.2	C15H41O6PSi4	15.628	66,443,619
36	L-Glutamine, 3TMS derivative	70591-28-5	362.2	C14H34N2O3Si3	15.797	37,273,988
37	3,5-Dioxa-4-phospha-2-silaoctan-8-oic acid, 2,2-dimethyl-4,7-bis [(trimethylsilyl)oxy]-, trimethylsilyl ester, 4-oxide	31038-13-8	474.2	C15H39O7PSi4	16.096	26,119,365
38	9H-Purin-6-ol, 2TMS derivative	17962-89-9	280.1	C11H20N4OSi2	16.15	10,571,541
39	DL-Ornithine, 4TMS derivative	1000333-13-9	420.2	C17H44N2O2Si4	16.239	12,409,558
40	β-D-Galactofuranose, 1,2,3,5,6-pentakis-*O*-(trimethylsilyl)-	7045-52-5	540.3	C21H52O6Si5	16.463	36,593,598
41	Methyl galactoside (1S,2R,3S,4R,5R)-, 4TMS derivative	2641-79-4	482.2	C19H46O6Si4	16.87	8,891,873
42	D-Ribose, 4TMS derivative	33648-69-0	438.2	C17H42O5Si4	16.904	4,541,821
43	β-D-Allopyranose, 5TMS derivative	1000380-16-2	540.3	C21H52O6Si5	17.114	4,235,219
44	L-Lysine, 4TMS derivative	55429-07-7	434.3	C18H46N2O2Si4	17.345	41,689,530
45	Hexadecanoic acid, methyl ester	112-39-0	270.3	C17H34O2	17.454	55,398,676
46	L-Tyrosine, 3TMS derivative	51220-73-6	397.2	C18H35NO3Si3	17.515	32,283,374
47	Talose, 5TMS derivative	56192-85-9	540.3	C21H52O6Si5	17.858	283,241,230
48	Pantothenic acid, 3TMS	1000331-78-2	435.2	C18H41NO5Si3	18.017	3,080,674
49	Uridine, 3TMS derivative	10457-16-6	460.2	C18H36N2O6Si3	18.201	3,874,622
50	7-(Trimethylsilyl)-2,6-bis [(trimethylsilyl)oxy]-7H-purine	1000332-89-5	368.2	C14H28N4O2Si3	18.289	29,581,436

(continued)

Table 3
(continued)

Cpd	Name	CAS no.	Mass (DB)	Formula	RT	Area
51	*N*-Acetylglucosamine, 4TMS derivative	55517-99-2	509.2	C20H47NO6Si4	18.757	3,015,552
52	D-(−)-Fructofuranose, pentakis (trimethylsilyl) ether (isomer 2)	1000380-16-7	540.3	C21H52O6Si5	18.954	14,118,485
53	Uric acid, *N,O,O′,O″*-tetrakis (trimethylsilyl)-	1000071-71-2	456.2	C17H36N4O3Si4	19.036	43,242,701
54	D-(−)-Fructofuranose, pentakis (trimethylsilyl) ether (isomer 1)	1000380-16-6	540.3	C21H52O6Si5	19.463	4,200,252
55	Homoserine, 3TMS derivative	1177129-58-6	335.2	C13H33NO3Si3	19.687	17,443,692
56	Cystathionine, 4TMS derivative	1000141-33-0	510.2	C19H46N2O4SSi4	20.081	149,547,570
57	L-Tryptophan, 3TMS derivative	55429-28-2	420.2	C20H36N2O2Si3	20.115	17,919,051
58	2,2-Dichloro-3,3-dimethyl-*N*-(1,1,3-trimethyl-2,3-dihydroinden-4-yl) cyclopropane-1-carboxamide, TMS derivative, peak 2	1000332-48-4	411.2	C21H31Cl2NOSi	20.176	3,526,896
59	β-D-Fructofuranose, 1,2,3,4-tetrakis-*O*-(trimethylsilyl)-, bis(trimethylsilyl) phosphate	28176-76-3	692.3	C24H61O9PSi6	20.38	5,051,343
60	D-(−)-Fructopyranose, 5TMS derivative (isomer 2)	1000380-16-5	540.3	C21H52O6Si5	20.414	7,687,391
61	Glyceryl-glycoside TMS ether	1000365-90-1	686.3	C27H66O8Si6	20.773	38,963,944
62	D-Glucopyranose, 1,2,3,4-tetrakis-*O*-(trimethylsilyl)-, bis(trimethylsilyl) phosphate	28176-73-0	692.3	C24H61O9PSi6	21.418	5,132,648
63	D-Mannose, 2,3,4,5-tetrakis-O-(trimethylsilyl)-, 6-[bis (trimethylsilyl) phosphate]	55520-82-6	692.3	C24H61O9PSi6	21.473	15,035,612
64	Maltose, octakis(trimethylsilyl) ether, methyloxime (isomer 2)	1000380-10-2	947.5	C37H89NO11Si8	22.6	3,926,294
65	Inosine, 4TMS derivative	1000332-74-7	556.2	C22H44N4O5Si4	23.027	15,797,515

(continued)

Table 3 (continued)

Cpd	Name	CAS no.	Mass (DB)	Formula	RT	Area
66	1-Monopalmitin, 2TMS derivative	1188-74-5	474.4	C25H54O4Si2	23.163	9,451,596
67	3-α-Mannobiose, octakis (trimethylsilyl) ether (isomer 1)	1000380-09-5	918.4	C36H86O11Si8	23.53	6,602,010
68	α-D-Galactopyranose, 2,3,4,6-tetrakis-*O*-(trimethylsilyl)-, bis(trimethylsilyl) phosphate	55530-81-9	692.3	C24H61O9PSi6	23.903	12,135,675
69	D-(+)-Cellobiose, (isomer 2), 8TMS derivative	1000380-08-0	918.4	C36H86O11Si8	24.215	6,399,082
70	D-(+)-Trehalose, octakis (trimethylsilyl) ether	1000380-43-8	918.4	C36H86O11Si8	24.487	306,455,609
71	Guanosine, 5TMS derivative	1000367-43-9	643.3	C25H53N5O5Si5	24.731	15,826,342
72	Glycerol monostearate, 2TMS derivative	1188-75-6	502.4	C27H58O4Si2	24.853	20,560,925
73	Lactulose, octakis(trimethylsilyl) ether, methyloxime (isomer 1)	1000380-09-7	947.5	C37H89NO11Si8	25.858	5,880,105
74	Adenosine, *N*-(trimethylsilyl)-2′,3′-bis-*O*-(trimethylsilyl)-, 5′-[bis(trimethylsilyl) phosphate]	32653-14-8	707.3	C25H54N5O7PSi5	27.284	45,486,155
75	D-Ribofuranose, 1,2,3-tris-*O*-(trimethylsilyl)-, bis (trimethylsilyl) phosphate	69744-64-5	590.2	C20H51O8PSi5	27.745	3,971,264

Extraction was performed using solvent B (Chloroform-MeOH-water = 2:1:1). Identification was performed using the NIST library as described in the method Subheading 3.5

18. Weigh the 2 mL microtube before harvesting nematodes in the tube. This helps calculate the net weight of the nematodes after freeze-drying.
19. The time needed for complete drying depends on the amount of samples and the efficacy of the freeze dryer. One overnight in the freeze dryer is sufficient to dry 10–50 microtube tubes in one flask.
20. To ensure the bead reaches the bottom of the microtube tube, it is preferable to use a 2 mL round-bottomed microtube tube from **step 3.3.2**.

21. The choice of extraction solvent is crucial, with a wide range of options based on polarity and solubility. Refer to the excellent article by Geier et al. [13], which screens various extraction solvents and techniques. Generally, aqueous-alcoholic solvents are commonly employed, taking into account the nature of the sample. In this context, we present two solvent systems as examples: a single-phase solvent of 80% MeOH and a two-phase solvent of chloroform-MeOH-water.
22. The non-polar phase requires 1–2 h for complete drying, while the polar phase takes 4–6 h due to the presence of water. The 80% MeOH phase typically takes around 2–3 h for complete drying. Confirm the drying status through visual examination to ensure thorough dryness.
23. This is to prevent the GC-MS injector needle from being unable to reach the samples when the quantity is insufficient.
24. GNPS stands for the Global Natural Products Social Molecular Networking [14]. For MZmine 3, please refer to the article by Schmid et al. [15]. Both are free web-based software for this purpose.
25. You expect to obtain a PCA score plot as shown in Fig. 1. Control groups (EtOH as a carrier) and treated groups (Naphthalene) are well separated. Other software options for metabolomics analysis include SIMCA, MATLAB, etc. For further information on this topic, refer to the excellent reviews [16–18].
26. In the case of OPLS-DA, a VIP value higher than 1.0 can be considered differential metabolites between the two groups.
27. You will obtain a heatmap result as depicted in Fig. 2. By utilizing distinct colors such as red and blue, you can readily visualize the metabolites that have either increased or decreased as a result of the treatment.

Acknowledgement

Part of this work was supported by the European grant "EFRO TOX FLOW."

References

1. Salzer L, Witting M (2021) Quo vadis *Caenorhabditis elegans* metabolomics-a review of current methods and applications to explore metabolism in the nematode. Metabolites 11. https://doi.org/10.3390/metabo11050284
2. Hastings J, Mains A, Virk B et al (2019) Multi-omics and genome-scale modeling reveal a metabolic shift during *C. elegans* aging. Front Mol Biosci 6:1–18. https://doi.org/10.3389/fmolb.2019.00002
3. The C. elegans Sequencing Consortium (1998) Genome sequence of the nematode *C. elegans*: a platform for investigating biology.

Science 282:2012–2018. https://doi.org/10.1126/science.282.5396.2012

4. Kim HM, Long NP, Yoon SJ et al (2019) Metabolomics and phenotype assessment reveal cellular toxicity of triclosan in *Caenorhabditis elegans*. Chemosphere 236:124306. https://doi.org/10.1016/j.chemosphere.2019.07.037
5. Butcher RA (2019) Natural products as chemical tools to dissect complex biology in *C. elegans*. Curr Opin Chem Biol 50:138–144. https://doi.org/10.1016/j.cbpa.2019.03.005
6. Williams PL, Anderson GL, Johnstone J et al (2000) *Caenorhabditis elegans* as an alternative animal species. J Toxicol Environ Health Part A 61:641–647. https://doi.org/10.1080/00984100050195125
7. Apfeld J, Alper S (2018) What can we learn about human disease from the nematode *C. elegans*? Methods Mol Biol 1706:53–75. https://doi.org/10.1007/978-1-4939-7471-9_4
8. Cole RD, Anderson GL, Williams PL (2004) The nematode *Caenorhabditis elegans* is a model of organophosphate-induced mammalian neurotoxicity. Toxicol Appl Pharmacol 194:248–256. https://doi.org/10.1016/j.taap.2003.09.013
9. Yin J, Hong X, Ma L et al (2020) Non-targeted metabolomic profiling of atrazine in *Caenorhabditis elegans* using UHPLC-QE Orbitrap/MS. Ecotoxicol Environ Saf 206:111170. https://doi.org/10.1016/j.ecoenv.2020.111170
10. Xia J, Mandal R, Sinelnikov I et al (2012) MetaboAnalyst 2.0 – a comprehensive server for metabolomic data analysis. Nucl Acids Res 40:W127–W133. https://doi.org/10.1093/nar/gks374
11. AlOkda A, van Raamsdonk JM (2022) Effect of DMSO on lifespan and physiology in *C. elegans*: implications for use of DMSO as a solvent for compound delivery. MicroPubl Biol. https://doi.org/10.17912/micropub.biology.000634
12. van Es-Remers M, Spadaro JA, Poppelaars E et al (2022) *C. elegans* as a test system to study relevant compounds that contribute to the specific health-related effects of different cannabis varieties. J Cannabis Res 4:53. https://doi.org/10.1186/s42238-022-00162-9
13. Geier FM, Wan EJ, Leroi AM et al (2011) Cross-platform comparison of *Caenorhabditis elegans* tissue extraction strategies for comprehensive metabolome coverage. Anal Chem 83:3730–3736. https://doi.org/10.1021/ac2001109
14. Wang M, Carver JJ, Phelan VV et al (2016) Sharing and community curation of mass spectrometry data with global natural products social molecular networking. Nat Biotechnol 34:828–837. https://doi.org/10.1038/nbt.3597
15. Schmid R, Heuckeroth S, Korf A et al (2023) Integrative analysis of multimodal mass spectrometry data in MZmine 3. Nat Biotechnol 41:447–449. https://doi.org/10.1038/s41587-023-01690-2
16. Guillocheau E (2020) OPLS vs PCA: explaining differences or grouping data? Sartorius Stedim Data Anal, August 25
17. Fiehn O (2016) Metabolomics by gas chromatography-mass spectrometry: combined targeted and untargeted profiling. Curr Protoc Mol Biol 114:30.4.1–30.4.32. https://doi.org/10.1002/0471142727.mb3004s114
18. Feizi N, Hashemi-Nasab FS, Golpelichi F et al (2021) Recent trends in application of chemometric methods for GC-MS and GC×GC-MS-based metabolomic studies. Trends Anal Chem 138:116239. https://doi.org/10.1016/j.trac.2021.116239

INDEX

Young Hae Choi et al. (eds.), *Natural Product Isolation and Identification: Methods and Protocols*, Methods in Molecular Biology, vol. 2895, https://doi.org/10.1007/978-1-0716-4350-1,

E

F

Q

R

S

T

MIX
Papier aus verantwortungsvollen Quellen
Paper from responsible sources
FSC® C105338

If you have any concerns about our products,
you can contact us on
ProductSafety@springernature.com

In case Publisher is established outside the EU,
the EU authorized representative is:
Springer Nature Customer Service Center GmbH
Europaplatz 3, 69115 Heidelberg, Germany

Printed by Libri Plureos GmbH
in Hamburg, Germany